TUNGSTEN IN PEACE AND WAR, 1918–1946

TUNGSTEN

IN PEACE AND WAR, 1918–1946

RONALD H. LIMBAUGH

UNIVERSITY OF NEVADA PRESS / RENO AND LAS VEGAS

University of Nevada Press, Reno, Nevada 89557 USA
Copyright © 2010 by University of Nevada Press
All rights reserved
Manufactured in the United States of America
Design by Kathleen Szawiola

Library of Congress Cataloging-in-Publication Data
Limbaugh, Ronald H.
Tungsten in peace and war, 1918–1946 / Ronald H. Limbaugh.
p. cm.
Includes bibliographical references and index.
ISBN 978-0-87417-820-3 (hardcover : alk. paper)
1. Tungsten industry—United States—History—20th century. 2. Pacific Tungsten
Company—History—20th century. 3. Nevada-Massachusetts Company—History—
20th century. 4. Tungsten mines and mining—United States—History—20th century.
5. Tungsten ores—United States—History—20th century. 6. Technological innovations—
United States—History—20th century. 7. Tungsten—Political aspects—United States—
History—20th century. 8. Strategic materials—United States—History—20th century.
9. United States—Commerce—History—20th century. 10. International trade—History—
20th century. I. Title.
HD9539.T82U475   2010
338.2'74649097309041—dc22   2010003782

First Printing
19   18   17   16   15   14   13   12   11   10
5   4   3   2   1

CONTENTS

ILLUSTRATIONS

"Leukemia Cluster: Tungsten Level High in Urine." A 2002 headline from the *Las Vegas Review-Journal* alarmed residents of Fallon, Nevada, sixty miles east of Reno, a community in Churchill County surrounded by military facilities, farms, and old mines. Further investigations by the federal Centers for Disease Control, however, found no clear evidence of a link between the community's high incidence of childhood leukemia and elevated levels of tungsten in the environment. That was good news for Kennametal, Inc., of Latrobe, Pennsylvania, whose founder, Philip McKenna, had been in the steel alloy business since World War I. Kennametal's Fallon plant has produced tungsten carbide and other processed metals since the 1950s. But it was not good news for Fallon residents, still seeking answers to explain the town's high cancer cluster.[1]

"Tons of Mercury Could Hit Market: U.S. Agency Considers Selling Toxic Stockpile," reported the *Chicago Tribune* late in November 2006. Following a series of stories on the impact of mercury contamination, the article said the Department of Energy had just announced that it was considering selling more than thirteen hundred tons of mercury—75 percent of the available U.S. total. The timing of the announcement had political implications. Earlier in the year Senator Barack Obama (D-IL), citing mounting scientific evidence of global mercury pollution in water and air, introduced legislation giving the president the authority to prohibit the export of elemental mercury after 2010. Obama's bill challenged the Bush administration's environmental policies. Fresh from their party's election sweep on November 7, Democrats called for a new approach to environmental issues. Reintroduced and amended in 2007, the bill gained momentum as Obama's star rose. It passed both houses and was signed into law two weeks before he was elected president.[2]

News stories like these are common features of postmodern journalism. Heavy metals today bear a stigma that was largely absent prior to the 1960s. Before the environmental movement, the developed world enthusiastically welcomed the extraction and diffusion of heavy metals. They were taken for granted as necessary building blocks of modern life.

This book is about one of those metals in its heyday, during the formative years of American industry. Tungsten—or wolfram, as it is known in Europe and Asia—is a heavy steel-gray metal with the highest melting point of any metallic element. Classified as a rare ferrous metal, it is about the same density as gold but much less valuable. On the global market, in 2009 the price of gold hovered around nine hundred dollars an ounce. An equivalent weight of tungsten was worth less than two dollars.

A few technical details are necessary to understand tungsten's importance to modern civilization. Producing metallic tungsten requires refining its ores, for the raw metal is never found in nature. Under different pressure and temperature conditions the molecular compound WO_3, tungsten trioxide or tungstite, combines with other elements to form two distinctive and gradational isomorphous series of minerals. The first series ranges from ferberite ($FeWO_4$) to hübnerite ($MnWO_4$), with wolframite ($Fe,MnWO_4$) the intermediate mineral. A separate calcium series includes the fluorescent minerals scheelite ($CaWO_4$) and powellite ($Ca,MoWO_4$). Although wolframite and scheelite are the principal ores of tungsten, the other three minerals in the two series have also been commercially important under favorable market conditions.[3]

Tungsten ores have been found in thirty-seven countries. The largest known deposits are in China, Russia, Kazakhstan, and Canada. These four together control more than three million metric tons, or 64 percent of the known commercial deposits in the world today. In the United States, with only about 8 percent of global reserves, the principal deposits are in Nevada, California, Montana, and Colorado, but much has already been consumed. As of 2008, Nevada tungsten properties reported aggregated reserves of some ten million units, ten times the Nevada-Massachusetts production during the senior Segerstrom era. Today, China accounts for almost 90 percent of new tungsten supplies coming on the market.[4]

Once valued mainly for its ability to harden steel and carry current in lightbulbs without melting, tungsten since the 1950s has been utilized in such industries as aerospace and electronics for superalloys and other applications where hardness, durability, and strength under extreme temperatures are necessary. Its primary use, however, is still in heavy industry for long-wearing, hot-running, and high-speed cutting and drilling tools, and in military hardware for armor and ammunition. At the beginning of the twenty-first century only 10 percent of tungsten production goes into lamp filaments and cathodes, but that ratio is expected to change as incandescent lighting gives way to new technology.[5]

As of 2008, the U.S. government maintained a national stockpile of twenty-seven different minerals. Stockpiling laws describe these materials as "strategic and critical," to be used only in case of war or other national emergencies. Tungsten has been stockpiled since 1939, but the strategic reserve is less important now than before World War II. Though conventional military operations still require heavy armor to protect "boots on the ground," postwar advances in military and industrial technology, as well as changing perceptions about America's wartime resource needs and how to manage them, have altered the views of strategic planners. All but two of the minerals in the current stockpile are currently for sale by the Defense National Stockpile Center, including eight million pounds of tungsten ore and concentrates.[6]

Most of the literature on tungsten is specialized and highly technical. It is also largely focused on the period after World War II. Many of the best studies are listed in the bibliography. As a historian, I have tried to broaden the context for understanding tungsten as a strategic metal. The ores of tungsten took centuries of experimentation and testing before they were chemically reduced to produce tungsten metal that could then be alloyed with steel. How tungsten became a "key" metal of modern industry and how it influenced strategic thinking and politics during and between the two great wars of the twentieth century are two of the main themes addressed in this book.

Within this larger context of tungsten geopolitics, I have also tried to tell a more focused story. The Segerstrom Collection at the University of the Pacific is a rich resource for analyzing the tungsten business from the standpoint of a mining company investor and manager. At the core of this book is a case study of business management and corporate survival in a turbulent era. The Nevada-Massachusetts Company, under the leadership of Charles H. Segerstrom, its founder and first president, emerged in the mid-1920s out of the financial collapse of its predecessor. By the early 1930s it had grown into one of America's most important tungsten mining and milling operations. With a background in law, real estate, banking, and mining investment, Segerstrom was superbly qualified to manage the business end of a productive mining and marketing complex. Aided by a staff of engineers, scientists, and field managers, Segerstrom successfully led Nevada-Massachusetts, as well as his own family enterprises, through two disquieting decades. How he responded to national and international events that directly affected the tungsten business, and at the same time kept his mines productive and his customers supplied with quality concentrates, is a central concern of this narrative.

These varied themes are intertwined within a single chronological framework. The first two background chapters follow tungsten through its formative era from earliest discoveries and experiments to commercial application as an alloy of steel, and then from peacetime use in autos and lightbulbs to military weapons in World War I. Chapter 3 begins with the economic aftermath of war on the tungsten market and the political impact of the metals lobby. It then describes the founding of the Nevada-Massachusetts Company and Segerstrom's emergence as a leader in the tungsten industry. The next two chapters discuss the company's marketing and management issues that arose during the Depression years, Segerstrom's efforts to build a tungsten cartel, and the political ramifications of the New Deal on the metals industry. Chapters 6, 7, and 8 cover the war years 1939–45. They address issues specific to tungsten and the metals industry, and analyze Segerstrom's role as industry spokesman and company strategist. The book ends with a general assessment and a look at tungsten's postwar status.

A narrative history confined to one metal, one company, and one individual has obvious limitations. In discussing tungsten's strategic implications I have steered clear of ongoing debates over geographic determinism and resource scarcity as a cause of war. Nor do I delve deeply into what Michael Shafer called "mineral myths" that informed—or misinformed—policy makers advocating strategic mineral reserves. Western historians may wish for some discussion of the mining West as a distinct section, and its influence on American foreign policy—a topic broadly explored by Peter Trubowitz in a 1998 book. My narrative also only touches on several topics in business history that might be more fully expanded—foreign influence on domestic markets, factional and sectional disputes in the metals industry, processing patent fights, multinationalism, and the effects of changing technology, among others. All these are important areas for further analysis but beyond the scope of this book.[7]

ACKNOWLEDGMENTS

The idea for this book originated more than a few years ago, when I was serving the dual role of professor of history and archivist at the University of Pacific. On a research trip to Sonora, California, I met Mrs. Mary Etta Segerstrom, daughter-in-law of Charles H. Segerstrom. She first introduced me to the monumental accumulation of corporate records, correspondence, and other documents that make up the central files of the Nevada-Massachusetts Company. Her love of history and her concern for the preservation and utilization of the Segerstrom papers led to the development of a family collection at the university. I am deeply grateful for her assistance and cooperation. Unless otherwise indicated, all correspondence to and from Charles H. Segerstrom, and all records of the Nevada-Massachusetts Company and its subsidiaries, are located in the Segerstrom Collection at the Holt-Atherton Library, University of the Pacific, Stockton, California, cited as UOPWA in the endnotes.

Historical research and writing is truly a collaborative effort. In the past five years I have profited from reading and reflecting on the work of dozens of scholars in history, geology, economics, political science, public policy, and other fields. Their contributions are listed in the endnotes and bibliography. The librarians and archivists who preserve these materials and assist scholars in locating them are also an integral part of the research process. I owe a special debt to Sean Sutton, Michael Hurst, and Trish Richards at the special collections branch of the main University of the Pacific library where the Segerstrom Collection is housed. As a visitor at the University of California-Davis, I was welcomed and ably assisted by the special collections staff under Daryl Morrison, and by the staff and students in the government documents section under Linda Kennedy. In Reno, Jacquelyn Sundstrand at the special collections branch of the University of Nevada, and Eric Moody and staff at the Nevada Historical Society, offered valuable suggestions and went out of their way to expedite my research requests. So did Susan Searcy and staff at the

Nevada State Archives in Carson City. At the California State Mines and Geology Library in Sacramento, Dale Stickney and his staff were generous with their time and resources. The reference library staffs at the Colorado School of Mines, the California State Library, and the Sonoma State Library in Rohnert Park, California, all provided generous assistance.

The comments and suggestions of many friends and associates have also greatly enhanced the organization and readability of this history. To the following individuals I owe a profound debt of gratitude: To Willard P. Fuller Jr. and Frank and Floralee Millsaps, who read the entire manuscript and made many valuable suggestions; Joseph V. Tingley, who knows more about tungsten in Nevada than any other geologist, and whose careful scrutiny helped improve the narrative's technical accuracy and analytical content; Charles Murphy, Roger Burt, and Johnny Johnsson, for their help in improving the style and substance of chapters 1 and 2; Peter Hahn, whose knowledge of Nevada tungsten resources was particularly helpful, and who critically reviewed chapter 2; James P. Fell and Laurence P. James, for information and help in locating resources relative to Colorado's tungsten and smelting industry; Jeremy Mouat for his assistance in exploring tungsten geopolitics during World Wars I and II; Daniel Edelstein, copper commodity specialist of the U.S. Geological Survey, for locating and providing heretofore unpublished government data on monthly copper output during and after World War II; Richard W. Haesler, for information on the Haesler family; Gene Gressley, for helping me understand the broader dimensions of Thurman Arnold's career; Richard Turner, whose artistic rendition of the Cold Spring mine shaft house in Nederland, Colorado, highlights the illustrations; Terry A. Reed and Sarah Hall of the Chambers Group, a consulting firm in Reno, for providing photos and other information regarding recent developments in and around Mill City; John Heizer, for help in understanding his grandfather's role in the Nevada-Massachusetts corporate story, and for his generosity in making family photos available for publication in this book; and finally to my wife, Marilyn, for her patience, equanimity, and loving indulgence.

TUNGSTEN IN PEACE AND WAR, 1918-1946

1

STEEL ALLOYS AND THE RISE OF MODERN INDUSTRY

If steel is a critical building block of modern industry, tungsten is an essential component of modern steel alloys. Steelmaking has an ancient history, but making steel harder by adding tungsten is a commercial process barely a century old. Introducing the evolving technologies and industries that led to tungsten's economic development provides historical context, and helps explain how and why one modest Nevada mining company rose to national prominence in the years between two world wars.

EARLY DISCOVERIES AND EXPERIMENTS

Tungsten was discovered late in the history of civilization, and even then it took another 150 years before it became commercially valuable. In his Pulitzer Prize-winning book, Jared Diamond wisely cautions us to beware of simplistic explanations for inventions that ultimately change the world. Necessity may or may not be the mother of invention, but clearly tungsten required both an advanced technology and an industrial necessity before its properties could be understood and utilized. Long before primates evolved and began using tools, tectonic movements in the earth's crust folded, fractured, and sometimes exposed on the surface hydrothermal veins of wolframite, the "black ore" of tungsten, along with minerals containing silica, gold, silver, tin, copper, molybdenum, and iron with which tungsten is often associated. Other deposits of wolframite and its "white ore" cousin, scheelite, could be found in the metamorphic contact zones in and around granitic intrusions (the so-called skarn ores), in low-grade "stockworks" containing thousands of tiny veins and veinlets, in sedimentary formations, and even in the solutions of brine lakes and hot springs. Where deposits cropped out on the surface, the corrosive effects of oxygen, the weathering effects of wind, water, and ice, and the pulling effects of gravity altered, eroded, fractured, washed, and carried minerals, along with gravel, sand, clay, and silt, far from their sources. Unlike other less widely diffused elements that went into ferrous alloys such as molybdenum, manganese, and vanadium, tungsten can be found in some thirty-seven countries around the globe, mostly in shallow, low-grade deposits.[1]

Probably the first humans to encounter tungsten were Neolithic placer miners and metalworkers looking for pliable pieces of native copper or gold. From streambeds or dry washes they may have picked up rounded "nuggets" of wolframite or scheelite, but tossed them back when heating or hammering failed to yield their secrets. There they lay for the next ten thousand years or so, unusable and therefore uninteresting until civilization crossed the intellectual threshold that separated the medieval and modern worlds.

Isolating and identifying the heavy gray metal were tributes to empirical observation and experiment, a crescendo in the symphony of science and reason we call the Enlightenment. A century after Newton and Galileo, natural philosophy—a premodern term for the physical sciences—was a popular and influential intellectual endeavor in Europe among the educated elite, rich or poor, amateur or professional. Academic institutions across the continent expanded their curricular offerings in chemistry, astronomy, physics, and other disciplines.

By the mid-eighteenth century Sweden had a thriving scientific community. In 1747 Johan Gottschalk Wallerius, a professor of mineralogy at the University of Uppsala, studied the properties of a black mineral that generated a muddy froth in the process of smelting tin ore. Two hundred years earlier the German physician and scientist Georg Bauer, better known as Agricola, in a technical paper published prior to his famous treatise on metals, *De Re Metallica*, had termed it *spuma lupi*, perhaps because of its tendency to hide or "eat up" cassiterite to the consternation of smelter workers. Wallerius translated Bauer's term into German. He thought *volf rahm* was an ore of tin, but was unable to reduce it.[2]

In the 1770s and '80s at the same school, Professor Torbern Olof Bergman experimented, lectured, and wrote on the properties of metallic compounds, influencing generations of scholars and students. His chemical analysis of iron provided the first scientific distinction between iron and steel. One of his students, Karl Wilhelm Scheele, an impoverished pharmacist with a modest pension from the Stockholm Academy of Science, studied the reduction of metals and other substances, working after hours with a few simple laboratory instruments. Over a lifetime of experimentation Scheele discovered more new chemical compounds than perhaps any other scientist before or since. In

1781, building upon the work of Wallerius and Bergman, he boiled in nitric acid a light-colored mineral that was thought to be an ore of tin or iron, and precipitated a spongy substance that dried to a yellow powder. His experiment proved that the mineral was calcium combined with a "peculiar" acid. His mentor Bergman suspected that further reduction of the insoluble acid would reveal a metallic element, but neither he nor Scheele was able to break it down. Scheele labeled the suspect metal *tungsten,* combining the Swedish words *tung* (heavy) and *sten* (stone). Forty years later a British mineralogist, C. C. Leonard, honored Scheele by applying his name to calcium tungstate ($CaWO_4$), thereafter known as scheelite. The term *tungsten* was then given to the metallic element.[3]

Unknown to the Swedish scientists, the work they had done had been carefully noted in the spring of 1782 by one of Bergman's students, who was in reality a chemist employed by the Spanish government to engage in industrial espionage for the benefit of homeland defense. Returning to the Seminario Patriótico in the Basque province of Vergara, Juan José de Elhuyar worked with his brother Fausto, a professor of mineralogy, to successfully reduce tungstic acid and isolate the powdery steel-white metal for the first time.[4]

THE IRON AND STEEL AGE BEFORE TUNGSTEN

As a technological by-product of the modern steel industry, tungsten became useful only after advancing science and technology in both Europe and America made mass production of steel compounds and alloys possible. The wars and imperial rivalries of the late nineteenth century spurred expanded steel use in both hemispheres. In Europe, steel became big business during the two decades between the Crimean War and the Franco-German war, a period marked by accelerating industrialization, nationalist uprisings and unification movements, and expansionist drives and colonial rivalries in Asia, Africa, and South America. In the United States, the end of the Civil War marked a watershed in the industrial growth of a nation that was resource rich, economically integrated, and culturally eager to exploit whatever raw materials lay in its path. On both sides of the Atlantic, with capitalism the primary engine of economic growth, steel was the industrial icon of the age.

Strengthening iron by adding carbon and removing impurities through heating, quenching, beating, and blowing were techniques nearly as old as the Iron Age itself. Before the first mass-production process was developed in the 1850s and 1860s, however, steelmaking was laboriously slow and inefficient.

For three thousand years practical ironworkers made steel without understanding the chemistry behind it. The charcoal they used to heat iron ore also oxidized, reduced, and fused the molten metal with very small amounts of carbon. Depending on the amount and quality of charcoal used as well as the heating time and temperature, the carbon content in early steels varied from under 0.02 percent to 3 percent and above. Oxidation and reduction removed impurities that floated to the top of the molten metal as slag. Carbonation through fusion created a molecular structure with a crystal lattice that could be altered by heating, hammering, and cooling. This added tensile strength and hardness to iron but also made it brittle. The properties of steel could also be changed by eliminating almost all carbon and adding other metals and chemical compounds to the molten mass. Yet not until the late nineteenth century did metallurgical science and technology advance far enough for those possibilities to be tested empirically and marketed as new products.[5]

By trial and error, early metalworkers learned to make malleable wrought iron by repeatedly heating, stirring, and hammering brittle cast-iron "pigs" in a refining forge with a strong air blast, thereby oxidizing some of the carbon and other impurities. They learned to homogenize and shape wrought iron into blooms or billets by hammering or rolling out slag and diffusing the carbon. They learned to make blister steel in a carburizing process called cementation, whereby wrought iron is slowly baked in a bath of carbon powder, creating blisters where the carbon was absorbed. They learned to improve blister steel by hammering, to soften and toughen it by annealing, and to balance hardness, toughness, and ductility by tempering.[6]

By the eighteenth century steelmaking was a venerable craft in its third millennium, yet still limited by small batch size and lack of quality control. Those limitations became increasingly apparent under the modernizing pressures of empirical science and emerging industrialization. The chemistry and metallurgy of iron and its alloys were just beginning to be understood in 1742 when Benjamin Huntsman, a "Quaker clockmaker" in Sheffield, England, took a stride toward improved quality by melting blister steel and wrought iron in a crucible and casting an ingot. By improving carbon diffusion and slag removal, crucible steel set a new standard of quality. It increased steel demand and established Sheffield as the center of crucible steel production. For 150 years Sheffield steel remained the favorite for makers of fine cutlery, precision tools, gun barrels, saws, axes, and other high-quality cast-metal products. But making crucible steel was time consuming and labor intensive, hard on both men and equipment. Cast steel was hard but also brittle. It tended to fracture under pressure and heavy use, and was thus unsuited for many industrial applications. By the mid-nineteenth century, with the pace of industrial growth accelerating and global demand rising for railroads, plows, pumps, mills, machine tools, and heavy equipment, the need was clearly apparent for a cheaper, faster, and more durable mass-produced iron alloy.[7]

As a world-renowned steelmaking center for more than a century, Sheffield brought the best minds and materials in the business together with the most advanced techniques. Out of this critical mass came the "puddling" process for producing cheaper and more ductile steel than the crucible method. Puddlers made essentially a "high-carbon wrought iron" by reducing pig iron in a reverbatory furnace, oxidizing impurities with lime and other fluxes, removing slag, and homogenizing the product by repeated hammering, rolling, and reheating. Though puddling may have earlier origins, Joseph Bennett Howell, a Sheffield metalworker, is given the principal credit for developing the process, largely based on patents he registered in the late 1850s. Before the advent of bulk steel production, puddling was the only way to make large quantities of high-quality carbon steel.[8]

IRON AND STEEL IN AMERICA

Steelmaking know-how crossed the Atlantic before the American Revolution, but iron was the only real choice for most industrial and commercial applications prior to the Civil War. A desultory colonial iron industry developed first near coastal settlements where bog ore was available, then inland as ironmasters tapped into shallow hematite and magnetite deposits in the carbonate terrain of Pennsylvania, New Jersey, and other regions. Some antebellum iron makers also made small batches of blister steel and experimented with the crucible process for making iron alloys, but high cost, poor transportation, and limited markets hindered development. Even after domestic bulk steels became available, they were hard to sell.[9]

Throughout most of the nineteenth century, American craftsmen preferred Sheffield steel for making cutlery, saw blades, and other high-quality metal products. They resisted the efforts of steel industry protectionists to thwart foreign competition with tariff barriers, primarily for economic reasons but partly for the sake of tradition. In steelmaking as in mining, Americans improved on older inventions and methods rather than creating something entirely new. Innovation, rather than inventiveness, characterized America's contribution to the advance of industrial technology.[10]

Iron making grew rapidly after 1800, spurred by the proximity of urban markets in the Northeast and rising domestic industrial demand. Along the Susquehanna and its tributaries, in the Juniata region of central Pennsylvania, and west of the Alleghenies into the Ohio Valley, dozens of furnaces, fineries and refineries, forges, foundries, machine shops, and mills sprouted near sources of ore, fuel, and water. Eastern woodlands provided abundant hardwoods for the production of charcoal, the chief power source before the coal era. A generation of ironworkers in the 1840s and '50s experimented with anthracite as furnace fuel, though it was tricky to use and hard to light. After the Civil War, when Connellsville coke proved the superiority of bituminous over anthracite, major iron production in Pennsylvania shifted westward. By relocating to Pittsburgh and the Allegheny Valley, iron and steel entrepreneurs gained better and cheaper sources of fuel. Using low-cost transport by rail and barge, they tapped abundant raw materials from the newly opened Lake Superior iron deposits. They also profited from the burgeoning inland market for bridges, rails, railcars, and tools for factory and farm.[11]

BULK STEELMAKING

Despite centuries of experience in metals development, steel remained an expensive commodity until the 1870s and the advent of successful pneumatic bulk processing. For twenty years or more prior to the widespread adoption of the Bessemer converter, practical metalworkers on both sides of the Atlantic had tried to make large batches of steel out of pig iron by injecting the charge in a closed furnace with blasts of superheated air. Henry Bessemer's process saved money by utilizing economies of scale in the construction of massive converters that pivoted on an axis to load and unload. Smelting heavier charges in a closed container improved efficiency and shortened the conversion time by eliminating slow and expensive intermediate steps. It also reduced both the number of workers needed and the skill level required. Though Bessemer by 1859 had perfected his process using a selective sample of low-phosphorus iron ore, the poor quality of Bessemer steel made from more common pig iron with high phosphorus content marred his reputation and left some consumers resistant to change. A lengthy patent dispute in the United States also delayed recognition of Bessemer's achievement and the widespread adoption of his process. In 1864 Alexander Holley, an American engineer, obtained rights to the Bessemer process and began building a converter at Troy, New York. Before he could blow in, however, he was slapped with a suit by a midwestern investment group that controlled the patents of William Kelly, a Kentucky metalworker who asserted an earlier claim to the same process. After a two-year fight the rival contenders settled their differences by pooling patents and organizing the Bessemer Association, a holding company that controlled the licensing of steel plants using Bessemer converters.[12]

As a wide-ranging consultant for the Bessemer Association, Holley designed nearly a dozen steel mills in the United States between 1865 and 1876. His most innovative was the Edgar Thompson works he built for Andrew Carnegie at Braddock, Pennsylvania, in 1875. Combining the latest technological advances with improvements in shop design and organization, Holley, along with Carnegie's brilliant mill superintendent William R. Jones, made Edgar Thompson the nation's most efficient—and most profitable—steel plant. For the next quarter century the Bessemer process dominated bulk steelmaking technology, and Carnegie mills dominated the market for steel rails and structural steel.[13]

Across the Atlantic, however, blast furnaces gradually gave way to a new mass-production technique. Early in the 1860s Carl Wilhelm Siemens, a German inventor living in England, built an open-hearth furnace based on reverbatory principles. In a process called regeneration, coal-fired and preheated exhaust fumes drawn back and forth through brick-lined chambers built up superheated gases that melted pigs and burned off carbon and impurities. The open-hearth process made slow headway at first. Brick linings sagged under the intense heat until hearths were strengthened by adding iron framing. Only low-phosphorus pig iron could be used, since silica in the firebricks raised the acidity of the charge and inhibited chemical reactions that removed phosphorus compounds. In 1878 two practical metallurgists in England solved the phosphorus problem by changing the pH value from acidic to basic, using firebrick linings composed of calcined dolomite or magnesite and adding more lime to the charge. The Gilchrist-Thomas process, adaptable to both blast and open-hearth furnaces and eventually controlled by the Bessemer Association, made feasible the commercial development of common high-phosphorus ores in Britain, Sweden, the Lorraine basin, and other parts of Europe. The more deliberative pace of open-hearth processing also improved the quality and consistency of bulk steels. By 1900 on both sides of the Atlantic it was the predominant method of converting pig iron into standard carbon steel.[14]

By the beginnings of the twentieth century, the steel mill was "the exemplary industry of the material age," in the words of economist George Gilder. Cheap carbon steel produced by conversion in blast furnaces or by regeneration in open hearths accelerated the pace of urban-industrial modernization, but there were limits to its quality and utility. The extreme heat and agitation of pneumatic-bulk processing fused metals into a molten mass. This made slag removal easier, but the process left oxygen pockets that weakened the refined product unless such "blow holes" were treated by adding expensive ferromanganese or other deoxidizing agents to the charge. Moreover, inexpensive bulk steel made from a mixture of crude pig iron, iron ore, and scrap steel was brittle and often contained undesirable metallic elements that lowered its quality and consistency. Despite the increasing price differential, American precision toolmakers still preferred more expensive crucible steels from Sheffield to the cheaper bulk steels. Even puddled steel, though increasingly harder to find after 1870, was more homogenous, less brittle, and had fewer impurities than bulk products.[15]

Bulk steel also lacked the versatility required in modern urban-industrial societies, where munitions, high-compression engines, electric motors, and other specialized machines and tools took an increasing share of the consumer dollar. Keeping up the pace of modern life required steel that was not only superior to earlier versions but better suited to special needs of government, home, and industry. Improving steel's quality and versatility without limiting production, or adding substantially to cost, and making the tough hardware necessary to speed up the cutting and machining of bulk steels were powerful incentives to metallurgists, producers, and consumers alike. The answers lay in the development of tungsten steel and other ferrous alloys.

THE EMERGENCE OF STEEL ALLOYS

As a steelmaking center, Sheffield's reputation for quality control made it a natural leader in steel alloy research and development. Henry Bessemer recognized Sheffield's leadership in 1858 when he built his first steel plant there. However, it was unsuccessful until he enlarged the tuyeres to increase the air blast and added spiegeleisen, an alloy of iron and manganese, to oxidize impurities. Adding spiegeleisen to "clean" molten pig iron was a process developed by Robert Forester Mushet, son of a prominent Scottish ironmaster. Mushet's experiments at Sheffield earned him widespread recognition, but his greatest achievement came from his work with tungsten. In 1868 he patented "RMS," otherwise known as Mushet Special Steel. Developed from prior work by Austrian chemist Franz Köller, RMS was a hard, tough alloy, ideal for machine-shop cutting tools. Mushet made it in small batches at the Osborn plant in Sheffield by charging crucible steel with 7 percent tungsten and allowing it to air dry. Improvements in tempering by Henry Gladwyn, another Sheffield metallurgist, and later in substituting chromium for manganese in self-hardening tool steels by Sanderson Brothers at Syracuse, New York, added even harder and tougher qualities to Mushet steel under the most extreme cutting conditions.[16]

Sheffield also produced another industrial giant in Robert Abbot Hadfield, whose 1882 patent for ferromanganese, an alloy with higher manganese content than spiegeleisen, is said to have begun "the age of alloy steels." A practical metallurgist and son of a foundry owner, he took over his deceased father's business in 1888 and built it into one of the largest steel foundries in the world. Commercial use of ferromanganese was slow to develop, however. One of the hardest alloys, low-carbon manganese steel could not be marked with ordinary chisels or cut with the shop tools of the day. By the 1890s it had begun to find commercial application in rails, switches, dredge buckets, ore-crushing machinery, and other heavy-stressed railroad and construction equipment. After World War I ferromanganese consumption mushroomed, both in defense and in peacetime industries. For a time in the twentieth century manganese steel was considered the "most useful alloy steel ever invented."[17]

Hadfield also developed a low-carbon silicon alloy known as silicon steel, valued for its superior electric conductivity and magnetic properties. The General Electric Company eventually acquired Hadfield's patent rights and used them in the production of high-voltage transformers, still one of silicon steel's principal uses.[18]

The route from lab to shop, from scientific theory to broad practical application, is often bumpy and slow. Technical problems as well as economic uncertainties can add many roadblocks and detours. The appearance of new technologies may threaten old habits, ideas, or livelihoods. All of these factors hindered the commercial success of specialty steels. Up to 1900 most blacksmiths and machine shops used carbon steel tools instead of more expensive ferrous alloys. Carbon steel was widely available and the cheapest steel on the market, but it contained impurities and lacked uniformity of hardness. Tools made from it had to be sharpened and changed frequently, and worked at slow cutting speeds. In the meantime, experiments continued to improve the qualities of specialized steels.[19]

In the 1890s, Frederick W. Taylor, an American shop engineer for Bethlehem Steel Company, tested the properties of both Mushet and Sanderson steels. This is the same Taylor whose time and motion studies aroused so much labor-management controversy over the next twenty-five years. Dedicated to improving performance and productivity, he applied the same principles of "scientific management" to materials as he tried to do with employees. With the assistance of Maunsel White, a Bethlehem metallurgist, Taylor discovered that low-carbon chromium-tungsten tool steel retained its cutting edge even while red-hot if it was first heated to nearly the melting point before quenching. This quality of "red-hardness" became the distinguishing characteristic of a long line of specialty products known thereafter as "high-speed steels."[20]

As an efficiency expert, Taylor demonstrated the revolutionary potential of high-speed steel at the Paris Exposition in 1900. Dr. Colin Fink, a distinguished chemical engineer at Columbia University, later described Taylor's Parisian triumph. Using Taylor-White-treated steel, he "first exhibited to the astonished gaze of incredulous machinists and metallurgists the spectacle of a tool cutting so fast and so deep that it delivered steel chips at a blue heat and in amazing quantities." Though the demonstration was intended to promote labor efficiency through technology, Bethlehem Steel, Taylor's sponsors, saw it as a commercial opportunity. Company officials anticipated a business boom in high-speed steel and a jump ahead of their rivals. They soon found themselves in a patent fight instead. Sheffield steel producers and their American subsidiaries claimed Taylor-White steel was not a new alloy but Mushet steel under a different name and a different heat treatment. After a two-year court battle, the trial judge sided with the Sheffield attorneys, leaving English steelmakers in command of the ferrous-alloy industry and the main suppliers of high-speed steel until World War I.[21]

The patent litigation was indicative of the economic stakes involved in high-speed steel production and distribution. Ironically, Taylor's exhibition did more to promote technological innovation than labor efficiency. As historian Nathan Rosenberg has observed, the introduction of high-speed steel caused a "technological imbalance" in the machine tool industry that became "an inducement for further innovation."[22] Adjusting to higher speeds required redesigning or replacing not only older, lighter, and slower carbon steel tools but also the shops that used them and the systems that powered and controlled them. Within a decade after Taylor's sensational demonstration, high-speed steel had transformed the tools and technical infrastructure of heavy industry.

Practical chemists and inventors like Hadfield and Taylor made important contributions to science and technology in the nineteenth century and continued to do so far into the twentieth, but the era of amateurs gradually gave way as science education expanded and new technical schools annually turned out a fresh supply of trained professionals for both business and academe. The University of Sheffield, formed after a merger of Firth College with Sir Frederick Mappin's Technical School, grew into a world-class research facility after 1905 with financial backing from Sheffield's steel industry. Leading the research in steel alloys at the university in its formative years was a practical chemist with limited formal training but a formidable reputation. John Oliver Arnold, son of a railway company manager in the British Midlands, joined the navy upon completing grammar school, and later gained experience by working for a series of private technical schools and corporations as engineer, chemist, manager, consultant, and lecturer. As chief adviser for the leading Sheffield companies, he experimented with molybdenum, nickel, tungsten, tantalum, chromium, and other low-carbon steel alloys. In the late 1890s, working with a sample of ferrovanadium provided by an Austrian colleague, Arnold added vanadium to steel and came up with the most resistant alloy yet developed against metal fatigue. Naylor Vickers Steel Company, one of Arnold's Sheffield clients, produced vanadium steel for automobile springs and shocks in England. Across the Atlantic, Henry Ford learned of the properties of vanadium in 1905 after testing a valve stem taken from the wreckage of a French racing car. Later that year he contracted with United Alloy Steel, an Ohio company that later merged with Republic Steel, to produce a vanadium steel chassis for his new Model T.[23]

Vickers and other Sheffield companies, cognizant of the threat posed by cheaper mass-produced steel, promoted alloy steel research in part to protect the Sheffield name as a quality steelmaker. Formulating a combination of metal alloys that resisted corrosion was a high priority. In the first decade of the twentieth century British, German, French, and American scientists all had been working on similar problems. In 1912 Harry Brearley, a practical metallurgist at Thomas Firth and Sons trying to develop a better alloy for rifle barrels, added a heavy charge of chromium to an experimental batch of nickel steel. The result was a hard and highly stain-resistant alloy, a forerunner of modern stainless steel. Brearley's English backers formed a syndicate to promote and produce the new product along with its American subsidiary, the Firth-Sterling Steel Company of McKeesport, Pennsylvania. In the process of patenting the new alloy, however, Brearley learned that an American, Elwood Haynes, an Indiana metallurgist and auto manufacturer, had applied for a similar patent. After two years of fruitless contention, Brearley and Haynes merged their interests and formed the American Stainless Steel Company to control licensing. Coming on the market just as the United States entered World War I, Haynes-Brearley stainless steel found ready use in both military and industrial applications.[24]

Unlike Harry Brearley, who distrusted formal education, Elwood Haynes was a college-trained metallurgist whose senior thesis at Worcester Polytechnic Institute was titled "The Effect of Tungsten Upon Iron and Steel." After a year in graduate research at Johns Hopkins and a few more years of teaching public school he joined the ranks of industrial entrepreneurs, first as an expert in natural gas, then as an auto manufacturer and co-owner of the Haynes-Apperson plant in Kokomo, Indiana. By 1905 he had made enough money to retire from active management and concentrate on metallurgical research. In his private laboratory over the next several years he formulated and tested various low-carbon steel alloys for hardness, tensile strength, elasticity, and resistance to corrosion. At the American Chemical Society convention in 1910 he announced the development of a cobalt-chromium alloy he called "stellite." Promoted as a tough, nontarnishing steel for cutting tools, Haynes made it even tougher and harder the next year by adding tungsten. With patents in his pocket he built a small plant next to his auto factory and began the production of stellite on a small scale in the fall of 1912. Stellite sales boomed with the coming of war, and continued to boom in the inflationary postwar rush to meet pent-up consumer demand. In 1920, after long negotiations, Haynes sold his stellite plant to Union Carbide, an industrial giant whose predecessor, Electro Metallurgical Company, had been one of Haynes' largest customers.[25]

TUNGSTEN AND ELECTRIC LIGHT

While most tungsten produced before World War I went into steel alloys, it had a much bigger consumer impact in the prewar period as filament wire, making possible the "white magic" of incandescent light. Cheaper and much harder than platinum or gold, generating less heat and more light at high temperatures than copper or aluminum, capable of being drawn into wire forty times smaller than the diameter of a human hair, and lasting longer than any other element without melting under the extreme heat of an electric current, tungsten filaments were essential tools of modern life. However, their development came late in the history of electric lighting. By the early 1880s Thomas Edison's Menlo Park laboratories, building upon a century of experimentation and achievement in Europe and America, had produced both a practical lightbulb and a systematic method of electric power generation and distribution. Backed by J. P. Morgan and the financial resources of Wall Street, Edison did more than any other entrepreneur to light American households and establish the basic structure of the modern electric industry. However, his incandescent system, using direct current and a low-voltage carbon filament, rested on technology that was already obsolete by the time Edison General Electric merged with a northeastern rival to form the General Electric Company in 1894.

With the stubborn "Wizard of Menlo Park" no longer in charge, over the next two decades GE amply demonstrated its financial power and technological resourcefulness. Using cutthroat marketing tactics, pooling arrangements and secret acquisitions to eliminate serious competition at home, and patent buyouts along with substantial investments in foreign rival companies to reduce threats from abroad, GE soon dominated the industry. Pooling patents with its arch rival, Westinghouse, after 1896 opened the door to the most advanced alternating current technology. In 1909, two years after Austrians Alexander Just and Fritz Hanaman developed a brittle tungsten wire filament that generated much more light with much less heat than Edison's carbon-alloy filament, GE bought the exclusive rights. The next year a company physicist, William David Coolidge, gave GE a "virtual lock" on filament technology and production by patenting a ductile tungsten filament that soon became the industry standard. By World War I the General Electric Company controlled more than 90 percent of the electric-lamp filaments produced and sold in the United States.[26]

Despite rising consumer demand, lightbulb filaments used very little raw material. Before World War I more than 90 percent of world tungsten production went into ferrous alloys, a ratio that has not changed much over the years. But commodity ratios do not indicate the increased volume of production necessary to meet the rising demand for raw material after 1900. World tungsten production rose from 4,025 tons of 60 percent tungsten trioxide concentrate in 1905 to 11,840 tons by 1915. In the United States, the 247 tons produced in 1900 was but a tenth of the amount mined, milled, and marketed by the second year of World War I.[27]

TUNGSTEN GEOPOLITICS AND INTERNATIONAL TRADE

Unpredictable and largely unforeseen in the presumptive innocence of the prewar years were fundamental geopolitical realities that became increasingly apparent as the market for tungsten products expanded. The most obvious were the geographic disparities affecting ore supply and demand. Tungsten ores are widely scattered across the globe, but the first to be exploited were the vein and placer deposits on the Iberian Peninsula, the skarn deposits in the mountain ranges of southeastern Australia, and the wolfram outcrops in the tropical forests of Southeast Asia. These three regions alone accounted for 55 percent of world tungsten production by 1913.[28] Without significant tungsten deposits of their own, the major steelmaking countries of Europe grew dependent on foreign sources, but those sources before 1914 were largely controlled by German corporations.

The Germans owed their prewar commercial dominance both to the technological leadership of the country's scientists and engineers in the reduction and processing of tungsten ores and to the marketing strategy of its leading metals dealers. Heading the list of dealers was Metallurgische Gesellschaft, an integrated international firm led by the Merton family of Frankfurt and linked by family ties and investments to both British and American metal merchants and markets. In the imperial years after 1870 German heavy industry surged forward, rapidly closing the industrial gap that had distinguished the agrarian, preindustrial German states at midcentury from their more urbanized and industrialized rivals on the Atlantic seaboard. By the 1880s and 1890s, as economic historian Alfred Chandler has written, Germany had developed an industrial "community of interest" based on a few giant corporations cartelized with binding contracts and specializing in metals, chemicals, and high-tech machinery. With an aggressive marketing system and with subsidiaries, branch offices, and agents around the world, German dealers in machinery and metals were fierce competitors.[29]

The strength of Germany's grip on international tungsten markets before World War I also depended on controlling the source and the beneficiation of ores. The German states had very little tungsten but did have a rich heritage of mining and metallurgy. With the world's leading technical institute at Freiburg in Saxony, with abundant deposits of copper, lead, zinc, and silver in Silesia, the Erzgebirge, and the Rhineland, along with coal and iron in the Ruhr basin, German science and technology made its presence felt in both the mining and the processing of metals. Geologists, engineers, and metallurgists trained in Germany staffed many mining operations around the world. German engineering expertise attracted German investment capital into foreign mining operations. Germany's drive for colonies in Africa proved disappointing for lack of mining possibilities, but German engineers found other opportunities in British, Dutch, and French possessions abroad. Major tungsten deposits in Malaysia, Australia, and Portugal came under German financial control. In the late nineteenth century, as industrialization advanced in the Rhineland and Silesia with the construction of blast furnaces, smelters, refineries, puddling mills, foundries, and the railways to connect them, metals brokers entered the tungsten trade in force. Looking to supply the domestic steel industry as well as serve the international market, dealers in Frankfurt, Hamburg, Nuremberg, Cologne, and other trading centers bought foreign tungsten concentrates and shipped them to the Fatherland for further processing and sale. Much of the final product, mostly in the form of ferrous tungsten and tungsten powder, went into domestic commodities, but foreign demand steadily increased in the prewar years. Sheffield steelmakers grew so dependent on German tungsten that by 1914 German traders were supplying up to 90 percent of Britain's tungsten needs.[30]

American industry was less dependent on foreign tungsten but still linked to the European market. World production averaged 5,759 tons of concentrate between 1905 and 1914, with American mines contributing almost 20 percent of that total each year. Prices from 1909 to 1914 averaged around eight dollars per unit, defined as 20 pounds of concentrate containing at least 60 percent tungsten trioxide (WO_3). Most domestic ore came from Colorado, where in 1904 the Wolf Tongue Mining Company, a subsidiary of Firth-Sterling Steel, started selectively mining high-grade ferberite at its Cold Springs mine in the Nederlands district near Boulder. By 1915 California took the lead, primarily by exploiting a high-grade scheelite deposit at Randsburg, discovered in 1905 and developed by the Atolia Mining Company. Both of these American firms hand-sorted ore before crushing and used conventional gravity-flow separating equipment to concentrate the metal values. Domestic mines supplied about half of the tungsten needed by American smelters, mills, and factories in the prewar period. The rest came from abroad, often through the marketing facilities of German traders tapping into British sources of supply.[31]

No less obvious in the prewar geopolitics of tungsten was the discrepancy between American and European trade policies. Since the mid-nineteenth century Western Europe had embraced economic development and free trade as solutions to problems inherent in a growing population with a limited continental resource base. German imperial policies injected a protectionist countertheme after 1879, especially in agriculture, but not enough to seriously erode the commercial interdependence of the industrialized powers up to World War I. Germany and France protected their ferroalloy steel manufacturers, for example, but Britain did not, and crude-ore imports were duty free throughout Europe. By 1914 imperial Germany and imperial Britain alone commanded almost 40 percent of the world's commerce, and much of that between themselves. Nearly one-fifth of German industry's prewar raw materials came from British imperial sources.[32]

In contrast, the United States during the same period turned decidedly protectionist, a trend inaugurated by the Whig-Republican coalition during the Civil War and reinforced by the Republican business leadership. Steel industry protection began with the Morrill Tariff of 1861 and continued long beyond the formative years of industrial development. Despite some downward adjustment of rates on steel rail during the two Cleveland administrations, protectionists dominated Congress and the presidency through much of the 1890s and into the new century.[33]

The domestic infighting that accompanied the American tariff debates revealed still other geopolitical forces that complicated the political economy of tungsten both at home and abroad. Independent companies and specialized steel producers faced increasing competition from the "steel trust" following the formation of United States Steel in 1901. Big combinations had the economic power to undercut smaller companies as well as the political leverage to elicit subsidies from Congress. As one independent factory owner bitterly complained after reviewing the impact of the Dingley Tariff of 1897, the artificially high floor on base steel prices gave trusts and combinations a 25 percent to 50 percent higher profit margin. He felt abused by a system of "tariff graft," which forced up prices all down the line from producer to consumer, with the biggest profits going to the largest combinations. His was a lonely voice among steel men, however. After the panic of 1907 and a long period of depressed business conditions nationwide, some steelmakers warned that abandoning high tariff rates would force many independents into bankruptcy.[34]

Business pressure and a stale economy prompted Republican "Standpatters" to insert a protectionist plank in the 1908 party platform, which claimed that the "true principle" of protection was to impose duties that reflected the differential between the domestic and foreign costs of production, plus a "reasonable profit" for home industry. The following year President Taft signed the Payne-Aldrich Tariff, touted as tariff "reform" but with few concessions to consumers. This was the first American legislation to impose ad valorem duties on steel alloys, including a 10 percent tax on imported tungsten ore and 20 percent on ferrotungsten. Four years later a bipartisan progressive coalition passed the more modest Underwood Tariff. It managed to lower some duties and eliminate others, including the tax on crude ores. The reductions mollified smelter companies, refineries, and other ore importers but alarmed domestic producers. They were the protagonists in the turbulent postwar years when protectionism reached new heights.[35]

The economic history of tungsten is inseparable from the history of modern steel development both at home and abroad. As the commercial use of low-carbon steel alloys expanded and their industrial importance grew after 1900, tungsten became a valuable commodity in the international metals trade. But tungsten's scarcity and unpredictable deposition around the world complicated its utility and distribution. Patent suits over tungsten processes or formulas revealed the intensity of economic competition and the monopolistic trend of big business in the industrial age. Tariff debates exposed the differences between competing interest groups, and drove a wedge between producer and consumer. As national rivalries intensified, the discrepancies between the production and distribution of scarce metals like tungsten became geopolitical talking points with significant implications for military strategy and government policy. The full impact of these complications, however, did not become apparent until the age of innocence ended with the coming of world war.

2

TUNGSTEN IN WORLD WAR I

In the two decades before the First World War, steel alloys had revolutionized the machine tool industry and altered the course of modern industrial development. After 1914 steel alloys changed the course of war. Copper, lead, zinc, aluminum, iron, and other metals were all enlisted as never before in the global struggle that followed the Sarajevo crisis, but to many contemporary observers the "key metal" of World War I was tungsten. From 1914 through the cold war, the enormous military demands for heavy equipment, tough steel tools and weapons, armor plate, and heat-resistant components of military hardware made tungsten one of the world's strategic metals, an element with political as well as economic importance.

THE ARMS RACE

Technological innovation is a commonplace feature of war, but it is easy to exaggerate the consequences. New military prototypes are often failures, and even the successes are long in the making. In World War I, the "critical technologies," as Alex Roland has argued, were not airplanes or tanks or other innovations, but older weapons that took years to perfect and fully utilize. Breech-loading rifles, cannons and machine guns, artillery shells, and even submarines, along with the technical and industrial infrastructure necessary to design, fabricate, test, and improve them, had been in various stages of development at least since the Crimean War. Other military historians give less importance to weaponry "as a shaping force" in war and more to nontechnical factors, including political decisions that influenced military strategy and tactics. But whether old or new, decisive or incidental, the killing machines of the First World War and the factories that made them depended on steel and steel alloys. Indeed, as some military historians have claimed, the rapid advance in steelmaking technology after 1880 can be explained largely by the demand for bigger, faster, and more durable armaments.[1]

As steel metallurgy advanced, so did experiments to improve arms and armor by using special alloys made of nickel, manganese, chrome, tungsten, and other metals. Silicon steel was used in the production of shrapnel bars. Tungsten's strength and hardness under extreme temperatures proved essential in the manufacture of airplane engine valves and other parts, both military and domestic. Long before 1914 nickel steel had replaced bronze, wrought iron, and carbon steel in the construction of rifles and heavy cannon barrels, but even barrels made of modern steel alloys eroded quickly and lost accuracy under the tremendous heat of rapid fire. During the war, however, metallurgists found a solution by lining the barrels with special alloys using molybdenum or tungsten.[2]

British leadership in the commercial production of specialty steels gave Brown and Cammel, Vickers, Armstrong, Firth, and other firms a head start in the arms race that preceded the outbreak of hostilities in 1914. Britain was the largest exporter of weapons in the two decades prior to war. From shell casings and projectiles to heavy guns and gun steel, British firms kept ahead of their competitors in both volume of production and advances in design. Even the Krupp company in Germany, Britain's biggest competitor, modeled some weapons along British lines.[3]

To the public and their political leaders in the industrialized world, the most visible part of the arms race was not guns or projectiles but big naval vessels. The great wooden-hulled sailing "ships-of-the-line" that gave Britain command of the seas in the first half of the nineteenth century had by the 1880s given way to steel-hulled, steam-powered battleships with a top speed of sixteen knots or more, armed with four twelve-inch guns in two turrets fore and aft, and a variety of secondary guns amidships. With the largest navy and the most to lose in an imperial conflict, Britain remained the dominant sea power over the next twenty-five years by outspending and outpacing all other countries in the number, armament, and size of its capital ships.[4]

Although the basic design of naval vessels changed little before the dreadnought era, advances during the 1890s in the development and processing of steel alloys set the stage for a technological revolution in naval armament and weaponry. Armor plate made prior to that time varied widely in thickness, hardness, and tensile strength. To improve it was a major concern of naval officials and a challenge to leading steel manufacturers. German and British firms led the way with experiments using hardened nickel steel, tungsten steel, and other alloys, but even the best alloy steel contained impurities and blow-holes that degraded quality.

European alloy technology was still in transition when an American inventor patented a treatment for armor plate that profoundly changed the industry. Hayward Augustus Harvey had studied engineering and metallurgy while apprenticed as a young man in a Hudson River factory. After experimenting for years with methods to improve the quality of Bessemer steel, he discovered that supercarburizing armor plate by slowly baking it in a special charcoal bath greatly increased its hardness and durability. The process transformed the crystal lattice into a dense molecular structure, producing a steel that could not be penetrated by the most advanced high-explosive projectiles of the day. Standard steel plate on completed battleships could be case-hardened by the Harvey process, as Harvey demonstrated to astonished American naval officials in 1891. Ballistic trials that year proved the superiority of Harvey-treated plate over all others. With government approval and a license from Harvey's company, Bethlehem Iron and Carnegie Steel, the only two American plate manufacturers at the time, began making Harvey-process armor plate the next year.[5]

The British arms industry at first resisted this transatlantic development, claiming that Harvey's process was little more than cementation, protected by British patent. An infringement suit followed, but before it was concluded British field tests proved Harvey-treated plate superior to the Sheffield variety. British steelmakers conceded after Harvey's patent was upheld late in 1892. German tests the following year were equally impressive. Krupp, the leading continental plate maker, adapted Harvey's process to its own patented method of making case-hardened nickel steel plate. The result was a superior product, licensed under Harvey's London subsidiary as Harvey-Krupp armor plate. By the mid-1890s ten of the largest steelmakers in Germany, Britain, France, and the United States had pooled their patent information and were producing armor plate at a fixed price under license from the London Harvey Company. The formation of an international steel cartel, a "true armor trust," raised the cost of rearmament but also generated intense public criticism, leading to government investigations, antitrust litigation, and competitive underbidding by independents. Under the weight of adversity and rising nationalist furor the London Harvey Company dissolved in 1912. However, as historian Thomas J. Misa recently observed, the armor plate cartel set a pattern for pooling and price-fixing in the steel industry that carried into World War I and beyond.[6]

The Harvey cartel was riding high during the naval race that heated up after the turn of the century. German ship construction accelerated rapidly after 1897 when Prussian militarists placed Adm. Alfred von Tirpitz in command of the imperial navy. German arrogance worried American imperialists, especially Theodore Roosevelt and Henry Cabot Lodge. The United States had been building ships at a faster pace after the "splendid little war" with Spain in 1898 and the Philippine annexation that followed, but Lodge told his Senate colleagues that hemispheric defense was a more immediate concern in view of Germany's aggressive *Weltpolitik*. Britain had more cause for worry, but had too much at stake commercially to risk directly challenging German imperial policy before 1914. Better to keep the Royal Navy dominant on the high seas! Between 1902 and 1908, the British built capital ships nearly twice as fast as Germany and 25 percent faster than the United States.[7]

Though the three navies added 158 ships of all kinds in these years, only capital ships mounting big guns of a single caliber really counted in the arms race after a revolution in long-range firepower made smaller naval vessels obsolete. Progress in ship design had been under development for decades, led by advances in the application of steel alloys to naval guns and other weaponry. Alloys strengthened shell casings, gun tubes, rifling, and breech-loading mechanisms, giving rise to larger calibers, higher muzzle velocities, faster firing sequences, and much greater range. Harvey-process plate, tougher and lighter than its predecessors, made larger vessels with bigger guns manageable under extreme conditions. The British led the way with the first dreadnought in 1906, but with every big navy rearming with essentially the same systems and virtually the same armor plate, the naval race was stalemated technologically almost before it began.[8]

The intensity of the naval race and the global competition among leading arms dealers in the years leading to the First World War tend to obscure the underlying interdependence of Europe's industrial powers in the prewar years. Academics disagree as to the origins and extent of European economic integration prior to 1950, but the metals trade is indicative. Before the war, Germany was dependent on British imperial trade for raw materials, just as Britain was dependent on Germany for processed tungsten, molybdenum, nickel, optical glass, chemicals, and electrical equipment. The outbreak of hostilities in 1914 disrupted an international stream of commerce that some regard as a natural development of modern European industrialization.[9]

WARTIME METALS SHORTAGES AND STRATEGIES

Awakened to its strategic vulnerability by the guns of August, Britain rushed to make amends. To prevent war material reaching Germany or its allies, the war office immediately froze German assets and prohibited all trading with the enemy. It slapped an embargo on all ore exports from British colonies and dependencies, reinforced by a naval blockade of likely routes to the Central powers. It also financed the construction of concentrators and refineries to handle raw ore imports and processing requirements. To offset steel shortages and stimulate production at home, the government prohibited iron and steel exports, and expanded metal imports from Canada and the United States. It placed restrictions on military conscription of metalworkers, financed domestic plant expansion, established an allocation system between military and domestic producers, and encouraged the use of lower-grade ores and steel scrap. It also created a new agency to encourage scientific research and development. In the meantime, warning of the "German octopus," British business interests and their conservative allies persuaded Parliament to pass legislation restricting metals trading to firms licensed by the government. The effect was to break the German commercial ties to British industry and finance, limit the competition, help bring prices under control, and place government strategic metals policy and practice in the hands of a newly formed state-sponsored agency, the British Metal Corporation. By the time the United States entered the war, Britain had imposed price ceilings on almost all steel products, alloys, and raw materials, including tungsten, but subsidized some war industries to make up for rising production costs.[10]

Germany was better prepared for war than Britain, but not for a long conflict. Anticipating shortages, German munitions makers in the prewar years had stockpiled vanadium and other critical materials. Despite the blockade in 1914, some material continued to come in from abroad, but how much and by what means are largely unknown. Throughout the war German industry received some shipments of Scandinavian nickel and aluminum. French nickel crossed enemy lines until 1915, and at least one shipment of nickel reached Germany from the United States in 1916 via the "merchant submarine" *Deutschland.* In the first two years of war Portuguese tungsten reached Germany by overland routes, but eventually dried up under increased pressure on the Portuguese government from Britain, Portugal's largest trading partner.[11]

In 1918 A. Mitchell Palmer, alien-property custodian and Wilson's point man on internal security matters, doggedly pursued American firms with German connections, seizing their assets after alleging a vast enemy conspiracy to control "most of the principal metal and smelting companies of this country." Based on the publicity surrounding Palmer's war work, Charles Segerstrom later claimed that before America entered the war, German metal traders bought "every particle of tungsten" mined in the United States regardless of price, and either smuggled it into Germany or stored it to keep it from enemy hands. Hard evidence on smuggling is obviously scarce, but American counterespionage agents did report foiling one tungsten smuggling plot in 1917. The "perps" each paid a one thousand-dollar fine and went to jail for a year. Palmer's aggressive tactics pleased conservative activists and prepared the country for his "red scare" campaign later as attorney general. It also gave the domestic high-speed steel industry access to patented German metallurgical processes formerly kept secret.[12]

To supplement foreign ore sources the German arms industry exploited every possible domestic deposit, however sparse and expensive to develop. Miners worked low-grade nickel deposits in Silesia, bauxite in Austria-Hungary, and copper schist in the Erzgebirge range along the Czech border in Saxony. Others found some wolfram in old tailings in the Erzgebirge, and a little vanadium in slag around old blast furnaces. By 1917 both sides were using metal scrap of every kind to meet growing shortages.[13]

THE METALLURGY OF WAR

For large industrial applications, standard carbon steelmaking after the 1880s rapidly advanced from slower puddling and crucible methods to bulk processing in blast or open-hearth furnaces. Until the second decade of the twentieth century, however, special steel alloys were still made in small batches primarily by the crucible process despite the advent of improved electric refining methods nearly forty years before. The electric furnace ultimately dominated the refining and fusion of various elements and molecular compounds to make steel alloys, but converting from older to newer metallurgical technology was a slow process before World War I.

In 1886, almost simultaneously, Paul Héroult in France and C. M. Hall in Ohio invented the electrolytic process for the production of metallic aluminum from calcined and chemically treated bauxite. Soon metallurgists in France, England, Germany, and the United States were experimenting with electric induction or arc furnaces to smelt and refine other metals. In Sheffield, the long-standing leader in high-speed steel production, experiments using electric furnaces to reprocess steel scrap and ferrotungsten alloys showed promising results, but the lack of quality control and high prewar electricity costs slowed the adoption of new furnaces. Until the war emergency forced a change, most Sheffield firms continued to use their crucibles, both for reducing imported concentrates as well as for producing high-quality specialty steels.[14]

The insatiable wartime demand for metals required changes in production technology as well as composition of metallic compounds. Electric furnace use rose on both sides, in part due to the rapid expansion of electric generation and distribution systems and the corresponding decline of power costs. Electrolysis also proved more efficient than older methods for the reduction of metallic ores, as well as for the smelting and refining of steel scrap. Before 1914 most tungsten alloys were made either by adding ore directly to a furnace charge without prior reduction, leaving whatever impurities in the ore to fuse with the alloy, or by reducing the ore with carbon using lampblack in a graphite or ceramic crucible. The latter produced a "crude tungsten powder" useful in some applications but unsuitable for alloy steel of the highest quality. Improving quality required purifying the ore by chemical or electrolytic reduction prior to further use. Adding this step increased production costs but offered alloy steelmakers the option of using pure metallic tungsten, tungstic acid (H_2WO_4), or ferrotungsten, an iron-tungsten alloy. The first two were sold in powder form, the latter as a fused ingot or "button," crushed and sized to buyers' specifications. All were nearly free of unwanted impurities. By 1918 secondary ore treatment had become a universal step in the alloy steel business.[15]

Changes in the content of steel alloys also helped keep the belligerents in essential metals. Before 1914 German specialty steels contained up to 30 percent tungsten, along with small percentages of chromium, vanadium, and sometimes cobalt, but metallurgists found they could cut the amount of tungsten by half without losing essential qualities. Tests using molybdenum as a substitute for tungsten in some applications began at the same time, but the shortage of "moly" during the first years of the war was almost as bad as the early tungsten scarcity, and the extent to which the two metals could be interchanged was left for future investigation. By war's end both sides were making steel alloys with greatly reduced tungsten content. The industry standard for Type I high-speed steel, the grade used for most cutting tools, was "18-4-1"—18 percent tungsten, 4 percent chromium, and 1 percent vanadium added to the charge of pig iron.[16]

For new "carbon-free" ferroalloys and high-speed steels that required extreme heat under carefully controlled conditions, the electric furnace was the "ideal instrument." Germany helped offset the effects of embargo by turning increasingly after 1915 to electrolytic processes. In Canada and the United States, alloy steelmakers tapped into hydroelectric generators at Niagara Falls for their expanding power needs. And in Sheffield, as historian Geoffrey Tweedale has asserted, the "high frequency (or coreless induction) furnace could perform virtually all the functions of the crucible with greater flexibility and economy, and its introduction in the interwar period condemned the Huntsman process to extinction in both Sheffield and America."[17]

THE AMERICAN BUSINESS RESPONSE TO WAR

Once fashionable among popular accounts of the causes of war, the "merchants of death" hypothesis has long been discredited as an explanation for America's entry into World War I. However, it is clear that American business interests, despite initial concerns, saw the war as an opportunity. In the summer of 1914, industry was in recession. Many commodity prices had slumped from previous highs, and exports were falling after Europeans began dumping American assets, fearing a devaluation of the dollar following implementation of the new Federal Reserve system.[18]

As the world's largest producer of farm commodities and the third-largest arms manufacturer after Britain and Germany, the United States stood to benefit most from war orders. For the first six months of the war the belligerents relied on domestic supplies and reserves, but as the attrition rate escalated, both sides turned to neutral nations to meet growing shortages of food, fiber, and munitions. American farmers and merchants reaped enormous profits from rising commodity prices, soaring production, and expanding international trade. Americans embraced President Wilson's declaration of neutrality at first, but the mood shifted as the German offensive stormed through Belgium and Allied propaganda mounted. American loans kept the Allies buying American goods, and Allied war bonds floated freely in American markets. British control of the seas helped turn American sympathies, but unleashing the U-boat was the decisive factor in tying the United States to the Allied cause. Although both Britain and Germany violated American neutral rights, in President Wilson's words, "Germany had been brutal in taking life and England only in taking property."[19]

For metals companies and dealers, the outbreak of hostilities offered enormous risks as well as opportunities. The steel industry remained in economic doldrums until war orders began to pick up after the first six months of war. American firms lagged behind the Germans in technological development and the British in quality control. For tinplate as well as processed ores and metals for ferroalloys, U.S. specialty steel companies relied mostly on foreign imports. The British embargo made imports more costly and much more cumbersome because of the red tape involved in satisfying the British government that imported material would not end up in enemy hands.[20]

Opportunity beckoned by 1915 as steel prices rose, foreign orders increased, and domestic demand accelerated. As the war cut off steel imports from Britain, Belgium, and other foreign suppliers, domestic firms had all the orders they could handle. Maximizing profits, however, required keeping costs under control and finding more reliable sources of raw materials. Indeed, as Gavin Wright has noted, two fears invariably appear during times of international crisis: fear that vital resources will run short and fear that scarcity will cause inflation. Though economists deny the validity of both fears, uncertainty added a measure of risk to any business venture. It also fed speculation and contributed to the volatility of commodity prices.[21]

On Wall Street, speculators anticipated scarcity by bidding up metals prices after war began. In the first three years of fighting, steel and mercury prices rose 500 percent, zinc tripled, and copper doubled. In spite of the blockade, German dealers or agents were active bidders in the American market, and Germany continued receiving some American material via Scandinavia or other collateral shippers at least through 1916. To thwart the "Hun," British officials published a blacklist of firms or individuals in neutral countries suspected of "trading with the enemy," and stubbornly stuck to their principles despite protests from blacklisted Americans and objections from State Department officials. The Admiralty did, however, show some flexibility in applying the rules, delisting those erroneously included or otherwise cleared for trading. That gradually eased the tension, allowing the Wilson administration to quietly bury retaliatory legislation against British shipping that the president had signed but never implemented.[22]

Bad war news shook investor confidence and contributed to the volatility of metal prices. Early in 1915 prices zoomed after the German admiralty

proclaimed a submarine blockade of Britain. The Allies responded by announcing search-and-seizure rules for all ships and contraband bound directly for
Germany or indirectly through neutral ports. In one two-week period of tense
diplomacy in the summer of 1915, prices quoted for high-speed steel nearly
doubled. As submarine warfare took its toll on Allied shipping, tungsten rose
an astonishing 700 percent above prewar prices, dropping back later in the
year after ore imports eased domestic shortages. By early 1917 the unit price
was down to $20. Lead jumped 250 percent in the first year of war, and then
dropped 28 percent in the last two weeks of July 1915 after the German government, under pressure following the *Lusitania* sinking, ordered submarine
commanders not to torpedo passenger ships. These wild price swings and supply uncertainties inevitably meant higher prices for manufacturers as well as
for consumers. Ferrotungsten, quoted at 60 cents per pound in 1914, was up to
$2.50 by July 1915. Pig iron rose steadily, from $13.75 a ton in mid-1915 to $34.05
by the end of the war. Mercury, vital for high explosives, jumped more than 30
percent when the United States declared war, and stayed at that level or higher
until after the Armistice.[23]

Despite rising consumer prices and some commodity shortages, Americans
prospered during the war years. Steel producers had more war orders than
they could handle. Farmers sold everything they could grow. Employment
was up, and so were profits. The few voices of caution worried more about a
postwar recession than wartime inflation. In 1915, after a lengthy study concluded that foreign cartels were having an adverse impact on America's export
trade, the Federal Trade Commission (FTC) recommended relaxing antitrust
legislation against trade associations organized solely to promote commodity
exports. Three years later Congress incorporated that recommendation into
the Webb-Pomerene Bill. In the meantime, Judge Elbert Gary, chairman of
United States Steel and chief spokesman of the steel lobby, warned that the
future "may be very dark and desperate" after the war if former belligerents
started dumping goods on American markets. His remedy was a protective
tariff. With steel prices at record levels, Gary's warning seemed disingenuous
at best and unpatriotic at worst. U.S. Steel Corporation had used the same
tactic, "systematic dumping," to build European business after the Big Steel
merger in 1901. Free-trade Democrats suspected the steel industry of violating
antitrust laws and colluding to hike prices artificially, but they supported the
Webb-Pomerene Bill after the president endorsed it in 1918.[24]

Winning a narrow victory in 1916 for a second term, Woodrow Wilson desperately hoped for peace but could not find a way to avoid the inevitable break with Berlin. In the interim between his "peace without victory" appeal to the belligerents and the opening of American hostilities, war planning slowly moved forward. Late in 1916 Congress established the Council of National Defense, consisting of six cabinet officers and an Advisory Commission of seven civilians. For advice on raw materials, minerals, and metals, the president turned to a Wall Street financier whose party affiliations belied his business interests. Bernard Baruch, a progressive Democrat from South Carolina, had amassed a fortune before the war from investments in the mining and processing of gold, copper, sulfur, and other metals and minerals. He knew tungsten well, having joined Frederick Worthen Bradley in financing the development of the Atolia mine in California. Like his younger contemporary Herbert Hoover, Baruch was a civilian volunteer, a "dollar-a-year man" in the Wilson administration, enlisting in public service after a spectacular career in private industry and finance. Though wearing different party labels, they held similar business views. Both championed the American system of free enterprise, with government encouraging rather than regimenting business activity. Both opposed antitrust laws and free traders, emphasizing business cooperation and efficiency instead of competition. When the war came, both accepted the need for centralized planning, and both enlisted businessmen as volunteers to help mobilize industry for the duration.[25]

Baruch was an outspoken advocate for war preparation. He directed an Advisory Commission study of industry's need for raw materials, minerals, and metals, and worked with chemical and copper executives to keep a lid on prices despite inflationary fires elsewhere. Judge Gary, however, distrusted progressive reformers and resisted patriotic appeals on behalf of national defense. Big Steel opposed piecemeal voluntary price reductions and rejected the navy's effort to buy armored plate at prewar prices. Instead, the industry offered a uniform price for armored plate high enough to ensure profits for all producers. Such arrogance outraged the president, who ordered the Federal Trade Commission to investigate steel prices.[26]

In the midst of this steel controversy the United States entered the conflict, and the government asserted sweeping emergency powers over industry, agriculture, transportation, labor, consumer affairs—practically every aspect of American economic life. In practice, however, as Baruch's role in dealing with Big Steel illustrates, centralized control was more symbolic than real. Neither the Wilson administration nor the American public wanted drastic structural changes in the economic system, even in wartime. Given greater responsibilities over the production and pricing of strategic materials after the Council on National Defense organized the War Industries Board in July 1917, Baruch played on industry fears of government intervention to renegotiate with steel executives. In September 1917, with FTC evidence of enormous steel profits in hand, the WIB and Big Steel reached an agreement that even Judge Gary supported. It capped steel prices without drastically cutting the profit incentive. With some mutually satisfactory adjustments after Baruch became chairman of the WIB in March 1918, the steel agreement lasted throughout the war. Though feared by some as an "economic dictator," Baruch demonstrated that wartime mobilization worked best through voluntary cooperation and compromise.[27]

THE DOMESTIC TUNGSTEN MINING BOOM

For domestic metal producers, the war boom furnished a rare but risky opportunity. With foreign metal supplies cut off or severely limited, with demand increasing and market prices zooming, any mine or prospect on American soil that might contain strategic metals seemed ripe for development. Old mines reopened, and existing mines expanded operations. Speculators poured money into new promotions despite the risks, as if trying to revive the glory days of America's mining heritage. Between 1910 and 1920 the number of domestic miners in the United States increased by 112,000. Almost all worked for wages in coal or base-metal mines, however. The glory days of gold mining had long passed.[28]

Though not classified as a "precious" metal, tungsten was the hottest commodity on the metals markets during the war. So feverish was the speculative interest that "everyone who had a tungsten supply thought he would eventually become a millionaire," writes military historian Stuart Brandes. In Colorado, the national leader of tungsten production at the beginning of the war, thousands flocked to Nederland in Boulder County. Established companies like the Wolf Tongue and Vascas mines hired some new employees, but most individuals worked alone or in small groups as in Gold Rush days, staking claims, trenching and sampling the ground near older ferberite deposits, hoping for rich new strikes. The Nederland district, known as a "poor man's camp," paid unskilled "diggers" an average daily wage of three dollars, and five dollars for experienced miners. By 1916 some 15,000 men were at work there. Prospectors also searched the San Juan Mountains in the southwestern corner of the state and reported good strikes, although production was very limited.[29]

Farther west, tungsten fever struck eastern California about the same time as Colorado. In the northwest corner of San Bernardino County, just east of the colorful old gold camp of Randsburg, the Atolia mine had been producing scheelite since 1905 from both alluvial deposits and shallow, high-grade veins in a faulted and fractured mineralized zone. Financed in part by Bernard Baruch and operated by the Atkins-Kroll Company of San Francisco, Atolia expanded operations in the boom years after 1915 but had to contend with a rush of newcomers. Most worked placer claims in the "Spud Patch" adjoining Atolia ground where scheelite "nuggets" resembling potatoes could be recovered by traditional gravity separation machines and methods. However, the Atolia Company controlled the best deposits and produced 95 percent of the district's tungsten total.[30]

Until the United States entered the war, Atolia's biggest customer was Krupp Arms—a fact that immediately aroused the suspicions of U.S. Attorney General A. Mitchell Palmer. The Justice Department launched a full investigation, but turned up nothing that suggested either collusion or conspiracy. As Glenn Emminger later remembered, Dave Atkins was "a cockney in speech and manners and Kroll would never be mistaken for a German."[31]

One hundred fifty miles north of Atolia, in the Tungsten Hills a few miles northwest of Bishop in Inyo County, new underground tungsten operations began in April 1916. The ore was scheelite, discovered in 1913 in a skarn zone, part of a fifteen-mile mineralized belt that yielded 2 percent WO_3 to the ton on average. After two months of frenetic work while tungsten prices were highest, the Tungsten Mines Company opened the deposit, completed roads to the railhead at Laws, brought in electric power, built a mill, and started production. Several other claimants opened adjacent mines at about the same time. Eventually, the most promising properties were acquired and consolidated by the U.S. Vanadium Corporation, later taken over by Union Carbide. Together with Atolia and other smaller operations, the output of the Tungsten Hills by 1917 made California the nation's leading tungsten producer.[32]

The tungsten bonanza stimulated mining activity across the West. With prices at "undreamed heights," the Bureau of Mines (BOM) in 1915 reported new workings in New Mexico, Arizona, South Dakota, Idaho, Alaska, and Washington. Sustained demand kept promoters and operators busy for the next two years. Across the country more than forty mills were operating by

1916, but investors often made dubious choices, as federal geologist Frank Hess reported: "Unfortunately, some [mills] had no ore on which to work, having been erected in advance of prospecting." High prices also tested the honesty of some workers, especially where unguarded ore lay about in stockpiles, ore cars, or shallow placer deposits. In the Atolia district, high-grade scheelite spuds brought good prices on the black market. One mine in Arizona was rumored to have had more ore stolen than sold. Despite thievery and financial boondoggles, the United States produced more tungsten than it used domestically in 1916 and more than a fifth of the world's supply the following year.[33]

Although the government negotiated with industry leaders to control the price of processed metals and many other commodities during the war years, the War Industries Board avoided mandatory price-fixing, and intervened only when government purchases were large enough to affect the market. Tungsten prices varied depending on the needs of individual buyers and sellers, the type of ore offered for sale, the amount of impurities it contained, and the prevailing market conditions at the time.[34]

Whether marketed indirectly through metal brokers or directly to smelter operators and other secondary processors, concentrates had to be reduced to metallic powder or ferrotungsten before they could be used in steel alloys. By 1915 ferrotungsten was the preferred choice of most steelmakers. Not only was it difficult to find high-quality tungsten powder during the war years, but the expanded use of electric furnaces made ferrotungsten easier to produce, if not cheaper. To avoid brokerage charges and other processing and handling fees, as well as to increase efficiency and throughput, many wartime high-speed steelmakers installed their own electric furnaces. The biggest names in high-speed steel during the period, all with electric furnaces, were independents operating in niche markets.[35]

THE FORMATIVE YEARS OF PACIFIC TUNGSTEN

Domestic tungsten bragging rights would eventually pass to Nevada, but significant production in the Silver State took a long time to develop. As early as 1865 the first recorded discovery of "tungstate of manganese," or hübnerite, came from the Mammoth (Ellsworth) district in Nye County. Nevada's main tungsten ore, however, is scheelite, first found in a Pershing County skarn deposit in 1907. In the following years prospectors located scheelite in dozens of other skarns and vein deposits, but only a few were commercially valuable, and production before 1917 was very limited.[36]

Nevada's tungsten industry took a significant step forward in 1917, when several new mines opened on what was destined to become one of the most productive scheelite deposits in the nation. Three years earlier, Emil Stank and a small party of prospectors out of Lovelock had explored the eastern slopes of the Eugene Mountains in Pershing County. About eight miles northwest of Mill City, an old mining town with an erratic record of silver and copper production, they panned some samples from an outcrop about 150 feet long in highly fractured ground on top of a hill. They were looking for signs of silver, but instead found some "heavy brownish-white sand" that they did not recognize. Leaving the ground unstaked, they packed a few specimens and decided to look elsewhere. After the tungsten boom and just before America's entry into the war, Stank took his specimens to Reno, where a geologist at the University of Nevada confirmed the presence of scheelite. In March 1917 he and some friends hurried back to file claims on "Stank Hill." The news prompted a typical rush, and claims soon blanketed the area.[37]

Mining began in earnest after the weather warmed in the late spring and summer of 1917. Though many claims had been staked, four principal groups emerged out of the scramble with enough investment capital to get under way. First in the field was Thomas Sutton, a Nevada miner who organized the Mill City Tungsten Mining Company and began work on Sutton No. 1 early in March. A half-mile west was Stank No. 1, named for the district discoverer who had leased his claims to a second investment group headed by George H. Copley and Harry E. Baker. On Springer Hill to the northeast the Copley group controlled five other claims, including a good prospect known as the Springer lease, named after Harry Springer, a well-known Nevada prospector. Less than a month after start-up Copley and Baker optioned their properties to E. S. Shanklin and Ira J. Coe, two San Francisco promoters with plenty of cash and connections. Hoping to tie up the best properties and sell them later for a good profit, Shanklin and Coe by April had at least thirty-three claims, options, or leases in their pocket. The third group of six claims, led by Thomas Durnen and John McCloskey, opened a shaft in March and completed about 500 feet of development work before Shanklin and Coe gave them an option they could not refuse. Farther north on Humboldt Hill, a fourth group staked claims and organized the Nevada Humboldt Mining Corporation under L. A. Friedman. A successful silver miner from nearby Rochester, Friedman had enough capital to ignore the blandishments of Shanklin and Coe. East of Stank and Springer hills, in a narrow valley accessible by a wagon road from

Mill City, lay the little mining community first known as Camp Sutton but eventually renamed Tungsten. It was hardly more than a store and a series of cabins to house mine workers.[38]

Before construction of a local mill, processing ore mined from all four operations was inefficient, slow, and expensive. Teamsters loaded raw ore on wagons and hauled it to Mill City, where it was reloaded on Southern Pacific railcars and shipped more than 50 miles to Lovelock. There it was unloaded and hauled to a custom mill at Toulon for concentrating. The process took weeks and cost mine operators nearly a third of their revenue from ore sales. At that rate only the high-grade ore was mined. Nearly all the concentrates were shipped in 120-pound sacks to the Atolia Mining Company in San Francisco, which agreed to distribute Nevada's smaller monthly production along with its own Mojave tonnage. Loaded railcars from San Francisco were consigned to Charles Hardy, a New York dealer in both domestic and foreign ores whose business suffered little from being blacklisted temporarily by the British in 1916. He found a ready market among specialty steelmakers for high-quality ore.[39]

"Picking the eyes out," as critics called corporate highgrading, was an inefficient practice, condemned by most professionals because it left behind low-grade ore that could no longer be economically mined. But wartime was no time for rational mine development. The government needed arms, and industry needed tungsten to produce them. Mining was a risky business, and patriotism alone could not produce ore. Entrepreneurs might fly the flag, figuratively or literally, to encourage congressional subsidies or public support, but the real motive was profit. The higher the market price for tungsten, the faster ore came out of the mine.

Despite the rush of expectations, mining took time and patience in a remote area under wartime conditions, with shortages of equipment, supplies, and men. Regular ore shipments from the Mill City district did not begin until November 1917. By the spring of the following year the mines had produced 1,000 tons of high-grade ore at $60 per ton. Subtracting operating costs of $22.40 per ton still left a substantial profit. Hoping for more while metal prices were high, Friedman and Sutton invested their capital in plant expansion and development, but Shanklin and his partners decided it was time to fish in bigger waters.[40]

They found a good prospect early in February 1918. Shanklin and Coe's main office was located in the San Francisco financial district, home of one of the oldest mining stock exchanges in the country, notorious for heady speculation and promotional frauds in Nevada boom days on the Comstock and later at Goldfield, but fairly quiet after 1906 until the wartime surge in metals. The financial district was also headquarters of some of the most prominent mining companies and engineers on the West Coast. Among these high flyers of mining finance was William J. Loring, a self-taught engineer who had risen from humble beginnings to international acclaim in mining circles. His reputation rested on his work as a gold mine manager, first at the famed Sons of Gwalia in western Australia under the tutelage of Herbert Hoover, then as Hoover's successor in Bewick Moreing Company, a British-based engineering and management firm, and finally as general manager for the Plymouth, Dutch-App, and Carson Hill gold mines, all in California's Mother Lode. When the two promoters caught up with him Loring was still at the height of his career, living in a plush suite of rooms in downtown San Francisco, borrowing and spending freely, investing heavily in mining properties and looking for more. Later he admitted that he knew nothing about tungsten when Shanklin came calling about some "wonderful tungsten claims in Nevada," but ignorance was no reason to miss out on the "present magnificent market."[41]

The strength of the market depended on the quoted price at the time of sale. For most metals, prices had stabilized somewhat after 1915 as supply began to catch up with demand. However, tungsten remained scarce—and the price volatile—through the first half of 1916, largely due to the effects of the British embargo. Spot prices during the period jumped briefly above $100 per unit before dropping to $80 or less. The scarcity prompted Crucible Steel, the nation's largest high-speed steel producer, to stop taking foreign orders until the shortage was relieved. Later that year, increased ore imports from Mexico and South America, combined with rising domestic production, eased shortages and lowered prices. Tungsten dropped below $20 per unit in 1917 despite America's entry into the war, but late that year began rising again as orders increased for arms and armor. By the time Shanklin decided to sell, the spot price was floating between $20 and $23 per unit.[42]

The deal Shanklin and Coe offered was an option on the Durnen group of claims for $40,000, provided Loring accepted within three days, paying $2,000 cash up front and the balance in six months. Shanklin knew his man. A more reflective investor might have balked at the short notice, but Loring thought enough of his own abilities to join the tungsten boom without falling victim to the financial pitfalls that had brought down so many others in the mining business. After a quick phone call to line up financial backing, he grabbed a seat on the first train headed east, rushed out to Mill City for a quick inspection, rushed back to San Francisco, and closed the deal.[43]

The phone call was to Charles H. Segerstrom, a banker and businessman from Sonora, California, in the heart of the southern Mother Lode. Born in Sweden and raised in Minnesota, Segerstrom migrated to California in the 1890s, earned a law degree, and married the daughter of a prominent Mother Lode mining investor. During the "golden age" of Mother Lode mining before World War I, he found more opportunities in banking, mining, lumber, hotel investments, and property acquisitions than legal work. By the time America entered the war he was head of a title company and cashier of the First National Bank of Sonora.[44]

A voracious reader with a nose for business, Segerstrom subscribed to dozens of newspapers and trade journals, and over the years built an extensive personal library. As the war progressed he was well aware of tungsten's growing importance and the government's strategic interests. He also trusted Loring's reputation and business instincts, and had hired him as general manager of the Dutch-App gold mine partly financed by Segerstrom's bank. When Loring phoned with a hurried offer to join in a deal for some "wonderful tungsten claims in Nevada," he did not hesitate.

Both Loring and Segerstrom later attributed their sudden interest in tungsten to the government's call for more domestic metal production, but the Shanklin deal was concluded before a federal stimulus package had been approved and announced. Rumors of possible subsidies had surfaced in 1917, however, and were hinted at in Secretary of the Interior Franklin K. Lane's annual report when he referred to the Bureau of Mines' investigation of "methods for increasing available supplies of necessary minerals and metals hitherto largely imported." By the time Loring was ready to sign the Mill City contract, Lane's hint had grown into a proposal to Congress, announced late in February 1918. In a *Wall Street Journal* article cited by Loring in his appeal for federal relief after the war, Lane described ten "vital war minerals," including tungsten, that American mines could supply "if they are given the proper opportunity and assistance by the Government."[45]

Whether the government would subsidize the domestic mining industry depended on the strength and speed of the national security lobby in Congress. As chairman of the War Industries Board, early in 1918 Bernard Baruch recommended a $50 million metals appropriation bill. Supported by his allies on the House Mines Committee, the measure authorized the president to purchase strategic minerals and metals over a two-year period, and gave him sweeping power to fix metals prices, adjust tariff duties, even take over and operate mines if necessary for the public good. Forsaking the marketplace entirely was too bold for most congressmen, however. Instead of price-fixing, the House version authorized the Department of the Interior to control prices by revoking the license of any producer of war material who tried to make an "unjust or unreasonable profit." Shorn of its price-fixing clause, the bill passed both houses and was signed into law on October 5. In effect, it was the nation's first attempt to create a strategic stockpile. Before it could be implemented, however, the war was over.[46]

Congress was just beginning to debate the mineral control bill in 1918 when Loring and Segerstrom accepted the Shanklin offer. Having joined the profit scramble, they could not afford to wait for news from Washington. Loring took immediate charge, relying on his reputation to corral the resources needed to begin operations. To confirm his own quick field observations and build financial credibility he hired a reputable mining geologist, Oscar H. Hershey, to investigate the properties and write a report. He also found a local superintendent, C. W. Terry, an engineer who had been recommended to him by a neighboring mine owner.[47]

Financing the operation was much more complicated, however. Loring and Segerstrom invested $10,000 of their own money securing the Shanklin-Coe option, but that was only a drop in the bucket. To spread the liability and promote fund-raising, early in March Loring organized the Pacific Tungsten Company. Incorporated under Nevada law, capitalized initially with 125,000 shares at $1 each par value, PTC was a typical start-up, more show than substance. The five directors were token officials who purchased just enough heavily discounted shares to qualify. Loring held 30,000 shares as trustee, and the rest he and Segerstrom split between them as collateral for their initial investments.[48]

The first real operating capital came from Edward A. Clark, a Boston financier who was already committed to several California gold mining operations under Loring's management. On Loring's cable, Clark promised $50,000 spread over several installments, but upped the ante considerably when Loring came to Boston late in April to talk things over. Market conditions demanded quick action, but Pacific Tungsten would go nowhere without a vast infusion of capital to tie up all the major claims, rush ahead with mine development and production, and build a sizable mill on site to process ore. With Clark's influence and the hired help of a Washington, D.C., lobbyist, Loring sought approval from the Capital Issues Committee of the War Industries Board to increase PTC capitalization to $1 million and market shares publicly. The company did not want direct government aid, Loring wrote disarmingly, but "merely asks to be afforded the opportunity to find the money itself." In a quick show of support to industry, the government approved the request that same day. Upon hearing the good news the old PTC directors immediately reorganized and recapitalized, installed Loring as president and general manager, made Segerstrom vice president and treasurer, and issued 125,000 shares to Clark. The three major stakeholders now held 75 percent of outstanding stock. Before he returned home, Loring signed contracts with several eastern brokerages to peddle the rest on the New York curb market.[49]

Raising money and making it were two different propositions. Loring was confident the Mill City investment "will make money for us there can be no question," he wrote Segerstrom, "unless tungsten goes to the Devil altogether." In a hot market they could make money two ways, by selling ore and by selling stock. One fed off the other. "A market will be made in the shares of the company . . . [but] the best way to keep the market alive is to produce payable ore." It all boiled down to a matter of production, getting the most ore out of the deposit with the greatest possible speed.[50]

To ensure that Pacific Tungsten controlled the deposit, Loring began consolidating as soon as he had the financial resources in place or at least ensured. He needed a lot more than he had raised thus far, but two Boston banks, the First National and the National Shawmut, obliged by loaning him $160,000, largely on the strength of Clark's connections. The money Loring tossed around was a quick windfall to prospectors and speculators alike. His effort to soften up the market by trying to spread local rumors that tungsten prices were falling fooled no one. For $20,000 and an option to buy, he took over Shanklin's control of the Stank-Forge lease. Another $75,000 went to Shanklin and Coe to acquire the Copley-Baker group. Buying half of the outstanding shares of Sutton's Mill City Tungsten Company and taking an option on the rest cost him $125,000. He exercised the option two months later, absorbing Sutton's operation into Pacific Tungsten. By the time Loring was through, PTC controlled 2,000 acres and had added at least fifty more tungsten claims in the vicinity "at prices varying from $1,000 to $15,000." The only holdout was Friedman's Nevada Humboldt Corporation. It remained financially independent until 1924, when PTC took over all its assets and liabilities, long after both had shut down.[51]

Speed was the prevailing theme in Pacific Tungsten operations, but how it was applied revealed significant differences in management style. The bottom line for Loring, the peripatetic engineer and capitalist, was quick development and a fast payoff. Expecting "handsome profits, as much as $30,000 or $40,000 monthly within the next two or three months," he prodded Terry, the mine superintendent, to hire all the skilled labor in the area, select the best ore available, and push production as much as possible. In the meantime use good judgment, he said, and "do not feel that we are going to put you in jail if you make a mistake." He reinforced the point to his banker friend in Sonora: "I can tell you that this business will require pounding on the back if we are going to get any action out of it and make the money that is staring us in the face." But Terry was not the go-getter his boss demanded. After five months Loring fired him and put in his place the mill superintendent, John Henry Bell, a mining engineer from Stanford. To Segerstrom, who assumed general management responsibilities when Loring was absent, speed was important but conditioned by cost, circumstances, and long-range consequences. Rapid development must not break the bank or ruin the mine.[52]

The new mill was a case in point. Loring and the impatient Clark, expecting stock sales to provide capital for improvements and the marketing of concentrates to far exceed operating expenses, induced PTC directors to press ahead with construction of a new one hundred-ton-per-day mill regardless of costs. Segerstrom was more expedient. He told the site manager not to bother with concrete foundations, but "set . . . [the floor joists] on wood sills so as to get immediate action, and we can arrange to set them on a substantial base at a later date if we desire." The Sonora banker was also realistic, hoping for quick construction but prepared for at least a four-month building schedule. Loring wanted it completed in ninety days and was furious over unexpected delays. But this was not 1916, when the Tungsten Mines Company at Bishop could put in a new mill in two months. With a war on there were endless impediments despite PTC's priority authorization from the War Industries Board. Even appealing to patriotism could not advance the work any faster, as Loring found out in dealing with local labor leaders. He finally hired a detective agency to root out any Wobblies or other radicals, but ran out of money before company spies found any agitators. PTC problems were more financial, logistical, and technical than work related. Under the circumstances it took ten months to get the mill fully operational. In the meantime Pacific Tungsten shipped seven thousand tons of high-grade scheelite ore to the Toulon mill for concentration, then on to Charles Hardy in New York for sale. Loring thought it best to use just one broker rather than "mess around all over the country trying to find buyers."[53]

Across the canyon from Pacific Tungsten, the Nevada Humboldt Corporation had similar ambitions but also similar problems. Friedman started construction of his own one hundred-ton plant about the same time as Loring's, but neither could afford to operate them without sharing water and power facilities. The result was a joint subsidiary, the Mill City Development Company, by which each company agreed to split the costs of building a thirty-three-mile electric power line from Rochester and a pipeline with two pumping stations to carry water seven miles from the Humboldt River and lift it eleven hundred feet to the mills. State and federal officials cheerfully acquiesced in permitting these developments, for national security trumped all other considerations. Loring had even asked the Bureau of Mines to study the possibilities of "lower[ing] the level of Lake Tahoe" in order to generate hydroelectric power for the mills, but the Rochester line proved more cost efficient. Environmental consequences notwithstanding, completing the infrastructure and financing all the labor, machinery, and supplies for two separate plants took much more time and money than either company had anticipated. Both were staggering under heavy debt loads when their mills finally started up within days of each other. Unfortunately, it was now November, and the Armistice had changed everything.[54]

WAR'S END AND THE MARKET COLLAPSE

For the American tungsten industry, 1918 was a landmark year that started well and ended in disaster. Just six months earlier both producers and manufacturers had anticipated "the golden possibilities of war orders," but when the fighting suddenly ended, so did the orders. The War Production Board (WPB) immediately canceled munitions contracts, and steel mills cut back on ore delivery schedules. The collapse of demand stopped the high-flying tungsten producers like a rifle bullet striking Harvey armor plate.[55]

On the supply side domestic prospects were even worse, thanks primarily to China's looming presence in the international tungsten trade. K. C. Li, a metallurgist and trader in the interwar years for the Wah Chang Company in New York, claimed credit for discovering high-grade placer deposits in southwestern China in 1911. The surface geology there resembled wolfram deposition he had studied in Cornwall while a student at the Royal School of

Mines. Handpicked and sorted by farmers and day laborers from shallow alluvial zones, Chinese ore flooded the American market by the summer of 1918. It sold for about one-fourth the going domestic rate but still higher than the price ceilings imposed by Allied governments on tungsten and other strategic metals. Month after month Chinese brokers dumped seven-dollar ore in the States until stockpiles ballooned and the domestic market shriveled. American producers, beginning a theme heard frequently over the next quarter century, cast aspersions at "coolie labor" and demanded protection. They were still crying when the war ended, and war-related industries virtually shut down overnight. With peacetime tungsten consumption less than 40 percent of war-time use, with a two-year supply of WO_3 on hand at war's end and still more coming from China, and with postwar producers facing rising costs for equipment, transport, and supplies, the domestic tungsten industry collapsed.[56]

Despite these homeland difficulties, some optimistic forecasters looked abroad. Postwar Europe seemed to offer a window of opportunity to expand American foreign trade. Prostrate after four years of horrendous slaughter and devastation, Europeans needed the kinds of commodities Americans could best produce—textiles, foodstuffs, chemicals, machinery, tools, and metals. But the window—or to use America's favorite Asian trade metaphor, the open door—would not remain open for long. American businesses seemed all too vulnerable to foreign trade discrimination once European agriculture and industry had been rehabilitated and the foreign cartels restored.[57]

Even if there were no immediate competition or cartels to contend with, the trade door could be opened or slammed shut by government policy decisions, either at home or abroad. During the war American business interests, as we have seen, had been frustrated by British and French efforts to prevent neutrals from trading with the enemy. The restrictions became even stronger after America entered the war. With the tentative agreement of Wilson's war cabinet, the Allied powers imposed a unified military command to tighten the blockade and "deal directly with virtually all materials and commodities for the prosecution of the war." But the blacklist and the postwar implications of a restrictive European trade policy made some Americans, both in business and in government, uneasy. Suspicions grew after the Armistice and the treaty fight that followed. Britain got nowhere with efforts to continue the wartime international trade consortium, although it effectively pursued a postwar domestic policy of industrial "consolidation and combination" for "national safety." Elsewhere, the political discord in the last years of Wilson's second

term divided the Allies and weakened the advocates of international cooperation. Instead of working to lower trade barriers, American business leaders
and their allies in government now worked to promote unilateral commercial expansion. Where foreign trade was concerned, government and business
united to fight fire with fire.[58]

A preview of the American postwar position came in the aftermath of the
blacklist flap in 1916. To negotiate directly with the Entente powers and cut
through the red tape, textile manufacturers and exporters in several other
industries, with encouragement from the State Department and the Federal
Trade Commission, had built ad hoc wartime trade associations. In 1918 Congress gave official sanction to these efforts in the Webb-Pomerene Act, in
effect setting aside antitrust laws in order to encourage the formation of
American cartels to better compete overseas.[59]

Among American metal producers, however, in the immediate postwar
world only members of the copper industry were able to develop an effective
export association. In a period of rising demand for electrical power, machinery, and appliances both at home and abroad, the Copper Export Association
controlled major copper deposits both in the United States and in South America. By uniting its membership, pooling resources, cutting costs, and lowering
prices, the CEA helped American copper producers expand their overseas markets and hold their own against a restored European cartel, at least until 1926.
In that year, after a steady decline of international copper prices, the American
and European cartels joined together to control prices worldwide.[60]

The steel industry followed a slightly different path. Wartime demand
accelerated steel use both at home and overseas, but postwar economic recession and overcapacity in Europe led to declining prices. European steelmakers
responded by seeking government assistance, strengthening the cartel, dumping excess steel, cutting the workforce, and making themselves leaner and
meaner through modernization and technological change. In contrast, Big
Steel in the United States, instead of modernizing, abandoned foreign markets
and turned inward, relying on a strong domestic market in the 1920s for structural steel, autos, and appliances.[61]

For other domestic metal producers, neither cartelization nor monopo-
lization was a viable option when the war ended. Unlike copper and steel,
postwar military and industrial demand for tungsten and other specialty
metals declined or at least leveled off, and they were too widely distributed
to be controlled or manipulated by a single geopolitical entity, either foreign

or domestic. Lacking the economic strength of Big Steel or the international copper cartel, miners and their political lobbyists in tungsten, lead, zinc, mercury, and antimony sought help from government to stay competitive. Thus, while the copper and steel industries enjoyed an interlude of prosperity and presumptive self-reliance, less organized and more vulnerable metal producers clamored for government relief and tariff protection. However, as internal competition increased in the steel industry and cheaper foreign steel eroded the steel trust's market share, Big Steel soon took up the cry of the "war babies" in tungsten and other specialty metals to lobby for higher protective tariffs. During the Depression decade Big Copper also joined the protectionist clamor after the collapse of copper prices and the rise of stiff competition from new mines outside the American economic orbit.[62]

As one of several steel alloys that added to the killing potential of weaponry in World War I, tungsten was important but hardly the philosopher's stone that one pretentious Wall Street journalist had predicted. "It may someday well be said that tungsten made democracy possible," he declared.[63] The hyperbole was indicative of the times, however. From the strategic viewpoint, tungsten and other "rare" metals were essential in both war and war's preparation, meaning they had to be available at all times to ensure the vitality and strength of the nation's military hardware. From the financial viewpoint, tungsten was a speculative investment, a catalytic market force that could generate great wealth or poverty, depending on the investor's "luck and pluck," to use the contemporary bootstrap philosophy of Horatio Alger. In broader economic terms, however, tungsten was deceptive, promising more during the war years than it delivered either in military security or in financial gain. Instead of making democracy possible, it contributed to a military stalemate and a deadly attrition rate on the battlefields of sea and land, where each side, armed with essentially the same weapons systems, slugged it out until both were exhausted. War's end brought a measure of reality to military planners as well as mining investors. Both had to adjust their expectations in light of postwar economic conditions.

3

FROM PACIFIC TUNGSTEN TO NEVADA-MASSACHUSETTS

American tungsten producers did not share in the corporate gains that characterized much of the economy during the Roaring Twenties. Except for a few operators still working to fulfill old orders, tungsten mining in the United States virtually died in 1919 and did not revive for nearly five years. First of the larger operators to close was Pacific Tungsten. With a mismanaged mill, no operating capital, and a mountain of debt, it folded in January. Next to go was the Atolia mine in California, once the largest U.S. producer. It shut down the first of March. By June all of Colorado's ferberite and wolfram mines had closed as well as the twenty-one mills that serviced them. Pacific Tungsten's neighbor, the Nevada Humboldt mine, ran until July before it "was closed down by its creditors," as an industry spokesman reported. Two months later the last major American firm, Tungsten Mines in the Bishop district, suspended operations after completing its contracts. Overseas the situation was not much better. Markets everywhere suffered from "over-production and accumulation of ore stocks."[1]

Other metal producers revived after a brief recession in the early 1920s. Urban growth and rising demand for electricity, oil, steel, copper, aluminum, and other commodities stimulated markets worldwide and led to profound changes in the productivity and corporate structures of the largest metal and power companies. For small businesses, however, and especially for American miners and farmers outside the urban-industrial Northeast, the "decade of prosperity" was more chimerical than real.[2]

Most of the blame for the tungsten malaise fell on Chinese producers and traders who kept the American market saturated until the mid-1920s with wolfram concentrates produced at a third of the cost of American ore. Asian ore ruined the American tungsten industry, declared the *Mining and Scientific Press* in 1920. It urged a protective tariff to prevent the United States from falling to the "mercy of hostile nations, which could instantly cut off supplies."

Resurgent calls for protection appealed to a postwar nation disillusioned by Wilsonian idealism and the cynical aftermath of war in Europe. As a corollary of isolationism, protectionism gained considerable momentum during the recession years of the early twenties.[3]

RELIEF LEGISLATION FOR STRATEGIC METAL PRODUCERS

To financially strapped mining investors in the immediate postwar years, relief came before protection. W. J. Loring typified their distress in a frank remark to a Pacific Tungsten stockholder in 1919: "Absolutely no business is being done in tungsten, and we are moving heaven and earth to get some relief from the Government." In March, while President Wilson was still in France wearing himself out with treaty negotiations, Congress passed the War Minerals Relief Act, otherwise known as the Dent Act after the Alabama Democrat who was its prime sponsor. It established the War Minerals Relief Commission with power to investigate claims and make recommendations to the secretary of the interior for distribution of $8.5 million provided under the act. As part of a bill to settle "informal army contracts," the relief measure implicitly acknowl-edged the government's liability for net losses suffered by producers of pyrites, chrome, manganese, and tungsten because of the abrupt cancellation of gov-ernment contracts in November 1918. Advocates of these four "war babies" hailed it as a great victory. In the three months allowed for filing, 1,287 claims totaling more than $17 million were recorded, raising "some question as to the policy of payment," as the American Mining Congress (AMC) secretary dryly noted. Attorney General Palmer, however, drastically reduced the number by narrowly defining the act's intent to apply only to those companies with spe-cific contracts or written orders. The opinion caused an outcry, reflected in a *Wall Street Journal* editorial claiming it would cause the loss of "millions of dollars of relief to the small miner who is the backbone of the mining indus-try." But the order stood, and the Sixty-sixth Congress, distracted by foreign affairs, deflected pressure to amend the law. More than half of the initial claims were tossed out, but the unhappy petitioners bided their time until a more amenable Congress could be persuaded to revisit the issue. They did not have long to wait. In the meantime the Relief Commission conducted hearings to gather evidence, called in investigators to check the accuracy of claims that passed the initial screening, and audited each claim before making recom-mendations. With nearly 600 claims to review, it was a laborious process that kept claimants on financial tenterhooks until awards were announced.[4]

Pacific Tungsten's $521,000 claim and its case history exemplified the complexities and frustrations of the lengthy procedure. A Bureau of Mines official set the initial tone by warning Loring that "it would be an impossible task to work exact and equal justice and preserve everybody from loss." But gloomy predictions would not sit well with investors. The company boss told E. A. Clark, PTC's principal backer, that he should expect "a substantial sum in relief because of the justice of our claim." To present their cases in Washington, Loring and Atolia president E. C. Voorheis hired, on a joint contract, a retired California judge, John F. Davis. Perky and presumptive as the proceedings began, Loring wrote Segerstrom that Davis "could fill the position most admirably, because he is not thin-skinned and would without being announced, butt into the sanctum sanctorum of the most private officials and domiciles of the Powers That Be in Washington, not barring President Wilson himself."[5]

Davis did not impress the government commissioners. In July 1920 they completed their final review of Pacific Tungsten's petition. The chief accountant disallowed $311,350 claimed for property acquisition and consolidation, and discounted the figure for actual operating costs by 25 percent, leaving an award of only $92,000, or less than 20 percent of the initial claim. Loring accepted it with "reservations as to future consideration for relief of losses arising out of purchase of land or leases." Davis's cut was 10 percent, leaving little left to pay an accumulated debt of more than $325,000.[6]

Loring hoped for much more. In 1921, after a Republican election sweep the previous November, Congress seemed about to oblige. It amended the Dent Act, loosening the eligibility requirements to include claims based on strategic mining encouraged by "published and printed solicitation," and permitting "additional awards based on rulings contrary to the amendment or 'through miscalculation.'" President Harding's signature on the bill set off another round of hearings by a new commission appointed by Albert B. Fall, the secretary of the interior whose shady deals with unscrupulous mining and mineral company executives would soon force his resignation and later land him in jail for accepting bribes. Judge Davis immediately filed a new claim on behalf of his client, but three years passed before the commission got around to secondary appeals. In the interim it approved many that Wilson's attorney general initially had ruled ineligible. By the end of 1925 Hubert Work, who replaced the disgraced Fall, had paid out all but $9,000 of the $8.5 million appropriated for the purpose.[7]

Pacific Tungsten was not among the recipients, however. Work refused a rehearing on the PTC petition, questioning damage awards relating to leased properties, whereupon Judge Davis petitioned a federal district court for a writ of mandamus to force the secretary's hand. While the matter was pending, the U.S. Supreme Court, in two separate cases, ruled that mining companies could not appeal adverse decisions of the secretary of the interior on claims filed under the Dent Act.[8]

The high-bench rulings in 1925 ended round two, but there were still other rounds to come. If the courts would not cooperate, litigants still could petition lawmakers for redress. Early in 1929, the lame-duck Congress in the last days of the Coolidge administration passed another amendment to the Dent Act. It permitted new court filings to review unresolved legal questions. The amendment was the opening bell in a series of cases relating to monetary losses due to interest payments on loans, property acquisition, and other indirect expenses incurred in trying to meet government needs during the emergency, all of which Secretary Work and his successor, Ray Lyman Wilbur, had denied. The wheels of justice moved slowly, but in 1931 the D.C. Court of Appeals reversed a lower court decision upholding the Interior Department's position, and the U.S. Supreme Court affirmed the next year. It was a pyrrhic victory for the claimants, however. The decision amounted to what later came to be called an "unfunded mandate," since Congress had not appropriated any more money for the purpose. The plaintiffs had to wait another five years for Congress to authorize an additional $1.25 million to settle the remaining ninety-one discounted claims, including Pacific Tungsten's second award of $50,942.[9]

For some litigants there was one final round, triggered by the sudden imposition of a statute of limitations. In 1940 a conservative coalition in Congress approved legislation barring further prosecution of old war claims under the Dent Act after July 1, 1941. The result was an immediate rush to file new claims before the deadline. C. H. Segerstrom, still hoping to cover personal losses as a principal investor in Pacific Tungsten, hired Joseph A. Trimble, a D.C. lobbyist and attorney who had argued one of the relief cases before the U.S. Supreme Court in 1931. Trimble filed with the Interior Department to have the PTC case "reopened and reconsidered," but the petition was denied and the federal district court sustained the decision. By that time the average relief payout was only 10 percent of the original claim. Trimble stood ready to appeal and even to promote new amendments to the old relief law, but in 1943, after a quarter century of fighting, Segerstrom finally laid down the gloves.[10]

RESURGENT PROTECTIONISM

Relief laws, though difficult to administer and costly to litigate when disputes arose, derived from universal principles of equity and the rule of law. Few questioned the legitimacy of efforts to compensate individuals or corporations financially damaged by arbitrary government policies or actions. The protectionist upsurge, on the other hand, was much more complex and partisan, at times more emotional than rational, both in concept and in administration. American tariff legislation in the late teens and twenties embodied the xenophobic nationalism of distressed and sometimes disillusioned domestic producers and their political allies.

To consumer-oriented progressives and free-trade Democrats, the postwar clamor for protection sounded hollow and hypocritical. It made no sense to discourage foreign trade at a time when American loans remained unpaid and Europe was financially prostrate. Neither were tariffs needed for revenue after the adoption of the federal income tax amendment in 1913. Moreover, high tariffs benefited special interests, not national interests. Instead of propping up established industries with more government subsidies, progressives wanted to level the economic playing field by preventing unfair competition both at home and abroad. Four measures implemented during the Wilson years formed the basis of progressive trade policies: the Underwood Tariff to lower tariff rates, the Federal Trade Commission to regulate trade practices, the Tariff Commission to administer tariff laws and investigate disputes, and the Webb-Pomerene Act, which relaxed antitrust laws for firms trying to expand exports and compete with foreign cartels. To the progressive mind, any further government intervention in the American economy would be paternalistic, and do more harm than good.[11]

Progressive idealism, however sensible to statesman, economists, and sociologists, fared badly in the postwar economic climate. If "fear is the mother of foresight," in novelist Thomas Hardy's words, protectionists were anxious prophets. The "dark and desperate future" that worried Judge Gary in 1916 seemed ever more apparent in a resurgent and cynical Europe, resentful of wartime profiteering and eager to rebuild at American expense. To hard-line protectionists there could be no prosperity without economic security, no international interest superior to national self-interest. As Senator Smoot later remarked, free traders were as hopelessly romantic as the advocates of "political internationalism" who voted for the League of Nations; both believed "the American people [should] underwrite the salvation of the world."[12]

To many domestic producers with falling profits, the first priority was preventing resurgent European competitors from dumping cheap commodities on American shores. Antidumping provisions had been written into previous laws, but proving intent to undercut prices in international markets was a major legal obstacle until Congress decriminalized the enforcement procedure in 1921. The popular response to unfair competition was to raise import duties, despite warnings from trade experts that protectionism at home invited retaliation abroad. Within a decade after World War I, as tariff barriers on cotton, steel, and other commodities went up worldwide, "trade protection lost all value," in the words of economic historian Mary A. Yeager.[13]

For producers of strategic metals, national economic stability and security were more compelling arguments for trade protection than corporate profits. Whether at war or at peace, domestic producers warned that America's "key industries" must remain viable. The security argument predominated in hearings before the U.S. Tariff Commission in the summer of 1918. In Denver, Colorado tungsten producers claimed that higher tariffs were the only means to stabilize domestic production and keep "other branches of the industry, namely . . . refining [and] . . . alloy steel making," in business. In San Francisco, witnesses insisted that tariff protection was the only way to avoid a recurrence of prewar conditions, when Germany dominated tungsten production and marketing. A prepared statement reduced their argument to a single sentence: "If the industry which was developed as a war time necessity, from the sole standpoint of guaranteeing our National security, is not worth [a $9 tariff] . . . we have no reason to ask for the legislation."[14]

To patriotic Americans discouraged by the "calamitous end of our trust in international harmony," as T. A. Rickard editorialized in the *Mining and Scientific Press,* higher tariffs were essential to secure our "economic self-sufficiency." With industry spokesmen clamoring for protection and their lobbyists in Washington hard at work, a legislative response was soon forthcoming. Early in 1919 Colorado congressman Charles B. Timberlake introduced a bill to tax imported tungsten concentrates $10 a unit and ferrotungsten $1.25 a pound. Oddly, the measure exempted imports from the Philippine Islands and the islands of Guam and Tutuila. The reason for the waiver remained a mystery until Judge Davis invited the elderly congressman to dinner and asked for an explanation. The reply was "almost comical," Davis later wrote Edward

Clark. Timberlake "knew next to nothing about the subject," but had lifted the tariff provisions "word for word" from a bill introduced in 1917 on behalf of "his home people in Colorado," and then borrowed the remaining sentences from a recent tax bill introduced by another congressman. When Davis pointed out that importers could avoid any duty by transshipping through the excepted ports, the embarrassed legislator readily agreed to "eliminate the objectionable phrase."[15]

With Judge Davis and fellow lobbyists helping Timberlake "doctor up" his bill, the substitute was reintroduced in June and sent to the House Ways and Means Committee for a hearing. A contemporary letter from Jesse H. Holmes Jr., a Colorado lobbyist, to the president of the Atolia Mining Company provides an insider's look into the tungsten producers' strategy. The "star witness," Vanadium Alloys president Roy C. McKenna, impressed on committee members his conviction "that if [the] tungsten mining industry of America is allowed to cease entirely, in cases of any national emergency it will be impossible to re-open tungsten mines and secure production within a period of less than one or two years, which would probably be too late." Holmes suggested following up by first having the American Mining Congress send briefs to "every Congressman," then wire the "chief western producers, requesting telegrams to be sent to Senators and Representatives to give [the] necessary final 'punch,'" and finally forwarding a graphic illustration for the knockout blow: "The cartoon you sent me cut from Denver News of June 11th showing the fat Chinaman collecting American millions for oriental tungsten ore and the poor little suffering Colorado miner in one edge of the picture, is a classic. . . . It is a splendid cartoon and tells the story more effectively than a three column article, and any Congressman can grasp the full significance of it at a glance."[16]

The Timberlake bill exemplified the protectionist upsurge after the 1918 election restored conservative control of the House. With Chairman Joseph W. Fordney inviting tariff proposals "in accordance with campaign promises of the Republican majority in Congress," homeland producers of metals, glassware, surgical instruments, chemical dyes, and other commodities joined the parade before the House Ways and Means Committee in the summer of 1919. Cheap foreign labor was the common complaint, and higher tariffs the proposed remedy.[17] However, despite the heady rhetoric on behalf of economic prosperity, national security, and other worthy causes, the hearings exposed deep political and economic divisions. Incompatible belief systems drove a deep wedge between diehard protectionists and uncompromising free traders.

Pragmatic policy makers looked more to facts than faith, but in the metals industry during the interwar years facts were elusive. Technical experts in science and industry disagreed widely about the extent of American strategic metal deposits and how best to utilize them.

In the tungsten business, most western producers were unbounded optimists, assuming they could meet every domestic need now or in the foreseeable future. They wanted a protective tariff to drive ore prices above the cost of production so they could get back to profitably mining and milling ore. For tungsten dealers and specialty steelmakers, however, high ore costs raised consumer prices and were a drag on business. They sided with free traders, but protectionists questioned their motives, as Judge Davis revealed during the tariff hearings in a surly letter to Edward A. Clark:

The importing people have spread a certain peculiar propaganda . . . namely, the suggestion that we have no assurance but that America is to have her next great war with Great Britain, and that we had better for the time being exhaust the deposits in foreign countries and not disturb them in our own so that when the clash shall finally come and America have [sic] to rely on domestic resources we will know just where to put our hands on the home materials. That is the characteristic military point of view and is equal to anything that might have been thought of in Germany under the Hun regime. That is why the ordinary importer and importing broker in these days parades himself as such a patriot.[18]

Davis's crystal ball, though cloudy, did foresee the political divide over strategic materials that complicated military and domestic policy formation for the next forty years.

For other executives in the metals business, more compelling than ideology or patriotism was the hard reality of economic self-interest. When tungsten prices dropped after the Armistice, many importers continued to buy ore at bargain-basement prices, creating a huge stockpile that had to be absorbed before domestic prices would rise. Just before the House passed the Timberlake bill, Frank L. Greene, a Vermont Republican, added an amendment taxing ore previously imported. Importers objected not to the higher tariffs demanded by western producers but to the ex-post facto Greene Amendment, which "practically stopped all buying" when importers heard about it. Since it had little practical support, removing the objectionable tax was an easy compromise to reconcile most dealers and producers.[19]

The complexity of the metals trade made some conflicts difficult to resolve.

Republican congressman Isaac V. McPherson of Missouri, champion of zinc producers in his district, complained that smelter companies paid nearly three times less for imported ore than the domestic product. Part of the reason, however—as a market analyst later pointed out—was the internal secrecy that kept zinc companies from taking advantage of the Webb-Pomerene Act to share information, cooperate to lower costs, or work together to compete in foreign markets. In the highly competitive ferrotungsten industry, the U.S. Tariff Commission found that most of the cost of producing domestic ferro-tungsten in 1919 came from the high cost of raw material. The Underwood Tariff had eliminated the tariff on raw tungsten and lowered the duty on ferro-tungsten from 20 percent ad valorem to 15 percent, but American tungsten producers wanted protection from all types of imports. They backed the Timberlake bill and anticipated victory when the House passed it in August after a highly partisan floor debate in which Democrats led by former Speaker Champ Clark bitterly denounced the measure as an edict from Wall Street. To Fordney's claim that the bill would "protect" American wage earners, Clark responded that lowering the cost of living and jailing profiteers would be a better way to "do something for the people." Later his colleague Cordell Hull, an uncompromising free trader, issued a patronizing press statement claiming that "every intelligent person" understands the need for trade expansion and reciprocity "except possibly the blind, stand-pat, Chinese-wall protectionist."[20]

Despite some preliminary victories, the ideological divide separating free traders and protectionists prevented any general tariff revision during the Sixty-sixth Congress. For months after the House had approved new duties on chemical glassware, tungsten, magnesite, and other commodities, tariff legislation stalled while the Senate debated the ill-fated Treaty of Versailles. After the treaty was voted down in November, the Finance Committee resumed business and opened tariff hearings. With so much at stake personally, Loring raised money for the tungsten lobby and traveled to Washington himself to shore up support. He came home "very much pleased with the Tariff situation," but his optimism was premature.[21]

Experienced insiders were more cautious, as Frank W. Griffin, a San Francisco consultant working for the tungsten lobby, reported after learning the navy opposed the Timberlake bill. To learn why, he went to see Wilson's naval secretary, Franklin D. Roosevelt. A brief visit convinced Griffin that FDR knew "no more of tungsten than a rabbit," but for more specifics he was shunted off to a naval officer in charge of the Bureau of Steam Engineering. According to this bluecoated bureaucrat, the heart of the problem was the navy's current budget restrictions. Already hurting because of a congressional cutback in naval appropriations, "anything that looked like an increase in the cost of materials meant the abandonment of certain necessary work." Griffin remained skeptical. He blamed the opposition on political infighting and the shady manipulations of that "old dog" Boies Penrose, Republican boss of the Senate Finance Committee. "There is so much crooked underhand work here," he wrote a friend, "that when you see fairness you must welcome it as a tired, dirty body welcomes a refreshing bath."[22]

The legislative impasse seemed to confirm Griffin's cynical assessment. After several weeks of deliberation the hearings bogged down in conflicting testimony and partisan rancor. By the spring of 1920, with politicians and the public distracted by lingering treaty issues and a national election campaign, all prospects for tariff revision had died. Protectionists had to await a more favorable legislative climate.

The tariff temperature rose quickly again after the Republican landslide of 1920. All three branches of government were now in sympathetic hands, and protectionists were confident of victory. Big Steel led the call to raise duties to protect the American "manufacturer and worker." Lobbyists swarmed in the nation's capital, promoting tariff revision as a legislative priority. Still hoping to revive Pacific Tungsten, Loring distributed a printed appeal to all tungsten producers asking financial support to keep the lobbyists at work. "We must surely stand together in expenses as well as in profits to come by the passage of the bill," he wrote. The new president heard the call. Soon after his inauguration, Harding urged Congress to protect home industry. In May 1921 the Emergency Tariff Act was enacted, strengthening antidumping rules and raising interim rates on both farm and industrial commodities until more thorough revisions could be adopted.[23]

Over the next year, as proposals to increase duties on raw iron, tungsten, lead, zinc, manganese, aluminum, and other metals stacked up in Congress, and as metal prices rose in anticipation of higher rates, opponents rallied for another battle. The Tariff Commission compiled sixteen volumes of testimony in hearings and statements, representing a broad spectrum of views from around the country and abroad. Protectionists claimed high tariffs would help "infant industries," but the proposed rate schedules on pig iron, steel rails, wheat, wool, and other traditional products of American fields and factories belied that notion. Government geologists, citing the need for further research, questioned whether the United States should use up all its resources. They favored importing foreign ores and saving domestic raw materials for future emergencies. The steel industry remained divided between the trust companies and the independents, with company size less of an issue than sectionalism and quality control. Tool steelmakers and other low-volume, high-quality manufacturers, almost all located in the industrial Northeast, protested high rates on ores and ferroalloys. In a highly competitive business where skilled labor made up 90 percent of total costs, they could not pass higher rates on to consumers without losing market share. The high-volume trust companies had no such problem.[24]

The steel industry split facilitated a three-way alliance in Congress. Late in 1922, representatives of Big Steel joined those favoring metal producers and farmers in the West and South to enact the Fordney-McCumber Tariff. Average rates rose 35 to 50 percent above those established in the 1913 tariff, and made permanent the emergency agricultural duties imposed in 1921. Economists deplored the manipulative power of western metal producers, who exploited national security fears to gain nearly prohibitive rates on scarce or limited raw materials that formerly were admitted free of duty. A market analyst calculated that the new tariff on ferrotungsten alone was equivalent to an ad valorem increase of 200–225 percent compared to the 1922 world market price. Steel alloy manufacturers in the East also resented the higher costs, but the industry profited from the booming domestic market for automobiles, appliances, and other specialized steel commodities. Clearly, the tariff of 1922 hurt American foreign trade, but despite the high duty, Chinese tungsten continued to enter U.S. markets, primarily by way of German and British importers, until domestic demand stalled after the crash of 1929.[25]

PACIFIC TUNGSTEN IN THE 1920S: FINANCIAL CRISIS AND COLLAPSE

The 1922 tariff revisions came too late to help Pacific Tungsten and other struggling "war babies." Shut down since early 1919 and saddled with an outstanding debt of $325,500, the company's "situation is hopeless," Loring confided to an investor. Even the company watchman, W. T. Connors, a crusty old bachelor, quit after three years of lonely life in the Nevada desert "with no one here to talk to" and little to do but drink his own moonshine. He left with a word of advice to his boss: "I would suggest that the man that takes my place not have any rough neck kids for this is no place to turn children loose that do not mind and about 9 kids out of ten boss their parents now days." Loring replaced him with Harry Baker, an old prospector, one of the original locators of the Stank mine.[26]

Keeping the company's surface plant and equipment safe and the unpatented claims protected by completing the annual assessment work was only a small part of Loring's postwar management worries. The financial burden was much more serious. A major investor in the company he started and headed, Loring had to save Pacific Tungsten from bankruptcy in order to save himself. For five years after the Armistice he maneuvered his tattered company through one fiscal crisis after another, desperately trying to stay afloat in a sea of debt. Clearly, it would take more than rising tungsten prices to reopen the mine. The new tariff protected only solvent companies or those that had enough financial strength to attract investors and raise development capital. Unless the government offered more relief or he could find a way to refinance, he and his company would sink into fiscal oblivion.

To survive in troubled times Loring devised a strategy of fiscal triage, making promises he could not keep to creditors and investors alike, but finding ways to avoid the most persistent creditors until he could raise enough cash to make a payment. One of his many complicated financial headaches grew out of a wartime deal with E. C. Voorheis, president of the Atolia Mining Company. Atolia had loaned Pacific Tungsten $100,000 just before the Armistice to complete the PTC mill, taking 100 tons of concentrate containing 7,125 units of WO_3 as collateral along with Loring's promise to pay a bonus of $1 a unit. After prices collapsed and Pacific Tungsten defaulted, Atolia held the ore until prices began to rise during the tariff talks in 1921, then decided to sell. Loring tried to renegotiate, offering to take back the ore if Atolia would waive the bonus, but Voorheis would agree to the waiver only if Pacific Tungsten paid the balance due in sixty days. With tungsten, in Loring's words, still "about as valueless these days as a bucket of water," and with Voorheis refusing to extend the loan, Loring was stymied, and the new deal fell through. Atolia eventually sold the ore but realized only $81,000, leaving an outstanding debt of $19,000 on the principal and $7,125 on the bonus. To complicate matters, Atolia later borrowed $10,000 from Segerstrom's bank, the First National of Sonora, assigning the $19,000 claim against Pacific Tungsten as security. The debt was finally liquidated in 1925 by PTC's successor, the Nevada-Massachusetts Company.[27]

Loring's business partners had troubles of their own. Segerstrom and Clark were both suffering from ill-timed investments in Mother Lode gold mining. In the prewar boom years, with gold at a fixed price of $20.67 an ounce, they had invested heavily in properties Loring had either recommended or managed. Wartime restrictions and labor shortages, however, followed by the destabilizing zigzag of inflation and recession, crippled or closed all but a handful of well-run gold mines that had plenty of good ore in sight.

Clark had the most to lose. In addition to his Pacific Tungsten investments he had supplied most of the operating capital for the Carson Hill and Dutch-App gold mines, all under Loring's general management. In the spring of 1919 Clark and several other investment bankers, along with the U.S. Smelting, Mining, and Refining Company, had purchased the defunct Nevada Humboldt mine, adjacent to Pacific Tungsten, hoping to merge the two properties and dominate the district. However, the dismal business climate delayed those plans, leaving Nevada Humboldt until 1924 under Loring's management. He reported to a separate board of directors representing the principal stockholders. The failure of all these mines in the early twenties left the Boston capitalist holding more than a million dollars in worthless stocks and bonds. Loring was still after him to refinance the gold properties and underwrite the reorganization of Pacific Tungsten when Clark died in 1922. His legacy was a mountain of debt and an insolvent estate.[28]

Segerstrom also had a substantial financial stake in the same properties. In 1909 he had purchased the Dutch mine in Tuolumne County and combined it with the Sweeney, a neighboring mine. A production crew took out $500,000 in gold over the next six years before Segerstrom turned it over to Loring's general management company in 1915. Riding high on his reputation as a hard-driving engineer and developer, Loring invested his own money and convinced his financial backers to expand the Dutch-Sweeney into a major operation along the lines of Carson Hill. The next year their investment syndicate, Pacific Coast Gold Mining Corporation, took an option on the adjacent App mine, borrowing money from regional banks for operating capital. But Loring's crew failed to find major new ore bodies, and rising production costs soon brought a crisis. The mine shut down in 1920, a heavy blow to investors. Loring claimed he lost "at least $50,000" in the deal; Clark and Segerstrom lost even more. Laboring under a debt of $300,000 in bank loans and bonds, the syndicate was hard-pressed. The Sonora banker managed to retain control of the property by refinancing and putting up tungsten stock as collateral, but it took years to pay off all the Pacific Coast Gold obligations.[29]

Loring had less of a financial stake to protect, but a bigger reputation. Relying on past laurels and the indulgence of business friends, he borrowed heavily and spent money recklessly. Frequently, his lifestyle ran ahead of his pocket-book, as a jaunty letter he wrote Segerstrom on an eastern business trip in 1918 suggests:

If Mr. Clark had not come to my rescue with $700 I would have been walking the streets of New York in a hopeless condition. This money did not last long enough to scarcely get home to my Hotel, and it was necessary to refinance myself when I tele-graphed you to send me $1,000. . . . I must say that I have at times felt like the Elephant that leads Barnum's Circus parade through a country village, very proud at times, because I have been investigated as to my mining reputation[,] experience, success or failure, age, family, birthplace, with whom I have done business, and I have been inter-viewed by bankers, brokers, and all sorts of business people, coming through it all with flying colors, so much so that bank presidents and their immediate subordinates treat me with a courtesy that I am sure is shown to most any respectable man who has been investigated and has had a favorable report.[30]

The haughty mood did not last long. As the debts piled up and the tung-sten market withered, Loring had few options. With the three largest stock-holders in financial trouble, Pacific Tungsten could not raise money through assessments — the traditional route for overleveraged and undercapitalized mining ventures. Selling the company assets was also not possible in a reces-sion, as he discovered in 1920 when he tried to interest Henry Ford in tung-sten. The eccentric auto mogul did not even bother replying to Loring's obse-quious letter.[31]

To save the company from bankruptcy, when tungsten prices began to rise the harassed manager took the only option left. With the blessing of Seger-strom and the remaining stockholders, he leased the mine in 1924 to Claude W. Poole, a Nevada miner who promised to rehabilitate the property and resume production. Pacific Tungsten's share was 20 percent of Poole's gross proceeds, barely enough to keep up annual claim assessments and debt payments. To help with start-up costs Loring negotiated a contract with Charles Hardy, under which the New York broker advanced Poole $20,000 in return for exclusive marketing rights to the concentrates and a 1 percent brokerage fee on all sales. Poole grossed $200,000 on the deal, earning a neat profit after a year of work. But he gutted the mine in the process, taking only the best ore in sight and spending nothing on development. When his lease expired in 1925 the mine was in shambles — an all too frequent outcome of contract mining.[32]

Poole left under a cloud, and so did Loring. His career had taken a nosedive, with his management company moribund, his investments in jeopardy, his wife filing for divorce, and his creditors circling like wolves around a tethered goat. He resigned as general manager late in 1924. As a parting gesture he proposed a familiar but dubious financial strategy for Pacific Tungsten: liquidate the liabilities by forming a new company that would acquire the old company's mining properties through foreclosure, and then raise operating capital through bond issues and stock sales. For all his "sweat equity" and personal loans he thought he deserved a 15 percent stake in the new company. He also suggested turning over management of the new enterprise to C. W. Poole. His former associates, however, had other ideas. By the fall of 1924 they thought he was "acting contemptible," and were glad to get rid of him.[33]

Loring never recovered from these professional and personal disasters. Destitute and desperate after 1925, he repeatedly called on his old friend Segerstrom for enough to tide him over until the next crisis, always promising to pay him back but never succeeding. A suit he filed against the Humboldt directors for unpaid management fees was dismissed in 1927 after the Sonora banker wrote off Loring's personal debts in return for all his tungsten stock and claims against the two defunct tungsten companies. For nearly three more decades the aging engineer took on various management and consultant jobs, trying to earn a living, but dubious investments and bad luck dogged him the rest of his life. He died penniless in Nevada at age eighty-three.[34]

THE EARLY YEARS OF NEVADA-MASSACHUSETTS

The leadership turnover at Pacific Tungsten in 1924 was not accidental. It came at the cusp of a significant upward swing of the American business cycle as recession and scandal under Harding gave way to "Coolidge prosperity." Rising incomes, low taxes, and mass marketing geared to the urban middle class spurred demand for automobiles, appliances, and other consumer goods. Increased consumer demand and lower production costs through cheaper power and greater worker productivity all contributed to industrial growth during the years of "normalcy." As early as 1922 the postwar recession in the metals industry was beginning to lift. Commodity prices had risen significantly from 1921 levels, though in 1924 they had dropped a bit from 1923 highs. The index of basic materials showed iron and steel production rising from less than 70 percent of prewar production to 115 percent early in 1924. The domestic mining index did even better, jumping from 30 percent to 110 percent during the same period.[35]

The friendly business climate and the decline of Progressive antitrust enforcement, however, accelerated the trend toward business concentration and oligopoly in many industries. Business consolidation was the norm for metal producers in the twenties, but the reasons have more to do with efficiency than greed. As Alfred Chandler explained in his seminal study of modern business trends, economies of scale made size an advantage in primary metals. The heavy initial investment and high operating costs over a long period made it necessary to achieve high-volume production in order to reduce unit costs. The high initial costs reduced the number of new firms entering the business, and the lower unit costs made it harder for small-volume competitors to stay in business. The result, as economist Robert Sobel has noted, was "concentration and vertical integration in almost all raw materials industries."[36]

The trend toward oligopoly in American metals was most noticeable in steel, copper, aluminum, and vanadium, where a few established corporations dominated their industries by absorbing smaller competitors and organizing holding companies to control ore supplies, transportation, production, distribution, and marketing. Through jointly owned subsidiaries, industry leaders were even more closely allied, although the lack of published financial data makes it difficult to determine the full extent of these "quasi-mergers," or their impact on markets and prices. Despite new competition in the twenties from upstarts at home and abroad, U.S. Steel remained the giant of the steel industry with 38 percent of total steel ingot capacity, compared to 13 percent for its nearest rival, Bethlehem Steel. Similarly, before the Depression, the four largest copper producers—Anaconda, Kennecott, Phelps Dodge, and Calumet and Arizona—determined domestic prices and supplies in the copper cartel. Among ferrous metal producers, the vanadium business was even more concentrated than copper. By the 1930s two friendly giants, Vanadium Corporation of America, affiliated with Bethlehem Steel, and U.S. Vanadium, a subsidiary of Union Carbide, controlled 70 to 90 percent of vanadium production, and 100 percent of its marketing and distribution in the United States. Aluminum was in a class of its own under the monopolistic control of Alcoa, the Aluminum Corporation of America.[37]

In contrast with the metal giants, the tungsten industry in the twenties struggled to emerge from postwar chaos and collapse. Tungsten mining, milling, and marketing were still in formative stages, with too many variables to consolidate and control. Unlike either copper or steel, tungsten resources in the United States were too thinly scattered in isolated deposits to be easily contained under one or two large corporate umbrellas. Technologies were still evolving for reducing tungsten ore, purifying tungstic acid, producing tungsten powder, and manufacturing intermediate products such as ductile tungsten wire and tungsten carbide. One study found more than seventy international patents registered in the twenties for tungsten smelting and sintering processes alone. Consumer products were also changing as government and business found new uses for tungsten alloys in automobiles, airplane engines, farm equipment, tank and tractor treads, cutting tools, electronics, and household appliances. Finally, tungsten price volatility in the transition from war to peace had a chilling effect on investment, leaving the industry unstable and disorganized. Even the armaments industry was on the defensive, with disarmament talks and antiwar activists crowding the headlines. The 1922 tariff offered a measure of stability by enabling domestic producers to raise the floor on consumer prices, but until the midtwenties the market was glutted with a backlog of foreign ore imported before the new duties went into effect. Clearly, tungsten was a risky business.[38]

Starting anew in the face of these risks might seem overly venturesome, if not foolhardy, but the hint of an uptick heightened the expectations of Segerstrom and his business colleagues. As Charles Hardy pointed out, the steel situation was "bad," but the tariff had curtailed tungsten imports and steel mill stocks were declining, so "any turn in the business cannot but develop in all quarters a strong demand for materials." Election-year politics was the real joker, with the Harding scandals still hot in the headlines, the Ku Klux Klan rising in the Midwest, the European economy in turmoil, and protectionist assumptions under attack. Democratic contenders clamored for change, but the party was badly divided. At their convention in June a dark-horse New York lawyer, John W. Davis, won the nomination after a long deadlock, in contrast to Coolidge's shoo-in when the Republicans met two weeks earlier. Market analysts saw no reason to panic. Hardy assured Segerstrom that the Nevada tungsten company would make a profit "even in the case of a Democratic Administration, which does not seem likely, and a tariff revision." He was certain that with ore stocks diminishing, if the Republicans won, "your position will become stronger and stronger."[39]

In reality the principal stakeholders in Pacific Tungsten had little choice. With bank loans overextended and a debt load already at $350,000 by 1924, they could not afford to wait for better times. Yet refinancing was out of the question. Reorganization was the only way to relieve the financial burden and salvage something from the company's remaining assets.

The corporate restructuring took more than six months to complete, with Segerstrom acting on behalf of western creditors and Philip E. Coyle, a Boston attorney, assuming similar responsibilities in the East. Early in November 1924 Coyle took the first step by filing incorporation papers under Maine law establishing the Nevada-Massachusetts Company, with five directors representing the three principal stakeholders. Of the 100,000 shares outstanding, Segerstrom and his Sonora First National Bank had controlling interest with the largest block of stock, 49,999 shares. The First National Bank of Boston held 37,396 shares, and the National Shawmut Bank of Boston 12,599 shares. Following essentially the same strategy Loring had outlined earlier, the new directors assigned their old claims to the new company, and then acquired the Nevada mine by suing Pacific Tungsten and winning a judgment that was executed by a sheriff's sale on June 3, 1925. In the meantime, Segerstrom, as trustee, assumed custody of the Nevada property on October 1 and began rehabilitation but could not resume mining until the day Poole's lease expired. After mining and milling more than eight thousand tons of high-grade tungsten, the contractor had shut down in September because of slack ore sales. He still had concentrates stockpiled for marketing when he finally vacated the property when his lease expired on March 18, 1925.[40]

Although the new company lacked full control until the spring of 1925, its new manager took charge of operational policy and planning even before Loring's departure in the fall of 1924. As lawyer, banker, financier, and principal stockholder, active in the company's formative years and well versed in tungsten mining and marketing, Charles H. Segerstrom was ideally suited for the job. Fortunately for Segerstrom, his career never turned sour, even though some of his investments did. Unlike Loring, he was pragmatic, learning from experience and hedging his bets like a riverboat gambler. A consummate professional and skilled negotiator, he was always polite but could be ruthless when dealing with adversaries. He ran Nevada-Massachusetts for the next twenty years with assertive self-confidence, yielding only to physical adversity at the very end.

With corporate restructuring well in hand, mine development and marketing strategy were the highest priorities as Charles Segerstrom took charge late in 1924. The business plan anticipated paying operating expenses with revenue from ore sales, but neither ore nor revenue could be generated without mine rehabilitation and development of new ore bodies. Even though he had approved Poole's lease when it was first proposed, the results taught him a lesson. "I consider the gravest error was committed," he wrote later, "in that Mr. Poole took every pound of available ore, and left the mine and mill in such condition, that it will cost us far more than we received in royalties to repair the mill and develop the same tonnage in new ore, therefore, we are not going to make any more mistakes of this kind." With operating capital raised through bank loans and assessments, Segerstrom's crews finished the restoration and development work in six months.[41]

To supervise the work in Nevada while he directed overall operations from the company's head office in California, Segerstrom hired "a well known mining man" with "a world of practical experience." Ott F. Heizer, a Nevada native and credentialed engineer, had grown up in the Great Basin mining business. Like Herbert Hoover an uncompromising field boss, he drove his men as hard as he drove himself. For eighteen years Heizer remained Segerstrom's right-hand man, first as superintendent, then as general manager in charge of all technical operations.[42]

By the spring of 1925 the property was ready for active mining. After concluding a marketing contract with Charles Hardy, Segerstrom saw a promising future. He predicted unit prices of $10 or more for tungsten over the next three years. Basing operating plans on ore supply and demand as reflected in spot market prices was standard practice for both producers and consumers in the tungsten trade.[43]

That was not the case with Big Steel, however. In his probing analysis of the formative years of the modern steel industry, Thomas Misa emphasizes the importance of "user-producer interactions" in shaping technological growth and change. Between the Civil War and the midtwenties, the structure and technology of the steel industry had been influenced by consumer demand in the form of railroads, armored ships, and automobiles. After that period, he argues, the size and leverage of the steel cartel stabilized the industry but at the same time isolated it from market forces that elsewhere contributed to structural reorganization and technological innovation. Thus, losing touch with consumers and market conditions was a prescription for stagnation and obsolescence.[44]

Tungsten producers and consumers did not follow the same ominous trend. Most domestic tungsten producers were small, independent mining companies that relied on brokers like Hardy who negotiated ore contracts based on current market conditions. The brokers stockpiled ore for future sales, or sold ore to secondary processors, usually smelting or refining companies. The secondary processors either sold their products directly to alloy steel manufacturers or chemical plants or made alloy steel in their own furnaces for sale to machine shops, toolmakers, and other fabricators. The linkage between these various producers and processors was not always direct, nor even widely understood or known to the public. Some producers and processors eliminated brokers by negotiating direct contracts; others integrated backward or forward through mergers or stock purchases. Yet throughout the steel specialty business, producers and consumers alike were sensitive to market forces. Downstream user needs and specifications had a direct impact on business decisions affecting production volume, quality control, milling technology, marketing strategy and structural change. The user-producer nexus remained close and interactive, as Segerstrom's management decisions for the Nevada-Massachusetts Company demonstrate.

In the summer of 1924, with tungsten stockpiles still high, world market prices at $2.50 a unit, and domestic production costs $10 or more, Segerstrom's optimism was a leap of faith. He justified the decision to resume active mining on two grounds. First, the 1922 tariff of $7.14 per unit equalized the differential between world and domestic prices. Demand would rise after supplies ran out, giving the most efficient mines a chance to revive. Second, Nevada-Massachusetts had a singular advantage over other domestic competitors. Its best ore was high-grade scheelite, averaging 1.5 to 3 percent tungsten per ton. Some "hot spots" ran as high as 7 percent. It also contained fewer impurities than lower-grade deposits, making it easier and cheaper to process. Colorado ferberite and hübnerite ore averaged less than 1 percent tungsten per ton. Atolia scheelite in California was higher in grade, but the deposit was small and nearly exhausted by the time ore prices began to rise. Even Nevada Humboldt ore at the mine next door assayed less than 1 percent—a disappointment that left its directors ready to unload the property when Segerstrom came calling a few years later. European alloy steelmakers had paid a premium for scheelite until the flood of cheap Chinese wolfram made the extra cost of purification worth the price differential. With the tariff in place and the competition weak, Segerstrom and his broker thought American tungsten consumers would pay at least a $1 premium per unit for high-quality ore.[45]

Continual advances in technology, however, undermined the premium argument. Since the war, electric reduction furnaces had evolved from small firebrick units using graphite electrodes to large steel furnaces with dolomite and other heat-resistant linings. Constant improvements in furnace design, lining composition, slag control, and purification procedures had cut manufacturing costs and increased quality and quantity of production. By the late 1920s, high-grade ferrotungsten could be made from any type of tungsten ore, concentrates, or steel scrap, with yields between 70 and 90 percent of the tungsten contained in the charge. When Segerstrom tried to interest a Union Carbide subsidiary, Electro Metallurgical Company (Electromet) of Niagara Falls, New York, in premium ore, the sales representative said his company had just spent $100,000 on a new process to recover tungsten from wolframite for the same price as scheelite, so charging a "premium of more than 30c per unit on Scheelite over Wolframite is almost out of the question."[46]

As Electromet's position illustrates, progress in technology had a significant impact on ore marketing and handling. Some of the largest alloy steelmakers continued to buy unrefined tungsten ore and upgrade it in their own mills, giving high-grade producers like Nevada-Massachusetts a direct consumer outlet for ore sales. Many smaller steel firms, however, found it more efficient, if not less expensive, to obtain processed tungsten from smelting companies and refineries. Preferring complex ores containing a variety of metals that could be sold as by-products, secondary processors either bought low-quality concentrates directly from producers, importers, or dealers at bargain prices or contracted with producers and consumers to upgrade "dirty" ores for a fee. With a staff of trained and highly skilled engineers, chemists, and metallurgists, these high-tech firms employed all available technologies and tools for reducing ores and cleaning metals, including electromagnetic separators, blast and electric induction furnaces, electrolytic converters, and chemical treatment plants.[47]

BUILDING A CUSTOMER BASE

The head of Nevada-Massachusetts returned to the premium argument time and again over the years, eventually securing some contracts from alloy steelmakers willing to pay slightly higher prices for exceptionally clean ore that could be used in direct smelting. But selling ore at a premium was a long-range goal. For a new company in a tentative ore market, Segerstrom's immediate concern was financial stability, or keeping an "even keel," as he explained to the directors. To balance operating costs and revenue meant building slowly, producing no more ore than the market could bear at a reasonable price. It also meant finding steady customers that would buy ore on contract rather than abandon regular suppliers whenever market prices dipped. To secure those contracts in hard times occasionally required offering ore at a discount, but that was cheaper than depending on the open market, and much better than shutting down.[48]

Charles Segerstrom kept abreast of the market through direct contacts with customers, but for routine market matters he relied on metals brokers, the marketing specialists in the metals trade. All of Pacific Tungsten's ore sales had been handled by Hardy and Ruperti, one of the oldest and most experienced eastern brokerage houses. The standard commission was 1 percent of net sales charged to the seller. Segerstrom saw no reason to change either dealers or fees, although he refused to continue Loring's policy of giving Hardy and Company exclusive rights. After Ruperti retired, his secretary, Joseph John Haesler, moved to another brokerage firm, Metal and Ore Corporation, taking the Nevada-Massachusetts account with him. A New Jersey native of German and Irish immigrant stock, Haesler while still a boy took odd jobs to help support his family after his father died in an industrial accident. He managed to complete high school and a year of commercial training before entering the metals industry as a clerk for Irvington Smelting, a New Jersey plant just ten miles from home. Proud of his practical experience, in later years he often referred to his "alma mater" as "The School of Hard Knocks." Haesler's hard work on behalf of his clients earned Segerstrom's respect, and the two became close business associates over the years.[49]

Hardy and Haesler both operated out of New York City, the marketing center for foreign and domestic metals. With so much of the steel alloy business concentrated west of the Alleghenies, however, Nevada-Massachusetts needed a regional agent. On Haesler's recommendation Segerstrom hired A. C. Daft, a Pittsburgh dealer experienced in marketing tungsten and other rare metals. Daft's commission was the same as Haesler's, paid periodically out of the joint account Nevada-Massachusetts established with Metal and Ore Corporation in New York.

Commission agents were particularly important in the early years, when Nevada-Massachusetts needed eastern help to build its customer base. Segerstrom, while preferring direct sales to steady customers, accepted the role of intermediaries and used them to supplement his own marketing expeditions to New York and Pittsburgh. He was especially anxious to retain clients who had purchased Pacific Tungsten ore, or at least had been interested during the halcyon days of World War I. From these earlier marketing efforts came two of Nevada-Massachusetts Company's best customers.

One of the earliest buyers of Mill City concentrates was Vanadium Alloys Steel Company of Latrobe, Pennsylvania. Organized in 1910 to capitalize on rising demand for steel alloys, by World War I the company had converted to electric furnaces and was shifting into high-speed tool steels under its hard-driving president, Roy C. McKenna, an electrical engineer from a Pittsburgh family of brass makers. For more than thirty years under his leadership, Vanadium Alloys remained keenly competitive.[50]

Though steel alloy clients were important, Nevada-Massachusetts sold more ore to the Molybdenum Corporation of America than any other firm. A smelting and refining company that specialized in rare metals, Molycorp had been established in 1920 after the merger of a Pennsylvania smelter and a western metals producer. For the first few years its principal founder was also its chief technical adviser, Dr. Alcan Hirsch, a chemical engineer and consultant who operated two private research labs in New York. Company management soon passed to Hirsch's younger brother Marx, a chemical engineer with a law degree as well. Under the younger Hirsch, Molycorp expanded aggressively both ways—backward to tie down ore supplies, and forward to stabilize distribution and sales. With plants in Washington and York, Pennsylvania, it was well positioned to supply metal products to steel companies of all sizes.[51]

Both Molycorp and Vanadium Alloys had been aware of Mill City ore long before Charles Segerstrom took the reins. During World War I McKenna's firm had been a steady customer, buying Nevada scheelite through Charles Hardy for conversion to ferrotungsten. In the early twenties, Loring's financial problems led him to propose a deal with the Hirsch brothers by which Molycorp would purchase Pacific Tungsten ore in advance of delivery, providing enough negotiable trade acceptances to resume mining. Loring left before the deal could be consummated, but Molycorp need a reliable tungsten supply and kept abreast of Mill City developments. In 1925 Hirsch proposed an arrangement with the newly formed Nevada-Massachusetts Company to jointly lease and operate the adjacent Nevada Humboldt property. Segerstrom was cool to the idea, but the Hirsch interests remained a good prospect for ore sales when Nevada-Massachusetts completed rehabilitation and began active mining in May 1925.[52]

PRODUCTION WITHOUT PROFIT

Unlike the frenetic style of his predecessor, Charles Segerstrom was a cautious manager, planning for slow and steady growth but always keeping an eye out for contingencies. As he patiently explained to creditors and investors, he recommended developing the ore bodies systematically, keeping production abreast with demand. It was timely advice. The economy in the Coolidge years made shallow but steady gains, with low taxes and a friendly business climate adding to the "illusion of prosperity," as one economist has characterized the late 1920s. With autos and skyscrapers leading the way, rising steel demand pushed up metal prices all along the line. Average domestic tungsten prices rose gradually, from $8.47 in 1924 to $13.13 by 1929. Hoover's election in 1928 reinforced the illusion and even won over some skeptics. Just before the stock market crashed in 1929, Lincoln Steffens, the muckraking journalist, conceded that "Big Business in America is producing what the Socialists held up as their goal: food, shelter, & clothing for all. You will see it during the Hoover Administration."[53]

Segerstrom welcomed Hoover's victory along with every other Republican loyalist, but hindsight made him cautious regardless of who ran the White House. Implementing the prudent business plan he devised for Nevada-Massachusetts meant careful management of on-site developments. Over a four-year period beginning in 1925 the company modernized and integrated mining and milling operations. Mine crews added electric hoists and a larger compressor. They deepened the shafts, exposed and followed new veins, and blocked out new ore bodies. Mill technicians added a sintering furnace to roast sulfides, cutting the cost of sending ore to an outside processor for purification by nearly 90 percent. With these improvements the company produced 10,381 units in 1925 at an average cost of $8.07 per unit. By 1929 the figures were 30,128 units at $9.85 per unit, with milling efficiency increasing from 58.4 percent of tungsten trioxide recovered for each ton of ore milled to 68.4 percent by 1929. Though labor costs for the same period rose, the ratio of labor costs to production dropped from 54.1 percent in 1926, the first full year of operations, to 43 percent by 1929, an indication of the emphasis on efficiency and productivity under the watchful eye of field boss Ott F. Heizer.[54]

While the work went on near Mill City, Heizer and Segerstrom took advantage of the slow tungsten market to consolidate and expand the company position in Nevada. In 1927, following a disappointing assay of ore samples, the

Nevada Humboldt directors and shareholders decided it was time to unload their heavy metal white elephant. They sold everything to their next-door neighbor, including the Humboldt mine and mill as well as all interests in the power and water facilities and other subsidiary companies operated singly or jointly in the Mill City district. Under Heizer's supervision the two mines were soon fully integrated and operating with two main shafts, the Stank and Humboldt.[55]

During the same period Segerstrom and his superintendent also explored other prospects. With company finances in better shape, in 1928 Nevada-Massachusetts directors approved the organization of a subsidiary company with ambitious plans to prospect, promote, develop, and sell mines, and "to carry on such business in any part of the world." Recognizing the need for quick action in the field if necessary, the terms of agreement placed Segerstrom and Heizer in complete control. They located the properties and transferred title by quitclaim deed from the exploration company to the main corporation.[56]

Under Segerstrom's supervision, Heizer's extensive search eventually took him all over the West. Whenever there was a lull in Mill City operations he investigated claims and prospects, old or new. Once in a boat on the lower Colorado he had to jump when the outboard motor caught fire, losing all his equipment and nearly his life. He found better tungsten locations, however, closer to home, including the Ragged Top mine in the Trinity Mountains a few miles west of Toulon, and the Oreana prospect along the western slope of the Humboldt Range, fifteen miles south of Mill City. With Ragged Top came the Toulon mill, which Pacific Tungsten and other mines had used for a time during World War I. A local operator had tried to use it to process arsenic after tungsten mining collapsed in 1919, but eventually abandoned the project. Acting on Heizer and Segerstrom's recommendation, Nevada-Massachusetts acquired both the mill and the unpatented claims on which it stood, but deferred development on all these properties until the need arose.[57]

In the spring of 1929, after negotiating a series of lease-purchase options with five separate parties, the exploration company obtained control of what was touted as Nevada's second-largest "tungsten-producing locality" in the Silver Star Mining District southeast of Hawthorne. Known collectively as Silver Dyke after an old nearby silver prospect, the properties extended more than four miles along a complex vein system in an altered sedimentary-volcanic contact zone below the northern crest of the Excelsior Mountains. All the surface signs looked good, even to the geologists who advised the company. Oscar Hershey was "favorably impressed" after a cursory examination in 1930. He told Segerstrom that there was "no mineralogical reason, and so far as I can now see no structural reason, why ore should not extend deep in the Silver Dyke vein." It took eight years and more than $300,000 to prove him wrong. The mine turned out to be a colossal failure, Segerstrom's worst mistake as president of the Nevada-Massachusetts Company.[58]

The Silver Dyke frustration still lay ahead, however, as bookkeepers prepared the company balance sheets in 1929. In four years Nevada-Massachusetts had stabilized and consolidated under Segerstrom's leadership. It was a propitious time, with tungsten carbide just beginning to enter the market for cutting tools. Rising demand for high-speed steel, along with declining stocks of tungsten ores in Europe and the United States, led to a price surge during the spring and summer. Chinese wolfram, once shallow, clean, and cheap, had declined in quality and risen in price. Exhausted Chinese surface deposits gave way to deeper, more expensive, and more complex ores with impurities such as arsenic and tin that added costs to purification, compared with the relatively clean scheelite from Nevada. By the fall of 1929 the Nevada-Massachusetts Company was the nation's leading tungsten producer. The gross income from ore sales had risen from $51,249 in 1925 to $414,447 four years later, but subtracting depletion and depreciation put the books in the red. Thus, despite improvements in mine development, quality control, labor efficiency, and overall productivity, the company made a profit only two years out of five in the "Roaring" Twenties.[59]

Profitability had not been the most urgent priority, however. Nevada-Massachusetts was a closed corporation owned by three banks and a handful of other investors, including Charles Segerstrom and his family. Their money had been tied up in tungsten since 1918. The Boston bankers had been ready to unload the property after Clark died and Loring grew "atrocious," but Segerstrom—a banker himself as well as an experienced mining financier— persuaded his eastern partners to reorganize under new management as a financial strategy designed to protect their capital investment and wait for better times. He was not foolish enough to guarantee long-term profitability, but by hands-on management and careful cost control he brought the company from the brink of bankruptcy to fiscal solvency and operational efficiency. With a mining engineer now as president of the United States, better times seemed just around the corner.

4

TUNGSTEN AND THE GREAT DEPRESSION

The formative years of the Nevada-Massachusetts Company ended just as America's greatest economic crisis began. Instead of better times, the tungsten business slid downhill along with the rest of American industry after the market crash in the fall of 1929. Through the decade of uncertainty and upheaval that followed, Charles Segerstrom managed his affairs with remarkable resilience and pragmatic adaptation to change. In the process his company not only survived but grew to national prominence as America's foremost tungsten producer.

As Nevada-Massachusetts gained greater stature, so did its president. Segerstrom found himself on a national stage during the early years of the New Deal as spokesman for domestic tungsten producers. The experience reinforced his convictions, common to conservative businessmen in the thirties, that government intervention was more hindrance than help in solving the nation's economic problems.

HOOVER THE "GREAT ENGINEER"

The American Institute of Mining and Metallurgical Engineers is the nation's oldest and largest society of mining professionals. When the Nevada branch met at Reno in December 1928, the audience was in a celebratory mood, for one of their own had just been elected president of the United States. On the speaker's platform hung a "handsome" portrait of Herbert Hoover, presented by fellow member N. H. Getchell, Hoover's Nevada campaign manager.[1]

The mining industry had reason to celebrate. Though Hoover had left mining behind upon entering public life in 1914, his reputation as a tough but fair businessman remained both a tribute to past successes and a promise of effective leadership. Engineers applauded his devotion to "systematic management," and his insistence on productivity and efficiency in the workforce. As secretary of commerce under both Harding and Coolidge, Hoover had enhanced his reputation in the business world. A capitalist with a progressive vision, Hoover posed no threat to the classical model of free enterprise. Business leaders welcomed his efforts to harmonize relations between government and industry through voluntary associations. He defended high tariffs and hard money and at the same time worked to improve labor relations and expand overseas markets for American goods. Promoted as both the "Great Engineer" and the "Great Humanitarian" for his earlier work on behalf of Belgian relief, Hoover seemed an ideal candidate to manage the nation's affairs.[2]

The principal speaker at the Reno meeting was Ott F. Heizer, general manager of the Nevada-Massachusetts Company. He carefully reviewed company operations and technical problems over the past three years. While admitting that Chinese ore imports had hurt domestic ore sales, he was encouraged by recent reports of the "increasing scarcity of Chinese placer." Rising demand for new products made with high-quality steel alloys also seemed promising. Safeguarding the tungsten industry's future, however, required political intervention, a conviction Heizer shared with Segerstrom, an ardent Hoover Republican on the California Central Committee. Both hoped the new president would fulfill the protectionist plank in the 1928 Republican platform by raising the tariff on tungsten.[3]

"SUPERPROTECTIONISM" AND THE SMOOT-HAWLEY TARIFF

Protection was high on the national agenda even before Hoover took office. With an agricultural recession still in effect and farm purchasing power lower than before the First World War, both parties endorsed protectionist planks during the 1928 campaign. For Democrats under pressure from southern agrarian interests the chief issue was "equalization" of farm income with other segments of the economy. At best they were reluctant protectionists, torn between wanting to help farmers and at the same time keeping a lid on consumer prices. Republican Party leaders, however, saw no reason to equivocate. In an op-ed piece released in January 1929, Senator Reed Smoot asked the American people to consider a simple question: "Shall tariff rates be high enough to promote and encourage our own citizens, or low enough to encourage foreigners?" His answer, of course, was to protect both agriculture and industry by raising the tax on imports to prohibitive levels.[4]

Pressured by calls for a special session on tariff revision even before Hoover took office, the lame-duck Congress began hearings in January 1929. Though farm problems had been the main issue, Willis Hawley, chairman of the House

Ways and Means Committee, restricted only the time and not the content of testimony. Nearly eleven hundred witnesses testified before the committee over the next two months, seeking higher rates on raw materials, processed commodities, and finished goods. Only in the steel and alloy trades was there much opposition to the "ardent" clamor of "fanatical" protectionists, as one observer described them. The division that had developed in 1921 resurfaced again between producers, on the one hand, and importers and consumers, on the other. Bethlehem Steel, for example, opposed raising rates on foreign manganese, claiming domestic production was too small to satisfy the demand. Importers of tool steel hotly disputed the need to raise the tariff on high-quality rolled steel. A cast-iron pipe importer said any rise in the pipe duty would put him out of business.[5]

Late in January the tungsten industry took its turn in the protectionist parade. Chairman Hawley called Nelson Franklin to testify. A Denver attorney and professional lobbyist, Franklin had helped draft the Timberlake bill passed by the House in 1921 and incorporated into the Fordney-McCumber Tariff a year later. For his tungsten clients he produced a short typewritten newsletter with the latest market quotes, political scuttlebutt, and news culled from the *Daily Metal Trade* and other trade publications. Loring had praised him highly, and Segerstrom retained him as company representative in Washington, D.C.[6]

Franklin's testimony before the House committee in 1929 repeated a familiar refrain: cheap Chinese tungsten was once again flooding the market. Figures supplied by the U.S. Tariff Commission and the Bureau of Mines bolstered his argument. Held in check for a few years by the 1922 tariff, Chinese imports had risen from almost nothing in 1924 to 755 metric tons in 1928, or 60 percent of the total tungsten imported. In the same period American production had peaked in 1926 at 1,382 metric tons but had fallen to 830 tons by 1929, nearly a third of which had come from the Nevada-Massachusetts Company. Even with the 1922 tariff adding $7.20 to Chinese unit costs of $3.00, domestic producers with higher unit costs could not compete and were shutting down. Adding to their plight was the loophole in the 1922 law that allowed scrap steel with a tungsten content equivalent to ferrotungsten to come in almost duty free. To equalize the competition, Franklin recommended reclassifying scrap and increasing the tungsten duty 50 percent higher than previous levels.[7]

With the committee controlled by staunch protectionists, the tungsten proposal became part of the "tariff readjustment" bill that Hawley sponsored at the special session of Congress called by Hoover soon after his inauguration. The president asked for a "limited revision" that primarily addressed agricultural inequities. As the debate dragged on, however, opponents said the proposed revision did "too little" for farmers and "too much" for everyone else. Tennessee congressman Cordell Hull, for example, complained that the proposed tungsten increase would add 58 percent more to the current ad valorem rate of 192 percent. The manganese recommendation was even worse. It would have the effect, he insisted, of excluding any further imports of a metal vital to the steel industry but scarce in the United States. Agricultural duties were equally exclusionary. Despite efforts by the Tariff Commission to reduce the 1922 sugar duty, the beet sugar trust, representing only 3 percent of American farmers, secured a 12 percent increase. The rates on cattle rose by 50 percent, butter by 42 percent, cream by 65 percent. A *New York Times* editor summarized the so-called benefits as "few for the many, but many for the few." As he saw it, protection had become "superprotection," with rates on some commodities reaching prohibitive levels.[8]

The "Hawley tariff" passed the House in a partisan showdown late in May, but over the summer the debates grew ever more contentious as the Senate struggled with its own tariff package. With Senator Smoot, chairman of the Finance Committee, leading the protectionists, and with Hoover threatening a veto of any bill he thought excessive, the sectional wrangling crossed party lines. Senators representing the industrial Northeast threatened to cut back support for any agricultural increases if the farm lobby did not accept a quid pro quo. The special session ended in a stalemate after remnants of the old Progressive coalition led by Senator William Borah introduced a resolution on the Senate floor that would cut out all proposed rate revisions except those on agriculture. It lost by only one vote.[9]

To salvage something from the legislative wreckage before the midterm elections in 1930, the Republican old guard beat back disgruntled insurgents during the regular session of Congress that began in December 1929. Six months later, as the economic fallout of the stock market crash was just beginning to be felt, the legislature passed the Smoot-Hawley bill, a compromise package that pleased no one. Farmers felt "betrayed," and industrialists were disgusted by last-minute changes that lowered rates they had demanded. Pennsylvania senators David Reed and Joseph Grundy, the latter known as the "king of lobbyists" because of his business connections, denounced the bill but voted for it anyway to end the "tariff agitation." Senator Robert La Follette, Wisconsin's flamboyant progressive, said the bill surpassed in infamy the 1828 Tariff of Abominations that had led to the Civil War. While the bill was still on the president's desk a thousand economists from across the nation signed a petition urging Hoover to veto, warning it would raise the cost of living, hurt foreign trade, harm American investments abroad, and invite retaliation. Hoover signed it anyway, lamely declaring that no tariff was perfect. As consolation he vowed to adjust any inequitable rates using the flexible provision of the 1922 tariff bill that gave him authority to alter existing rates by 50 percent up or down.[10]

Reducing the political influence over tariff decisions was one of many issues that reformers failed to resolve while protectionists dominated the government. The Depression soon changed the political climate and gave the critics another chance. By 1932 only the most die-hard protectionists still defended the Smoot-Hawley Tariff. President Hoover never seemed to grasp the geopolitical fallout caused by protectionism. Even before he signed the bill into law, foreign governments began retaliating by raising their own tariff barriers. Despite the precipitous decline of American foreign trade after 1930, the president stuck to his high-tariff and hard-money policies, and so did a dwindling set of admirers. In the 1932 campaign, Charles Segerstrom tried to rally a Sonora crowd by linking protectionism to employment. A high tariff, he claimed, protects American workers from having to compete with cheap foreign labor. "Therefore, Herbert Hoover is making a fight for humanity to maintain a Protective Tariff, and humanity is stronger than passion." Despite this appeal from one of its most prominent citizens, Tuolumne County backed Franklin Delano Roosevelt that year by a thirty-two-vote margin.[11]

NEVADA-MASSACHUSETTS DURING THE HOOVER YEARS

Instead of protecting the jobs of American workers, the Smoot-Hawley Tariff added to the industrial malaise that began even before the Wall Street crash. Symptomatic of broader economic troubles, the wholesale price index peaked in the summer of 1929 at about 73 percent above prewar levels, and then started a gradual decline that accelerated after February 1930. During the same period carbon steel production fell 28 percent, and alloy steel output fell even faster and further. As automobile sales plummeted, tungsten steel and other ferrous alloys dropped 38 percent below their 1929 production record of nearly four million tons.[12]

If Charles Segerstrom failed to see the connection between rising protectionism and declining markets at home and abroad, at least he recognized the dangers presented by falling commodity prices. Tungsten prices shrunk by 8 percent the first year of depression, another 9 percent the second, and a whopping 16.5 percent the third. Though production costs per unit remained steady until 1933, slumping sales and shrinking ore prices ate away at company revenue. Despite President Hoover's assurances in June 1930 that the Depression was "over," business did not begin to pick up until late 1933. Nevada-Massachusetts lost money all but one of the Hoover years.[13]

A loss of profitability was not the same as bankruptcy, however. Thanks to prudent management policies and the financial cooperation of both creditors and customers, the Nevada-Massachusetts Company avoided the fate of the more than 106,000 other businesses that folded between 1930 and 1933. By cutting costs wherever possible, Segerstrom maneuvered through these hard times without a complete shutdown, although the company suspended operations for the last eight months of 1932. Closing down entirely even for brief periods—the standard ploy of many factories during the Depression—was not an option for the Nevada firm because of hydrostatic conditions underground. Though not excessively wet compared to many western mines, the Mill City mines made enough water to require constant pumping. In 1929 electric pumps lifted 90 gallons of water a minute out of the Stank shaft alone. By 1937 the mine was still making 250 gallons per day.[14]

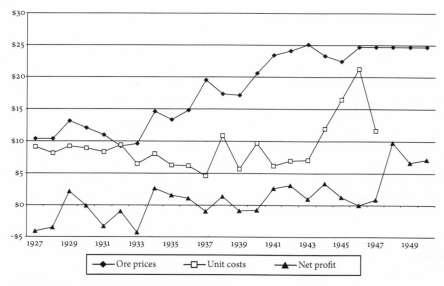

Nevada-Massachusetts profit and loss ($/u)

With a declining cash flow as ore prices fell faster than operating costs, how long Nevada-Massachusetts could remain in business depended on how much revenue it could generate through sales of its products. By the early 1930s bulk steel production had declined sharply as Big Steel retrenched and consolidated. As we have seen, however, Segerstrom's primary clients were not the giant firms but ore processors and alloy steelmakers. Of the latter, the largest was Crucible Steel Company of America, with six plants, two each in New York, New Jersey, and Pennsylvania. Formed at the turn of the century from a merger of several independent crucible steel producers in the Pittsburgh area, by World War I it was the foremost producer of high-speed steel in the country. Adding newer electric furnace technology to its stable of crucible and open-hearth furnaces, the Crucible Company made both standard and specialty steels. Until the late 1920s most of its tungsten for steel alloys came from imports, putting it at odds with producers and many smaller specialty firms that supported the protectionists. By 1929, however, Crucible had softened its tariff opposition, in part because of the declining quality of cheaper foreign ores. Chinese wolframite was still available at low market prices, but the Chinese had gone underground as surface placer deposits ran out. Deeper ores were more complex, containing sulfur, arsenic, and other impurities that had to be removed before concentrates could be refined. The added costs made foreign ore purchases less attractive. Moreover, political instability during the warlord era slowed ore production and delayed shipments from Shanghai and Hong Kong, the two largest ports for ore from the Chinese mainland.[15]

Faced with these uncertainties, encouraging more reliable domestic suppliers was a predictable strategy for American steel alloy companies that had to buy tungsten on the open market. Aside from Crucible Steel, most were small, independent firms that specialized in the production of a variety of alloys containing less than 1 percent to more than 20 percent tungsten. As John N. Ingham found in his study of steel companies clustered in the Greater Pittsburgh area, remaining small was a social as well as an economic decision. Owners had a "special mentality" to protect their elite status in the community by avoiding organizational changes beyond their control, keeping a paternalistic eye on employees, and finding a proper market niche for their products. To maintain profitability in a highly competitive field they had to be innovative and flexible, adaptive to the demands of steel consumers, just as ore suppliers had to adjust to the changing needs and specifications of steelmakers.[16]

Trying to interest specialty firms in Nevada-Massachusetts ore was a difficult task in the depths of depression. The slump in automobile sales after 1929 was particularly hard on alloy steelmakers. Autos consumed nearly 80 percent of steel alloy production, but the auto industry was drawn into the same downward spiral experienced by the rest of the steel business: inventory reduction, employee layoffs, production cutbacks, and plant closings. Some firms never survived these cutbacks; others consolidated or merged into larger operations. By 1931 annual production of bulk steel had fallen by more than half its 1929 tonnage. Alloy steel production over the same period dropped by almost two-thirds. Those tungsten consumers that remained in business operated on reduced schedules, buying ore on the spot market only as needed to fill existing orders. As Segerstrom complained in 1931 to an Electromet executive, "Everyone seems to only want to purchase sufficient for the shortest period of time and are willing to take chances on futures."[17]

Nevada-Massachusetts operated only part-time after the summer of 1930, producing just enough ore to meet the limited demand. To bolster production and attract buyers, Segerstrom considered various inducements. He let Molybdenum Corporation, an old customer, pay for contracted ore with non-interest-bearing trade acceptances, even though they were subject to discount when redeemed. When Ludlum Steel, a small alloy firm in Dunkirk, New York, ordered two cars and asked for deferred delivery and payment, he gladly accepted, adding that "at the price we are selling, there is absolutely no profit in this material for us." Early in 1931 he offered three carloads of high-grade scheelite to Crucible Steel on generous terms, but heard nothing in reply. A month later he gave J. J. Haesler, his eastern broker, permission to sell a single carload—thirty-five tons of concentrates—at any reasonable price, even if below cost. Haesler advised not selling for less than $11.50 a unit—just above the cost of Chinese ore plus the duty of $7.93, FOB New York—but Segerstrom needed revenue to keep the mine going. Crucible eventually bought a carload containing about twenty-five hundred units for $10 per unit—a dollar below the spot market—delivered free of charge to a subsidiary in Syracuse, New York. It was enough to tide the mining company over for a few more months.[18]

Segerstrom's eastern partners bowed to his leadership and experience, but kept a sharp eye on the bottom line. The alarming rise in bank failures across the nation in late 1930 cast a cloud over all of the nation's financial institutions, including the two Boston investment banks so deeply involved in the fortunes of the Nevada-Massachusetts Company. To ease cash-flow problems until ore sales picked up, the Boston bankers loaned the company an additional $340,000 between 1929 and 1933, taking unsecured promissory notes but insisting on prompt interest payments. Early in 1931 Segerstrom alerted them that Vanadium Corporation of America was in the area looking to purchase good tungsten properties. The Sonora banker seemed ambivalent to the possibilities of a buyout, but to worried investment bankers the chance to unload an uncertain asset was too good to miss. Setting the price at $1 million, the Nevada-Massachusetts board gave VCA a thirty-day option but extended it another month to give Vanadium engineers time to complete their examination. By May, however, the economy was in a tailspin, steel production was down more than 38 percent over the previous year, automobile factories were closing, and tungsten sales had completely dried up. Even President Hoover had changed his tune, blaming the Depression on global economic problems rather than Wall Street speculators. Because of these unfavorable "present conditions" VCA executives let the option expire.[19]

Segerstrom's voluminous papers do not reveal whether he was discouraged by the failure of the VCA negotiations or glad he still had a mine to run. The negative market conditions had clearly shaken his confidence, as his annual report to stockholders revealed. It was both unprofitable and unpleasant, he admitted, to have to work like a stock jobber trying to peddle discounted tungsten ore to "unwilling buyers who are operating their plants at approximately 25 percent capacity." Operating costs for the year exceeded gross income by $43,000, making 1931 the "most difficult year for operations in the history of our company." Yet he insisted the firm was better off than most mining companies whose losses he thought were much worse.[20]

If the past year was difficult, what of the immediate future? Market prospects for 1932 seemed hopeful at the beginning of the year. Though steel demand was down in three major industries—autos, railroads, and construction—declining inventories gave optimists some hope that business would pick up later in the spring or summer. One analyst in January told Segerstrom he expected steel plant output to reach 70 percent of capacity by October. Another thought tungsten prices might rise on the bad news from Asia, where Japanese adventurism on the mainland was disrupting ore shipments from Shanghai. On the other hand, all bets were off in an election year. The world was in a financial crisis, and the president's political capital was rapidly dwindling through futile efforts to stimulate the economy with a supply-side agenda that was too little and too late. Hoover's support for federal loans to business through the

Reconstruction Finance Corporation (RFC) was a major concession to government power, but most of the money went to a few banks, and little of it trickled down to small business or the hard-pressed populace.[21]

More worrisome to tungsten producers was the prospect of a Democratic sweep and renewed attacks on the tariff. Though protectionists still had powerful advocates in both government and industry, the Smoot-Hawley bill had been a political embarrassment, a "patchwork of shameful compromises." Reformers during the campaign were unhappy about the Democratic candidate's waffling on tariff matters, but given their party's history as the champion of free trade, clearly the odds favored downward revision if the Democrats won. Thus, long before the November election, the financial consequences of a lower duty had to be considered in any tungsten marketing negotiations.[22]

Tariff issues played a role in Nevada-Massachusetts ore sales during 1932 contract talks with Molybdenum Corporation. The tungsten processing business by the early 1930s had dipped as the steel trade declined, but Molycorp was still a major supplier of tungsten powder and ferrotungsten to alloy steelmakers and fabricators in the industrial Northeast. The company also sold ferrotungsten periodically to the fledgling Russian steel industry via Amtorg, the Soviet government's purchasing agent in the United States. Even though the U.S. government did not extend diplomatic recognition to the USSR until 1934, Soviet industrialization was modeled after the American emphasis on "mining, metallurgy, and machine building," as economic historian Kendalle E. Bailes has written. Amtorg, therefore, bought many more American industrial products than it sold Russian raw materials to Americans. Except for one anomalous year, 1933, the trade balance remained in America's favor throughout the twenties and thirties.[23]

Early in 1932 Segerstrom offered Molycorp three carloads of concentrates over a three-month period on what he thought were generous terms: $10 a unit, payable in non-interest-bearing trade acceptances due six months after each delivery. Molycorp's president, Marx Hirsch, a shrewd bargainer, demurred. With two Amtorg orders on the books and others under consideration he could use more ore, but three cars in three months at a fixed price was too risky to handle in an uncertain market, especially "if the Democrats get in." To conclude the deal Segerstrom had to agree to repurchase the ore if Molycorp had not sold it before the due date of payment, and then sell it back at the spot price of Chinese wolfram, including whatever duty applied at the time. As Hirsch told Segerstrom's agent, J. J. Haesler, this stipulation was "necessary . . . [to] avoid the risk of . . . not being able to dispose of the Ore prior to the elections in November."[24]

Hirsch had good reason to be cautious. Instead of improving as Segerstrom's advisers had predicted, business conditions grew worse over the spring and summer. The international financial crisis accelerated the rate of bank failures in spite of RFC efforts to bolster the credit system. Another round of wage cuts and factory closings added thousands to the unemployment rolls and drained the resources of relief agencies. By the fall of 1932 the Depression had forced nearly thirty-two thousand more American businesses to close and had left almost 24 percent of the workforce without any substantial means of support.[25]

One of those closed businesses was the Nevada-Massachusetts Company. Except for maintenance and pumping, it suspended all mining and milling operations after May 1, 1932. The Boston creditors insisted on cutbacks after Segerstrom reported he was "unable to sell a single pound" of ore following the spring negotiations with Hirsch. Scaling back in hard times was prudent fiscal strategy, a pragmatic adjustment rather than a permanent shift. Neither Segerstrom nor his backers lost faith in the mine's ability to generate positive cash flow under normal economic conditions. In the interim they kept the company afloat with commercial and personal loans. Their optimism was strengthened by two teams of postgraduate geology students led by Paul Kerr of Columbia and Theodore Hoover of Stanford. After several weeks of on-site study while the mine was shut down, they concluded that Mill City was "among the important tungsten producing districts of the world."[26]

During the shutdown Segerstrom crossed the country to line up customers and negotiate a better contract with Molybdenum Corporation. Hirsch had offered to buy all the ore Nevada-Massachusetts could produce at a reasonable profit, but the proposed price was lower than importers paid for Chinese ore. The Sonora banker was wary of any long-term deals that would tend to drive down market prices for tungsten. On the other hand, the Wah Chang Company, one of China's principal metal importers in New York, had been promoting the idea of an international tungsten cartel similar to those in the copper and steel industries, ostensibly to stabilize the market, but to Segerstrom the motives were unclear. He suspected the real purpose was to undercut domestic producers and force down the tariff.[27]

Faced with these uncertainties, and with no immediate revenue prospects,

Segerstrom was on the defensive in New York talks with Hirsch. The contract he signed in June sacrificed market flexibility and floating ore prices for a long-term relationship with a major ore buyer. The terms, effective the first of July 1932, gave Nevada-Massachusetts a base price of $7.93 per unit for ore delivered to Molycorp's Pennsylvania plants with no credit risk, plus a "contingent amount" based on a sliding scale for higher-than-average grades of ore. Although the base price was only 79 cents a unit above the tariff rate for imported tungsten, it was low enough to undercut foreign ore prices FOB New York, assuring Molycorp of a steady supply of good ore at the cheapest market price. In 1933 the Pennsylvania company refined only twenty-five hundred units of WO_3, a figure Nevada-Massachusetts could easily meet. If the tariff fell below $7.14 per unit, the contract was void. The amount of ore to be delivered was unspecified, but Segerstrom thought he could meet Molycorp's estimated needs with plenty to spare. If Nevada-Massachusetts fell short, the contract stipulated that Molycorp could buy ore elsewhere, and that Nevada-Massachusetts could meet the contract terms with purchased ore if necessary. An important concession was the clause allowing Nevada-Massachusetts to sell ore outside the contract to other customers for direct smelting, although the contract gave Molycorp a 15 percent cut of the net proceeds for any sales above the base price. The contract would remain in force five years, but either side could cancel after thirty months.[28]

Considering the circumstances Segerstrom thought he had driven a hard bargain. His report to Boston assured the bankers that Molycorp "will purchase *all* their ores from us and . . . *all* we can expect to contract for and with outside sales it looks like a wonderful arrangement to me." Other metal traders were less enthusiastic when details of the contract leaked out. Some importers, fearing that greater output from Nevada-Massachusetts would hurt their own businesses, tried to sabotage the deal by raising the specter of tariff opposition in an election year. David Taylor, the New York metals dealer who had opposed efforts to raise tungsten duties in 1930, told Hirsch bluntly that "he did everything he could to create ill will between Segerstrom and ourselves as it was to his interest to do this." Hirsch kept his cool, however, assuring Haesler that "he would not be a party to such a scheme."[29]

Even without outside opposition the two firms had trouble implementing the contract. The timing could not have been worse, with business at a standstill and no prospects for change until after the election. Confronted with an ever-shrinking economy and a looming social calamity, the Hoover administration vacillated. The incumbent president had done more than any predecessor in previous depressions, but he hesitated to abandon fundamental conservative ideals despite an unprecedented economic and social emergency. The political crisis left businesses like Nevada-Massachusetts in financial limbo. It needed to sell at least half a carload of concentrates per month to meet expenses, but with markets at a virtual standstill the steel business was moribund. Hirsch refused to make specific commitments as to volume and price, but without them the contract tied Segerstrom's hands without generating any revenue. Eventually, the refinery president agreed to better terms, but delayed ordering more ore until Molycorp's stockpile had been consumed. When orders did come in they took months to fill, and Molycorp's trade acceptances could not be cashed until six months after delivery.[30]

In the meantime, Segerstrom operated on borrowed money and advances from sales. The contract allowed ore sales for direct smelting, but customers were scarce and skittish in the waning months of 1932. As the company president told his investors in September, he had "scores of inquiries for ore, showing that everyone is in need but they do not want to finance the purchasers [sic] until the orders for steel are actually in their hands."[31] Even so, Hirsch grew concerned that the direct sales option might undercut Molycorp's ferro business, since both firms would essentially compete against each other for the same customers. The few orders Nevada-Massachusetts did receive were hard to fill. With his own mine closed, Segerstrom bought small lots of ore wherever he could, and then mixed them with stockpiled Nevada-Massachusetts ore and concentrated or otherwise treated the mix to make up a sales order. These were stop-gap measures that lost money, but Segerstrom thought it was better to lose a little while awaiting a rebound than shut down entirely and face a total loss.[32]

Most of the contract difficulties with Molybdenum Corporation were resolved through negotiation over several months, with J. J. Haesler acting as intermediary between the two principals. Haesler's objectivity might be questioned, however. He and A. C. Daft, the independent Pittsburgh broker, drew commissions from both Molycorp and Nevada-Massachusetts, but Haesler was beholden personally to the Sonora banker for loans that had helped rescue his own Metal and Ore Corporation from bankruptcy during the lowest months of depression. Segerstrom, in turn, relied on Haesler's eastern connections and counsel on financial matters. To avoid any complications over direct sales by the Nevada-Massachusetts Company, Haesler suggested that Segerstrom conduct some business on the side under the name of the Western Tungsten Company. This was a private family firm quietly organized in New York earlier in the year to buy and sell ore. Haesler and Daft, respectively president and treasurer, handled all sales, but Segerstrom and his wife controlled the company assets, including all the family stock in the Nevada-Massachusetts Company. By keeping the accounts separate and the records closed, for years Western Tungsten competed with both Molycorp and the Nevada-Massachusetts Company with no apparent concern for conflicts of interest.[33]

TUNGSTEN PRODUCERS AND THE NATIONAL RECOVERY ADMINISTRATION

Early in 1933, after the Molybdenum Corporation received a sizable order from Amtorg that required Nevada-Massachusetts ore to fill, the Mill City plant resumed operations. The closure had nearly brought down the company, but FDR's inauguration improved the business climate even if clouds of doubt remained. Segerstrom hedged his bet with a cautious assessment to stockholders in May 1933: "Considering that the year 1932 will probably go down in history as the lowest point of the depression, we believe that our operations at Mill City reflect better than average conditions."[34]

Unlike some Republicans who had abandoned Hoover by 1932, the boss of Nevada-Massachusetts remained loyal to his conservative roots and skeptical of the vague promises of the New Deal. Trade issues were especially worrisome in view of traditional Democratic opposition to protectionism. Segerstrom had welcomed Hoover's signature on legislation approved the day before he left office that added a new trade restriction on American imports. The "Buy American" Act, which Oklahoma senator Thomas Gore (unrelated to the later vice president) sneeringly labeled the "Goodbye American" Act, required government procurement agents to buy products of American "origin or manufacture" whenever possible. In the words of a Canadian economist, the measure had the combined effect of a "tariff, a quota, and discrimination against imports of a certain type." Segerstrom and other domestic producers thought it would stimulate business, but the law was incompatible with New Deal trade policy. Because of discretionary loopholes it was also unenforceable. Yet it remained on the books as an expression of American nationalist sentiment, and periodically revisited.[35]

However important for economic recovery, trade issues took a backseat to emergency relief during the frenetic "Hundred Days" following the March 4 inaugural of Franklin D. Roosevelt. The Great Depression had reached bottom during the last days of the Hoover presidency, not just in statistical terms showing production and price declines, bank failures, and foreclosures but also in the human consequences of unemployment, homelessness, hunger, and despair. With the Republican opposition discredited and demoralized, the new administration in a burst of energy proposed a series of bills that a compliant Congress rubber-stamped with almost no debate. Promising to aid people directly by reopening the banks, creating jobs for urban youth, recognizing the rights of labor, establishing a national relief system, expanding farm and home mortgages, even legalizing beer and light wine, FDR eased the grumbling discontent that threatened the very foundations of American capitalism. Some of the government's promises proved hard to keep, but at least they gave people in need some reason to hope. Even statisticians could sense the changing mood. After plotting the conjunction between quarterly business trends and the impact of Roosevelt's first month in office, one reported: "The quarter closed with faith and hope rapidly reviving."[36]

Lifting the economy proved longer and more difficult than lifting public spirits. Inflationists won early victories by taking the country off the international gold standard and devaluing the dollar, but opponents in both government and business warned of dire consequences. Weakening the dollar, however, had some beneficial effects on the tungsten market. A year after devaluation, Segerstrom had to admit that ore prices had risen as the foreign exchange rate had declined and imports became more expensive. That was good for domestic metal producers, but how long would it last? A cheaper dollar would mean little if new tariff legislation cut the rates on imported ore.[37]

Although tariff reform topped the personal agenda of the new secretary of state, Cordell Hull, the president put off that battle until more pressing domestic recovery legislation had cleared Congress. The farm problem was a Gordian knot of complex price, production, and trade issues. Proposals to raise farm prices by reducing the surpluses through federally designated crop allotments, price supports, and other measures had been considered in the twenties, and New Deal farm programs drew on those precedents. But the idea of using federal commerce and tax powers to manipulate the domestic market for farm commodities drew critics from both extremes—the Far Right for too much regimentation and the Far Left for too little. Internationalists like Hull and Henry A. Wallace also criticized short-term nationalist solutions to a long-range problem, arguing instead for lower tariffs and expanded trade. Nevertheless, import quotas and other subsidies provided substantial benefits to the nation's largest farmers, cutting production, lifting commodity prices, and raising farm income.[38]

Government intervention was also the initial solution proposed for industrial recovery, but policy makers were divided over both the means and the potential benefits. On June 16, 1933, the president signed the National Industrial Recovery Act (NIRA). He defined it as a public works plan to "put people back to work," but it was much more than a stop-gap relief program for the unemployed. Promoted on the assumption that economic recovery in a national emergency required cooperation instead of competition, the legislation endorsed incongruent principles and policies long held either by labor or by business leaders. In sanctioning employee rights to join unions, bargain collectively, and strike, Section 7(a) sought to strengthen the rights of labor to offset the power of management. In authorizing businesses to work together and create their own "codes of fair competition," the new law fostered industrial cartelization and undermined older populist-progressive antitrust laws. To balance these "countervailing" powers and oversee the program, the NIRA created a labyrinthine new federal agency, the National Recovery Administration (NRA), staffed with thousands of bureaucrats under an irascible and audacious chief, Gen. Hugh S. Johnson.[39]

More than any other legislation during the first months of the New Deal, the NIRA challenged the old economic system and offered a new vision of power sharing among government, business, and labor. But piecing together a hodgepodge of programs under a cumbersome bureaucratic structure that lacked adequate enforcement power was not a sustainable formula for economic recovery. W. M. Kiplinger summed up the fundamental problems in his newsletter that fall: "Too much was promised, too little delivered within the promised time."[40]

American metal producers responded to the NRA with a mixture of cautious optimism and alarm. They liked the provision delegating code-setting power to each industry, but opposed the wage and hour standards, and vowed to fight any changes in the tariff schedules. A week before the NIRA was signed into law, the American Mining Congress, an industry lobby anticipating the political struggle ahead, organized a new division of "Rare Metals and Non-Metals" and called delegates from that division to Washington to discuss the implications of the new legislation. Declaring a unity of interest in developing fair industry codes, equitable taxes, and "proper and adequate tariff protection," J. F. Callbreath, AMC national secretary and spokesman, appointed well-known business representatives to chair each component of the new division. The man he picked for tungsten was Charles H. Segerstrom.[41]

Now fifty-three years old, a family-oriented businessman, active in local church and charitable organizations when not engaged in banking and mining, Segerstrom had little time for national politics. Though careful to cultivate key lawmakers with campaign contributions and friendly advice, he preferred financing other lobbyists to getting personally involved himself. When called to the national stage, however, he could not resist becoming an active player. He responded with alacrity, traveling to Washington late in June 1933 to plan strategy with Callbreath and other industry leaders.

The AMC meeting was a reality check on the prospects of tripartite cooperation envisaged by the NIRA. Metal producers, like other business leaders, resisted any NRA standards that threatened industry profitability and independence. Assuming each branch of industry would develop a separate code, the tungsten delegates discussed the general code requirements and established the "Tungsten Industries Association," soon renamed "Tungsten Producers," with Segerstrom the chairman. They envisioned a producer's cartel, similar to the steel trust and the oligopoly of major copper companies, that would determine what constituted unfair competition and take appropriate action against violators. Asserting an unlimited mandate, the association proposed to oversee "wages, hours of labor, conditions of employment, number of employees, production, shipments, sales, stocks, prices, and other data pertinent to the purposes of this Code."[42]

Turning to tariff issues, the delegates complained that the tungsten industry was operating at "a substantial loss" and could not stay in business without a ban on foreign ores. All opposed tariff reduction but were divided on how to make that message clear. Those who saw advantages in using the NRA as a weapon wanted to invoke the "club" of retaliation given the president under Section 3(e), the NIRA import control and embargo provision that the administration had added as a bargaining chip to win industry support. The Colorado delegates, however, took a harder line. They wanted the members to refuse to join NRA "unless protection is given ... against low cost and low prices [of] foreign competition."[43] Left to govern themselves, tungsten producers equated public interest with self-interest.

How the government would respond to industry concerns was not immediately clear. NRA officials offered mixed signals, reflecting internal divisions over policy, organization, and implementation. In an era when the concept of limited government still resonated with the courts, the industrial leadership, and most of the citizenry, cooperation and voluntary compliance seemed better than compulsion. Thus, Gen. Hugh Johnson, the impulsive NRA boss, regardless of his statutory authority to regulate industry through a licensing system built into the NIRA, never implemented that provision, preferring moral suasion to legal threats. In a meeting with the assembled delegates he was full of patriotism and promises, as Segerstrom reported to the Atolia management: "Johnson the Administrator told the meeting . . . that if the various groups organized, they would protect them even to the extent of an embargo on foreign ores, so we are working hard on the code." Johnson's remarks were presumptive, ignoring the political impact of surging antibusiness public sentiment, and overlooking the State Department's foreign policy shift toward international trade and cooperation. Nevertheless, they led the assembled mining executives to believe the government was leaning their way.[44]

At least that was Segerstrom's initial reaction. He saw Johnson's appeal for cooperation as an opportunity to "stabilize" the tungsten industry with government-sanctioned production and price controls. To adopt the AMC's draft code and submit it for official endorsement, however, required the support of every major tungsten producer in the country. As chairman of the Tungsten Association he worked for six months to build a powerful industrial cartel, but internal dissent and external complications frustrated the effort.[45]

The first signs of trouble came from NRA officials after reviewing a draft preamble that took a hard line on tariff protection. They flatly rejected any quid pro quo on imports, insisting that industry codes first had to be presented and approved before the president would consider any protectionist appeals. They also questioned the labor and price clauses, the two sticking points for every industry trying to comply with NRA regulations, and the subsequent executive order establishing minimum-wage and -hour guidelines for industries without approved codes.[46]

The labor provisions of the draft code gave lip service to labor rights but set a different standard for wages and hours than the president's Re-employment Agreement, a set of guidelines promoted under the banner of the Blue Eagle, Johnson's symbol of industrial compliance. The draft established the wage rate at forty cents per hour, a dime higher than the president's minimums, but adopted the Blue Eagle workweek of forty hours only where "geographic conditions" warranted. For two months Segerstrom and other mining representatives negotiated with NRA officials, finally compromising on a provision that made the forty-hour week standard and a forty-eight-hour week the exception in remote areas of the West where it was hard to find skilled underground workers. The provision still had to be ratified by association members and approved by the president, but at least it was a step forward.[47]

Compromise also resolved the question of how to establish a fair market price for tungsten. Segerstrom, whose Nevada scheelite was less expensive to mine than Colorado ferberite and less costly to purify and refine into secondary products than Chinese wolfram, wanted a base price geared to the cost of producing ferrotungsten. Smaller operators with higher production costs, however, feared they might be priced out of the market by any index based on secondary cost factors. Since the NIRA prohibited selling ore below cost, the tungsten code required a price-setting standard that would satisfy all producers. After months of discussion, industry executives settled on a modification of the draft that omitted any reference to a base price. Instead, the preamble claimed that by agreeing to "shorter hours and higher wages," the government authorized tungsten producers to "cooperate as to secure a return of those required expenditures, plus a fair profit." Antitrust critics of the New Deal, however, equated price-setting with price-fixing—bad for consumers and, like high tariffs, a symptom of monopoly.[48]

During the summer of 1933, while still struggling over code language, Segerstrom and his associates learned that the organizational rules had changed. Instead of filing separate codes for each branch of the industry as originally suggested, NRA administrators now wanted a unified code for "as many of the industries as would work [together]." After completing the common code, each subdivision could then select its own officers and propose specific rules for working hours and wages. Every action or change, however, was subject to approval by an "Executive Emergency National Committee" consisting of representatives from both industry and government.[49]

Callbreath, the Mining Congress secretary, thought it was good strategy to adopt this simplified structure as soon as possible. He told Segerstrom that the AMC was "trying to have this appear to be the administration's code rather than our own, for purposes which will be clear to you."[50] But big and little producers from different geographic areas had divergent, sometimes incompatible, goals. Faced with a September deadline for code submission to the NRA, the tungsten chairman spent all summer trying to devise a workable formula that would satisfy all interested parties and still meet government guidelines.

Some companies were enlisted by offering monetary or other inducements.

The Atolia Company, for instance, agreed to sign the proposed code after Segerstrom offered to handle smaller lots of Atolia ore through the sales and distribution system of the Nevada-Massachusetts Company. Others simply needed more information, as Segerstrom learned after sending a draft to the Boriana Company, a small Arizona firm that had not been invited to the initial tungsten strategy session. A reply came back two weeks later. We "do not wish to convey the impression . . . that we are quarrelsome, obstinate or recalcitrant," said the indignant company vice president, but he threatened to file a protest if he did not get a full explanation "immediately." Privately, Segerstrom complained to his friend Haesler that Boriana was "on the warpath," but after a quick trip to their Los Angeles headquarters he used a combination of charm and intimidation to "whip them into line."[51]

More troublesome were two Colorado firms who felt threatened by foreign tungsten. The Wolf Tongue Mining Company and the Tungsten Production Company still produced significant amounts of commercial ore from the Nederland district near Boulder. However, they were struggling with diminished ore bodies and rising costs. Shipping high-grade ferberite concentrates by rail to the Firth-Sterling refinery in Pittsburgh, for example, cost the Wolf Tongue Company $14.50 per ton, compared to $4 per ton that Chinese ore importers paid for bulk shipments by sea from Hong Kong to New York. Labor in the Colorado mines was also expensive, amounting to 63 percent of Wolf Tongue's direct costs in 1931, compared to 51 percent for Nevada-Massachusetts.[52]

Faced with mounting problems, the Colorado producers wanted Segerstrom to take action without delay so they could petition the president for an embargo, even if it meant filing a separate code for tungsten rather than a generic one for all metal producers, as the new directives instructed. William Loach, the Wolf Tongue CEO, thought that if FDR rejected or declined to act on an embargo petition, AMC should lobby the U.S. Navy. Naval officials, he assured Callbreath, had great influence on Roosevelt, since he "is a strong Navy man and with the big program for increased Naval equipment, the need for Tungsten is apparent to all of us."[53] As Loach would soon discover, however, military planners had a different view. Where strategic minerals were concerned, they wanted tariff barriers lowered, not raised.

The tungsten industry was still struggling with code language and internal disputes when the NRA filing deadline passed early in September 1933. Hundreds of industry codes had been submitted by that time, some with questionable labor and price provisions, but General Johnson wanted all codes filed before starting the NRA review and revision process. Despite objections from Colorado, however, Segerstrom hesitated. Since FDR's inauguration metals prices had risen, undermining the argument that foreign imports were hurting American producers. It seemed also strategically wise to hold up on tungsten until the NRA took action on the proposed gold code. Segerstrom knew the gold business as well as he knew tungsten. Gold was a western commodity, produced, like tungsten, by a wide variety of independent operators under similar geological conditions. Miners of gold and tungsten were virtually interchangeable in terms of skills, wages, and expectations. The gold code, which Segerstrom had helped prepare, was a good model for tungsten, especially with its clause setting a forty-eight-hour workweek in remote areas. If the gold code passed NRA muster, a precedent would be set for any codes that followed.[54]

Procrastination may have had strategic merits, but to some producers the filing delay seemed more like a tactical blunder. After the NRA deadline passed unheeded, the complaints from Colorado grew louder and more ominous. Segerstrom needed the cooperation of all producers to build an effective cartel, but he refused to file a premature code. To avoid an open revolt, the chairman finally called the first formal meeting of the Tungsten Association. A handful of delegates, including key producers from Colorado, Nevada, and California, gathered at Salt Lake City on October 21, 1933. After installing officers and directors the participants adopted the code drafted four months before with only minor modifications, subject to ratification by the directors of each company represented. The meeting ended amicably, every delegate assuming that quick action would be taken to complete the code and get it approved.[55]

Every delegate but one, that is. Charles Segerstrom had quieted dissension in the ranks with the Salt Lake meeting, but rather than submit the code prematurely, he decided to wait. He still hoped to piggyback on the gold code, which had never gone beyond the drafting stage. Neither had a code been developed for the refining industry, which if completed would have a direct bearing on tungsten ore and products. Moreover, tungsten prices remained steady for the rest of 1933, reducing the demand for imported ore despite the continued calls from Colorado for an embargo. When the foreign market began to fall off in January 1934, J. J. Haesler suggested it was time to send the code forward. The tungsten chairman, however, thought differently. By that time external events cast a shadow over the whole NRA program, as Kiplinger earlier had predicted.[56]

The warning signs were obvious by the beginning of 1934. Code enforcement had bogged down. Antimonopolists had grown increasingly strident, led by Senator William E. Borah, the old progressive "lion of Idaho." Consumers had gained an official platform with the NRA-sponsored Consumers' Advisory Board. In December 1933 the Federal Trade Commission rejected the quicksilver industry's call for embargo—a portent for other groups with similar appeals. Three months later it was the steel industry's turn to squirm. Responding to criticism from consumers, small businessmen, and their political allies in Congress, the FTC condemned the steel code's pricing policies and forced a revision of its code provisions. Earlier Callbreath had warned Segerstrom that unless tungsten had a code of its own, the NRA might force the tungsten group under the steel code umbrella—in effect, putting tungsten producers at the mercy of their customers. Big Steel's bad image, however, made code consolidation highly unlikely.[57]

With the NRA under a cloud and the FTC on the offensive, Segerstrom saw little reason to persist in tungsten code writing. When an eastern company threatened to invoke steel code provisions against Nevada-Massachusetts after it failed to pay for some steel castings within the thirty-day billing limit, he acidly replied, "As you well know . . . it will not be long before both prices and terms will be a matter of mutual adjustment and not of code provisions." He was more prescient than he realized. Less than a year later the U.S. Supreme Court struck down the entire experiment in federally sanctioned industrial cartelization.[58]

For Segerstrom and other conservative businessmen, the NRA experience was a lesson in frustration that only reinforced their instinctive distrust in the New Deal. Business self-regulation through "codes of fair practice" had failed as a mechanism for industrial recovery. But the Blue Eagle emphasis on cooperation, and the growing influence of labor and consumers in federal planning, seemed to them an ominous trend. Instead of accepting their share of responsibility for the shortcomings of NRA code formulation and enforcement, they blamed the process and the philosophy behind it. Herbert Hoover, now a conservative elder statesman, warned that "cooperation" through "regimentation" was incompatible with "the primary concepts of American Liberty." Yet voluntary cooperation—Hoover's model in the 1920s—had failed miserably in the early Depression years as a formula for economic recovery. For most American industries, real growth remained anemic until war clouds began to gather a few years later.[59]

5

MINING AND MARKETING
DURING THE NEW DEAL

Franklin Delano Roosevelt is considered one of America's greatest presidents, yet history has been hard on New Deal domestic programs. Most scholars agree that despite all the governmental stimuli designed to raise production, prices, and wages, recovery remained elusive until after the rearmament rush as the world geared up for another war. Businessmen did not need the perspective of hindsight to reach the same conclusion. They had partnered with government during the national emergency, but by 1935 they grew increasingly skeptical of all the "cracked ideas in circulation," as Charles Segerstrom characterized the later years of the New Deal. Like other fiscal conservatives, he equated government intervention with government intrusion, a threat to free enterprise and the rights of private property.[1]

But Segerstrom was no libertarian ideologue. Pragmatic necessity, rather than ideology, should determine when and how to use the power of government for the common good—however narrowly defined. Like his friend Herbert Hoover, Segerstrom saw no inconsistency in providing federal loans to banks or protecting American business from cheap foreign competition, while at the same time opposing government efforts to raise wages or control prices. He did not think the "welfare of your fellow man," as a Nevada Works Progress Administration (WPA) official suggested in 1936, was the proper concern of the Nevada-Massachusetts Company or any other business.[2]

THE NEW DEAL AND THE TARIFF REVOLUTION

Segerstrom's dream of building a government-sponsored tungsten cartel faded along with the NRA and the climate of fear that had precipitated the emergency legislation of the Hundred Days. With the economy showing signs of life in 1934, the inchoate Tungsten Association fell apart in the market revival and the competition for customers. Regardless of their individual and corporate differences, however, tungsten producers were united in their opposition to free trade. They believed that tariff barriers were essential to keep American mines open and miners at work, despite the growing body of evidence that high duties made the Depression worse instead of better.[3]

For more than a decade protectionists had been able to ignore the carping criticism of Cordell Hull, the recognized champion of free trade. Now he was secretary of state, with a Democratic Congress and a president seemingly open to tariff reform. Though FDR's political instincts made him reluctant to take on international trade issues in the middle of a national emergency, Hull had no such reservations. With the Smoot-Hawley imbroglio in mind, in May 1933 he published a blistering polemic that blamed the Depression on a series of "utterly blind" economic nationalists. These American business and political leaders, he said, had deceived the public and themselves with phony arguments and self-serving propaganda. "The only remedy" for the "existing panic," he claimed, was "a system of liberal trade agreements pledging fair, equal, and friendly trade treatment and trade relations."[4]

Promoting trade through direct diplomacy required not just an action plan but a revolution in the process of establishing formal commercial relationships around the world. Hull was a visionary, a Wilsonian internationalist who believed in the principle of reciprocity. The practice of negotiating equal rates among trading partners had been abandoned as part of U.S. tariff policy between 1909 and 1934. In that period Congress gave the president unilateral power to impose higher rates of up to 50 percent on imports from countries found to be in any way discriminatory against American interests; after 1930 he could exclude such imports altogether. Hull's proposal, signed into law in June 1934 as the Reciprocal Trade Agreements Act (RTAA), restored the equity principle and removed Congress from the formative process. It delegated to the executive branch full authority to negotiate and monitor reciprocal agreements, subject only to congressional approval every three years. The change reduced the power of legislative lobbyists and gave consumers and export associations a larger voice in the formation of tariff policy. It did not take politics out of the process, however. Old theories and arguments were merely updated and sent back into battle from a different staging ground. Lobby tactics also shifted along with the transfer of power from Congress to the executive branch. Instead of writing bills for friendly lawmakers to introduce, interest groups now focused on undermining new laws before they could cause much harm to affected industries or constituencies.[5]

After the RTAA went into effect the State Department initiated the first in a series of bilateral trade talks. Punitive rate increases remained as a "stick" to threaten potential discriminators, but New Deal liberals preferred the most favored nation (MFN) principle as a "carrot" to bring tariff schedules down to the lowest rates approved by any one of America's trading partners. Reductions established under the RTAA automatically lowered the rates of earlier agreements.[6]

To protectionists, unconditional MFN was an alarming doctrine, adopted as a U.S. trade concession almost by accident during the Harding administration.[7] It meant little while Congress controlled trade policy, but tariff-dependent industries felt threatened by diplomats dedicated to applying the lowest rates equally across the board. With the MFN to worry about, every round of trade talks under the new rules posed risks for protected industries. Subsidized marginal producers contributing little to the national economy were especially vulnerable to the criticism of consumer-oriented free traders and their allies in the State Department.

For protectionists in the metals industry the best defense was national defense. In the early 1930s, with the nation still looking inward and focused on Depression issues, national security was a potent political foil to the macroeconomic arguments of liberals in the Roosevelt administration. The nationalist upsurge helps explain the political power of "small group interests," in the words of Abraham Berglund, a contemporary economist and longtime tariff critic. Prior to 1934, he said in an analysis of New Deal trade policy, American tariffs had promoted monopoly interests; after the RTAA "they are props of a vast number of small and often high-cost producers."[8]

The domestic manganese industry gave Berglund a prime example of "small group interests." Manganese producers, despite their weak economic position, demonstrated substantial political strength during trade talks with Brazil and Canada, both with significant manganese deposits. Though domestic ore supplied less than 15 percent of the manganese used in the steel industry, mining companies had a potent protagonist in J. Carson Adkerson, president of the American Manganese Producers Association. In July 1934 he delivered a petition to FDR seeking reassurance that domestic industry would not be hurt in pending trade talks with Brazil. It was signed by nearly half the Senate and 145 members of the House. However, it did not prevent diplomats behind closed doors from dropping the duty on raw manganese ore by 50 percent. In formal ceremonies at the White House early in February 1935, with FDR looking on, Cordell Hull hailed the Brazil agreement as the first big step away from "medieval merchantilism" [*sic*] and a model for the future. A home industry that produces only an "infinitesimal portion" of the nation's metal needs, he said, should not be subsidized at the expense of America's export trade.[9]

Blindsided by the diplomatic talks, the manganese spokesman struck back with a personal attack on Cordell Hull. In a press release, Adkerson warned that the secretary's reciprocity treaties were "crippling industries vital to the nation's defense." Taking their case to the public was a risky but widely used tactic by the lobby industry, a last resort if other options failed. Against Hull it backfired, as a sympathetic journalist reported. The "campaign of public criticism," he said, was "so virulent that even the gentile Secretary of State called in the reporters to protest against it." Publicity campaigns also invited public scrutiny of the organizations behind them. New disclosure efforts, if not substantial lobby reform, soon followed.[10]

By linking tariff policy to national security, Adkerson hoped to build a backlash that would pressure Congress to alter the provisions of the RTAA. However, as critics pointed out, domestic producers of the dark metal were still comfortably protected by a raw ore duty equivalent to 55 percent ad valorem, and a 25 percent duty on manganese chloride. The dip in U.S. manganese market share between 1935 and 1938 had as much to do with the lingering effects of the Depression as it did with the MFN impact of the Brazil treaty. Consequently, despite their concerns, American manganese producers were not seriously hurt by tariff reductions. In the decade prior to the RTAA, domestic miners produced, on average, 30.5 percent of the total manganese consumed in the United States. In the following decade the ratio was 27.5 percent.[11]

Despite similar reform pressures, tungsten protectionists managed to hold the barrier high until after World War II. With a larger market share of domestic consumption than their counterparts in manganese, tungsten producers used their economic power to gain political leverage. Their primary opponents were dealers in Chinese tungsten, led by the David Taylor Company of New York. Soon after the Democrats took command in 1933, traders renewed efforts to lower rates by filing complaints with the Tariff Commission and trying to enlist consumers in the reform campaign. To counter the downward pressure, some high-cost smaller producers tried to start a political backfire by seeking even higher duties. In this tug-of-war between economic and political arguments, Charles Segerstrom found a middle ground: use all possible political means to retain the existing tariff rates, but hedge your bets with economies of scale to remain competitive if the tariff goes down.[12]

As in previous tariff battles over metals, the politics of national security gave the edge to tungsten producers. By casting economic arguments in political terms they obfuscated free-trade rhetoric and derailed the opposition. Before the RTAA, protectionists had relied on the western congressional delegation as the first line of defense. Ott Heizer, Segerstrom's general manager, reminded his representatives that tungsten was "one of the most important of the war minerals . . . and in time of a national emergency we could be depended on to furnish a very considerable part of this country's needs." At the time, three-fourths of the tungsten produced in the United States came from Heizer's firm, but up to half of its production was stockpiled until the economy began to revive. By 1934 consumption exceeded annual production, adding pressure for tariff reform. Nevertheless, Senator Pat McCarran of Nevada, a staunch defender of western interests, reassured Heizer that "it will be my intention at all times to maintain a tariff on this commodity."[13]

After the passage of the RTAA the tungsten tariff advocates, like other protectionists, shifted focus. They tried to undermine the legislation itself by building political backfires in Congress, expand their political network, and defend the existing duties. During the fight with tungsten importers, Segerstrom described the effort in a terse telegram to his Boston backers: "We have used every influence in all directions to prevent any negotiations for reduction. . . . We are members of American Mining Congress[,] American Tariff League and have personally worked through United States Chamber of Commerce[,] California State Chamber of Commerce and various Senators and Congressmen from the West who favor retention of duty."[14]

The combined effort was an important counterforce in subsequent tariff negotiations, as Segerstrom admitted late in 1935 after dodging a bullet in the Canadian treaty. Duties were reduced on all ferroalloys except ferrotungsten. "I do not know how long we can continue to do so," he told a Boston creditor, "but our friends are surely doing their stuff for us in Washington, and all our work along this line has not been in vain."[15]

NEVADA-MASSACHUSETTS IN THE MID-1930S

Keeping an eye on trade negotiations that threatened the tungsten tariff was only one of the challenges Charles Segerstrom faced as president of the Nevada-Massachusetts Company in the middle years of the Depression. As industry began to revive, New Deal labor and tax policies imposed new restrictions on mining practices and financial opportunities. International instability and rising militarism increased the volatility of prices for tungsten ore and concentrates, complicating and often frustrating marketing strategies. In the manufacture of steel alloys the gray metal itself faced serious competition from its cheaper "sister metal," molybdenum, especially after Climax Molybdenum began mass-mining and -marketing ore and concentrates from its world-class deposit in Colorado. Segerstrom's biggest challenge, however, was a fundamental problem in economic geology: how to offset the steady decline in ore grade and volume at the main company property near Mill City, Nevada. Both the life of the mine and the health of the company depended on the amount and quality of ore produced. Through these troubled waters Segerstrom ably steered his firm, managing even to reach the highest prewar production and profit levels in the company's history. It was a remarkable achievement.

For tungsten producers after 1933, attempts to nullify the impact of New Deal labor policies were not nearly as successful as their tariff defense. Having reluctantly accepted the NIRA provisions that recognized the basic rights of labor, they grew increasingly resistant to subsequent legislation that implemented those rights. During congressional hearings early in 1935, for example, the American Mining Congress argued that the Wagner Act would place the federal government in the position of supporting a union monopoly under the American Federation of Labor. Mine owners also rejected the progressive argument that encouraging unionization with the help of the National Labor Relations Board would forestall worker radicalism. When organizing efforts began the next year at the Mill City mines, Segerstrom blamed communist agitators but stole their thunder by raising wages sixty cents per day. That was more practical than Haesler's advice to "run every last one of the Bolsheviks out of Mill City even if you have to go down there with a gun."[16]

Of greater concern than a few union organizers were the wage and hour stipulations of the Fair Labor Standards Act. This landmark bill, effective October 1938, imposed a forty-cent-per-hour minimum wage and a forty-four-hour week on industry in its first year, sliding to a forty-hour week over three years. A time-and-a-half overtime pay provision was included. To fiscal conservatives it was dangerously inflationary, a Keynesian experiment that would ruin the country. "Until the administration makes up its mind to stop trying to remake the world, and go back to old time business economics," Segerstrom told an industry lobbyist, "we are going to be in for a very hard time." This was unduly pessimistic, however. The Mill City boss had to reduce the workweek from forty-eight to forty-four hours but was already paying the minimum wage. When miners demanded the same pay for less work, he negotiated a compromise that gave them a Sunday off every other week. The net result of the concession, he told his Boston creditors, would cost the company only a 5 percent loss of production and no change in wage rates. Indeed, labor costs at Nevada-Massachusetts actually declined after 1937 and remained below 35 percent of operating expenses until America entered World War II.[17]

New Deal tax policies aroused the most vocal industry criticism. Deficit spending was no more popular in the 1930s than it is today. Fiscal conservatives questioned the wisdom of "gambling on the future" by burdening their grandchildren with a mushrooming national debt. When Congress passed new revenue bills to plug loopholes in the old tax code and reduce the deficit, critics called them "wealth taxes" that "soaked the rich" with surcharges and higher tax rates. To the economic elite, using federal tax power to "redistribute wealth" was "creeping socialism," and FDR was a "traitor to his class." The reality was quite different. Though the new laws frightened business leaders and increased their hostility, taxes aimed at the highest income brackets were not substantial sources of revenue. Mark H. Leff's revisionist study concluded that New Deal tax changes were actually short-lived symbolic victories that masked the heavy burden of taxes borne by the American middle class.[18]

History is replete with paradoxes. Just as the working masses in the 1930s failed to recognize the inequalities of a largely regressive tax system, businessmen were convinced that the new corporate taxes were "unduly heavy, violating the canons of equity, certainty, convenience, and economy." Segerstrom thought they were a form of "national suicide," but his comment was little more than political hyperbole—a conservative lament that belied his personal manipulation of the tax laws. In a confidential memo to his Boston bankers, for instance, he estimated 1934 profits "on the books will be about $180,000 [but] our actual about $240,000."[19]

Early in 1937 the Sonora banker hired a firm of tax accountants to ensure that "we are not overlooking anything that may be of advantage to us." By paying off the company's debt of more than $800,000 with capital reserves, he avoided the new surcharge on undistributed company profits that Congress had imposed in 1936 but repealed two years later after heavy lobbying from business interests. Segerstrom was proud of his financial feat, for Nevada-Massachusetts was the only Boston-financed "war baby" to pay its debt in full with interest. Federal officials, however, charged the Nevada firm with tax avoidance, and won the case in federal court. Eventually, the $22,000 assessed in extra taxes and interest was cut in half after the tax accountants negotiated a compromise with the Treasury Department. Late in 1937 Segerstrom, in a peevish mood, told his friend Haesler that "we must get the common people to feel that they are being sold down the river." Only "when the people see that we are killing the entire country then I think we will see action and someone will get out and put the heat on the whole New Deal."[20]

However effective pressure-group tactics were against New Deal labor, trade, and tax policies, they did not work against market forces beyond political control. Tungsten prices were pegged to the London metals market, which in turn was affected by daily changes in metals supply and demand around the world. Unlike copper and steel, the domestic tungsten business was small and diversified, with neither producers nor importers strong enough individually or in combination to dictate either production or price. U.S. tungsten mines produced on average only 9 percent of the world's supply between 1933 and 1939, but consumed 14 percent. Ore importers therefore had considerable market power. China remained the largest tungsten producer. By the mid-1930s the nationalists had monopolized the mines to reduce foreign exports and maintain stable prices, but Japanese adventurism on the Asian mainland after 1931 threatened both the supply and the control. Market volatility was the inevitable result.[21]

"A miner is a peculiar animal," Segerstrom observed early in 1939. "Men may go out into the hills to prospect and they are about the most independent set of men in America." The remark reflected his frustration with market uncertainties and the lack of unity among tungsten producers and dealers. Before American industry began to mobilize for war, domestic tungsten demand remained out of "sync" with foreign demand. While European tungsten consumption rose steadily, largely on the strength of Germany's not-so-secret rearmament program after 1934, America's isolationist policy kept the military at bay and industry overly dependent on automobiles and other domestic goods. Despite Roosevelt's open attacks on the corporate elite during the Second New Deal (1935–38), big business retained a powerful grip on wages and prices. Leon Henderson, a rising star among FDR's "brain trusters," blamed Big Steel's inflationary price increase early in 1937 for the two-year recession that followed, but Segerstrom and his conservative associates blamed administration policies. "I hope the political situation will be ironed out sufficiently that we can go back to real work again," he wrote, "although only God knows what is going to happen to us, with so many cracked ideas in circulation."[22]

After a decade of development, the Nevada-Massachusetts Company by 1935 was poised for prosperity once the economy revived and the market stabilized. But the "chaotic" market conditions for tungsten in the late 1930s hurt the big producers and drove many smaller ones out of business. Risks were high under such circumstances. Just the threat of lower tariffs after the RTAA passed in 1934 spurred domestic production as mining companies hoped to cash in before prices dropped. Overproduction invariably brought a price decline, forcing domestic mines to cut production and leaving regular consumers short unless they turned to foreign suppliers. In a declining market some dealers bought and stockpiled domestic tungsten at bargain prices from debt-laden smaller companies desperate to pay their bills. When shortages appeared some speculators tried to control the market, buying up mines and bidding up supplies. Haesler advised strengthening the producer's association to help stabilize prices, but the tungsten business had too many divergent interests and geopolitical uncertainties to be readily contained.[23]

For five years Segerstrom's long-term contract with the Molybdenum Corporation of America insulated the Nevada-Massachusetts Company from the vagaries of the tungsten market. Haesler thought of it as an insurance policy, providing an unlimited market for his client's ore at a guaranteed base price of $7.93 per unit, plus a "contingent" amount that varied with the extractable tungsten content. For the first three years Segerstrom never knew how much ore Molycorp wanted. That was clarified in 1935 by a quota system by which Molycorp agreed to purchase only 45 percent of its tungsten requirements from the Nevada-Massachusetts Company.[24]

In practice, Molycorp bought ore from the Nevada firm at $2-$3 per unit below the market price for Mill City scheelite. Segerstrom considered the discount a fair price to pay in exchange for steady income during the Depression's bottom years, but as business picked up his pact with the Pennsylvania refinery began to seem more like a Faustian bargain. Molycorp expanded rapidly after 1934, gaining market share by underbidding rival ferro and powder makers. Its success was reflected in its stock price on the New York Curb, which rose 60 percent in 1935 and another 29 percent the following year before starting a decline. In 1933 Molycorp's two plants in York and Washington, Pennsylvania, had refined 2,500 units of tungsten. Two years later the figure was above 72,000 units, a 2,800 percent gain! During the same period, production at the

Nevada-Massachusetts plant rose only 45 percent, from 32,774 units in 1933 to 72,640 units two years later. Had not Segerstrom insisted that year on capping shipments to Molycorp, Hirsch's company would have acquired the entire Mill City output at a substantial savings to his firm and a considerable loss in profits to Nevada-Massachusetts. But even with a quota, the Nevada plant could not keep up with Molycorp's insatiable ore demands. If the Nevada firm could not meet its quota and Molycorp had to buy ore on the spot market at more than the contract price, Hirsch charged the Nevada-Massachusetts account for the difference.[25]

The Molycorp discount was not the only problem with the 1932 contract. It allowed Segerstrom to sell ore to direct smelting customers, but Molycorp received a 15 percent "commission" on any amount above the base price regardless of whether Hirsch's company or the customer refined the ore. This clause cost the Nevada-Massachusetts Company thousands of dollars. On six carloads of scheelite sold to the Vanadium Alloys company in 1934 at $15 a unit, for example, the Nevada company owed Molycorp a $15,000 kickback. With steel production and prices on the rise the Sonora boss wanted to terminate the Molycorp deal and go it alone, but the Boston bankers thought it was too risky to rely entirely on direct marketing. Against his better judgment Segerstrom conceded, allowing the contract to run its course and adjusting differences with Hirsch personally whenever he could.[26]

The two executives remained cordial but highly competitive, each seeking advantage over the other whenever possible. Hirsch had a standard line in response to complaints about the Molycorp discount: ore prices had to be kept as low as possible to prevent steelmakers from turning to molybdenum substitutes or demanding lower tariffs. Segerstrom's response was to shift delivery priorities, especially as Mill City production declined. Reserving his best grade of scheelite for direct smelting customers, when production ran low he bought and shipped off-grade concentrates to Molycorp. At times even low-grade ores were unavailable, leaving the Molycorp boss fuming because of delayed or incomplete shipments from his chief supplier. Segerstrom also minimized the molybdenum threat even though by 1937 a new steel alloy, patented as "Mo-Max" by the Cleveland Twist Drill Company, was making some inroads in the high-speed steel business. After visiting the Cleveland plant and seeing production figures, he told George Emery, a Boston banker and board member, that he had seen nothing "that would really cause us any alarm."[27]

On the other hand, Hirsch was often accused of "not playing fair" in dealing with his main supplier. When product inventories were high he tried to win away Segerstrom's direct smelting customers by selling refined tungsten at cut-rate prices. Although Vanadium Alloys preferred to buy concentrates direct from the Nevada company, Crucible Steel and Latrobe Electric, the latter a small high-speed steel manufacturer with electric furnaces in Pittsburgh and Latrobe, Pennsylvania, signed contracts with Molycorp that gave Hirsch added leverage as broker. He bought ore from Segerstrom at bargain prices, then refined and sold the products at a discount. Claiming that his shrewd marketing practices helped Nevada-Massachusetts stay in business, Hirsch brazenly billed his supplier for half of the discount price! Unhappy with slow or delayed shipments from his Nevada supplier, he also frequently deferred payments on trade acceptances for ore already delivered, coyly promising to settle in cash when his supplier completed the shipments due. Then it was Segerstrom's turn to fume. He threatened to store ore promised to Molycorp and require the eastern firm to prepay in cash, but that would have strained relations to the breaking point, and Boston said no.[28]

In 1936 Congress unwittingly came to Segerstrom's aid with a legal reason to alter the contract. The Robinson-Patman Act outlawed discriminatory pricing. Even though it was directed against chain stores, a year later Segerstrom used it as an excuse to end the 15 percent kickbacks.[29]

Dealing with Segerstrom's contract changes was less worrisome to Marx Hirsch than keeping his refineries operating at maximum efficiency. To avoid supply shortages and production delays, Molycorp in the mid-1930s began a search for alternative tungsten supplies in the mining West. The move did not worry the Nevada-Massachusetts president, who had already sent Heizer scouring the country without turning up any new major deposits. Segerstrom reassured his board of directors that there was nothing to fear from the field investigations of Hirsch's vice president, Van Rensselaer Lansingh. An MIT graduate, Lansingh had been president of York Metal and Ore, a small alloy steel company in York, Pennsylvania. He came to Molycorp when that firm absorbed the York plant and its staff. He may have been a competent engineer, but Segerstrom thought little of his mining abilities. He wrote George Emery that "it is actually comical to see Mr. Lansingh scouting around the West for ore with men that we would not employ for a shift boss."[30]

In 1935, doubtless at Lansingh's urging, Hirsch took an option on the Atolia mine in California. Its owners had fallen on hard times and were looking for a buyer. Once the nation's largest tungsten producer, by the 1930s the

Atolia's main orebody was seriously depleted. What ore remained was high in phosphorus yet still usable for making ferrotungsten, Molycorp's primary product. The eastern refinery invested heavily in new development, hoping for ore extensions and discoveries under the old lode workings, but within a year Hirsch admitted the effort had been a "total loss."[31]

A year later Molycorp tried again, this time by leasing Boriana, the same mine whose obstreperous owners had given Segerstrom difficulties during the NRA days. It was an old Arizona property Lansingh had worked during World War I as head of York Metal and Ore. News releases promoted the Molycorp acquisition as the start of something big, but Segerstrom had scouted the mine before and thought it would "not be a serious competitor in the tungsten business." Despite extensive new development work the mine failed to return Molycorp's investment. It closed in 1943. Total production under Molycorp's control was about half of Nevada-Massachusetts' output during the same period.[32]

Molycorp's failure to find good alternative ore sources strengthened Segerstrom's position in subsequent dealings with both Hirsch and his own board of directors. In the fall of 1937 Lansingh approached him with an offer to option the Nevada company. The Boston creditors were still willing to unload, but Segerstrom refused to consider it. Having survived tough times and built a solid foundation, he told George Emery, "Now when conditions looked better, it was a very poor time to sell."[33]

Segerstrom soon found a way to avoid both the restrictions of the Molycorp contract and the inherent conservatism of his corporate board of directors by expanding his personal business at the expense of both. His chance came in 1935, when a rule change by the Interior Department added to the annual assessment work required to hold unpatented mining claims, including the mill site at Toulon and its associated tungsten claims at Ragged Top, a few miles to the west. The Nevada-Massachusetts Exploration Company had acquired both in the late twenties. Segerstrom thought the assets worth keeping, but the bankers on his board did not. R. G. Emerson wrote that it would be "hardly worthwhile to spend much money hanging on to claims which there is only vague prospect of our being able to use or sell." His response typified the board's bottom-line mentality. It came after a discouraging report that extensive new development would be needed to convince eastern capitalists that the Mill City holdings were worth the million-dollar price tag that company directors had placed on the property four years before.[34]

Segerstrom also watched the bottom line carefully, but had much more mining experience than his eastern associates. After years of investigating mine properties he knew that good tungsten deposits in the American West were few and far between. He was confident that the Toulon claims, along with a few others Heizer had located, could eventually become profitable if properly developed. But by whom? Rather than see them fall into the hands of Molycorp or other competitors, he organized a new syndicate, Rare Metals Corporation, headquartered at Lovelock, Nevada. It was ostensibly made up of "eastern capitalists" but financially controlled by the Segerstrom family firm, Western Tungsten. Ott Heizer at first joined in the investment, but soon bowed out, preferring operational management to high finance. It took a year to complete the transfer of Toulon property and arrange financing with several eastern steel alloy customers for mine rehabilitation and mill reconstruction.[35]

While Rare Metals was in the formative stages Segerstrom kept the real ownership secret, especially from Molycorp executives who doubtless would have thought their main supplier was trying to skirt provisions of the 1932 contract. All mail and inquiries went to New York in care of the Metal and Ore company, the marketing outlet for the new firm. At the same time, however, he directed Heizer to make certain Rare Metals acquired any good prospects that might become available.[36]

The deception did not last long. By 1937 Segerstrom was openly running both companies from his Sonora home, with Heizer managing field operations and Haesler marketing the pooled ore. The Rare Metals Corporation freed Segerstrom from outside interference, and he made the most of it. The Boston bankers, as board members, could still influence the policy and operations of the Nevada-Massachusetts Company, but neither they nor Molycorp had any legal or financial leverage over the Rare Metals Corporation. Rare Metals became known as a subsidiary of Nevada-Massachusetts, although ownership and control remained entirely separate.[37]

Coordinating the operations of two separate production companies under a unified management, each with a different stable of mines and mills but marketing to the same customers, took careful preparation and planning. It also required instant and reliable communication between managers and marketers. To clarify the relationship and keep the Nevada-Massachusetts directors happy, Segerstrom secured an "understanding" with Rare Metals to avoid conflicts of interest, pool staff and resources, market through a single agency, and control output to keep prices stable. At the same time he installed teletype machines at each field office, and linked them with his Sonora office and with Haesler's brokerage in New York. The surviving transcripts of these almost daily conversations document the phenomenal energy and interaction that Segerstrom demanded of himself and all those associated with the business.[38]

By expanding production beyond Mill City and providing a broader base of operations, the Rare Metals subsidiary strengthened the position of Nevada-Massachusetts just as the national tungsten market was gaining momentum. It also lessened the appeal of a long-term contract with Molycorp. Even though he still needed more ore, Hirsch had earlier conceded that their contract referred only to production from Mill City mines. In the fall of 1937, with strong demand from specialty steelmakers and with Boston's reluctant backing, Segerstrom told Hirsch that the 1932 contract would not be renewed. The two leaders parted on friendly terms, however. Nevada-Massachusetts still needed a market for off-grade ore, and Molycorp still lacked adequate alternative supplies to meet its expanding market share. Despite having to pay higher prices, Hirsch's company remained one of Segerstrom's best customers.[39]

ECONOMIC REALITIES OF THE MILL CITY MINES

Of all the issues facing the Nevada company's president after 1935, declining ore grade headed the list. In 1934 the yield from Mill City ore was more than twenty pounds of concentrates per ton. Three years later it was down to just over fifteen pounds per ton. In 1938 it dropped even further, and remained below fifteen pounds per ton during Segerstrom's remaining years as head of Nevada-Massachusetts. In the same period, despite all-out efforts to raise tonnages mined and milled at Mill City, ore production remained fairly static until 1940, when input from leased properties bolstered the Mill City total. From 1936 to Segerstrom's death ten years later, his company contributed, on average, only 22 percent annually to the domestic output.[40]

In the meantime tungsten mining elsewhere zoomed ahead of Nevada-Massachusetts. Events in the Far East spurred new developments in many western states. Ore prices shot upward during the Japanese advance on China in 1937. Union Carbide, Molycorp's chief competitor in the production of ferrotungsten and powder, expanded operations in California. Its subsidiary, the U.S. Vanadium Corporation, took over the Pine Creek mine near Bishop and built a new plant. Another new mine opened nearby at Benton Mills in California's Inyo County. Smaller mines started up about the same time in Arizona, Colorado, Idaho, Washington, and western Nevada.[41] The boom

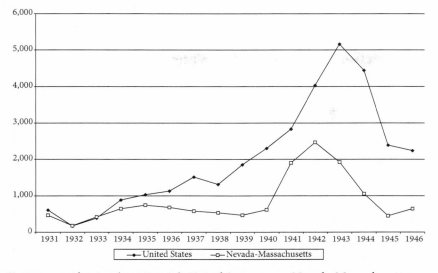

Tungsten production (metric tons), United States versus Nevada-Massachusetts

was premature, however. The Japanese invasion soon bogged down, allowing China to resume shipment of tungsten concentrates to Hong Kong for the foreign markets. At the same time American industrial demand, slowly climbing since 1933, turned downward late in 1937. The recession lasted nearly two years, closing many late starters and leaving even mature companies ambivalent about tungsten prospects right up to the start of World War II.

Segerstrom's extensive communications network kept him abreast of national or foreign events that might impact mining and markets. He tried to position his company so it could quickly respond to any contingencies, but without a reliable source of ore he could not even keep his regular customers supplied when they needed more concentrates.

The problem of declining ore grade had arisen first at the Mill City mines in 1935. As ore prices lifted Segerstrom ordered a full-scale underground assault. The decision reversed his earlier policy of systematic development and selective mining, but to the Boston bankers he justified the change as a pragmatic necessity in light of uncertain economic conditions. "I am going to take out every unit of tungsten that we can possibly extract as I do not believe the present situation will continue indefinitely, and we must make what money we can in the very near future." Diamond drill crews probed company ground and found promising leads, but the size and value of the orebodies could not be estimated without extensive—and expensive—new underground development. Anticipating the added load, Heizer expanded the Mill City plant for the second time in as many years, increasing milling capacity to 250 tons per day.[42]

Expanded development required more skilled miners—hard to find in the remote West. One inexperienced new man hired in 1936 died from a rock fall after just one day on the job. It was the company's first fatality. Employees had a group policy costing each worker $1.50 a month. It paid $10 a week for illness or disability, plus a $1,000 death benefit. Heizer had promoted it to his men the year before as an "exceptional opportunity." Although the company's mines had a low accident rate, questions of health and safety were union issues that Segerstrom wanted to avoid at all costs. To ease concerns and prepare for less serious incidents, after the fatal incident he bought a Chevrolet "touring car" and converted it into an ambulance. Later he expanded company insurance to cover "all our plants against riots, commotion and other public liability of every type."[43]

With expanded crews and great expectations, the Mill City mines and mills ran night and day after 1934. Segerstrom's highest priority was to wrest whatever ore remained from the company's two central mines, the Stank and the Humboldt. Drilling crews deepened the Stank shaft, installed new pumps, opened new levels, drifted along irregular veins, and crosscut toward others. By 1937 the Stank was below eleven hundred feet in very wet ground, but the core drill indicators proved misleading, and Stank production dropped precipitously after that.[44] The Humboldt mine also proved disappointing after some initial optimism. Work had proceeded slowly after takeover by Nevada-Massachusetts until 1933, when Heizer and his crews opened a new shaft and rehabilitated the old World War I mill. Miners encountered some good ore after sinking the shaft to eight hundred feet, but the volume and grade dropped quickly. Continued sinking and drifting over the next five years proved fruitless at lower levels. By 1940, with ore grade so low it did not pay to mine, the Humboldt seemed nearly played out.[45]

Simultaneously with work on the main orebodies, exploration continued on extensions of the Mill City holdings, but with limited results before all-out wartime development. Crosscuts and raises from various levels in the Humboldt to the upper Springer workings tapped into a narrow but high-grade vein system. By mixing Springer ore with lowgrade from other workings, Heizer's mill crews temporarily helped raise millhead values. By 1937, however, most of the richer ore had been mined out. The presence of underground faulting and hard granitic intrusions also complicated mining and slowed development of the Sutton ground. Although diamond drilling located "considerable new ore" early in 1939, the low-grade quality made further work problematical until the great push for production during World War II.[46]

The disappointing results of their Mill City campaign forced Segerstrom and his technical staff to look elsewhere for exploitable orebodies. In 1929 they had invested in the "Great Silver Dyke," high in the remote Excelsior Range of Esmeralda County. According to previous owners, the mine during World War I had produced 50,000 units from 10,000 tons of concentrates. Actually a scheelite-bearing quartz vein system, it proved as misleading as its name. Relying perhaps too much on optimistic geologic reports, Segerstrom in 1931 had sent a small crew under the supervision of W. G. Emminger to find more ore. But from the start of major development in 1934 to closure five years later, the net returns on the investment never met operating expenses. At nearly eight thousand feet elevation the men labored under unpredictable weather conditions, rebuilding the old gravity mill, laying an eleven-mile pipeline over two mountain ranges to bring in more water for milling, and drilling and drifting in a search for payable ore. Blizzards, hard freezes, and high winds caused frequent disruptions, especially when slides or drifts blocked their only access to Mina, the nearest supply center. An occasional earthquake caved some of the underground workings but fortunately spared the miners. What little ore his men found was contaminated with phosphorus, sulfur, and other impurities. After milling, the concentrates were sent to Toulon to be mixed with cleaner concentrates before shipping to eastern customers.[47]

Monthly reports from superintendent Emminger showed sporadic progress at Silver Dyke over a five-year period. The mill produced between 8 and 11 tons of concentrates per month at best, a tenth of its capacity. In 1936 the hardworking superintendent, hoping to speed ore shipments and supply deliveries, contracted with the Works Progress Administration to improve the dirt road to Mina. He soon regretted the decision, for the men sent to the job site were inexperienced and indifferent road builders. After a few weeks his complaints that the hired labor cost the company more than it was worth prompted an indignant reply from the WPA foreman, a local politician feeding off the popular antibusiness frenzy at the time: "To save the Nevada-Massachusetts Company the paltry sum of $2 per day in transportation costs you would ruthlessly, and without the slightest degree of compunction, relegate these men to a gruesome future—these unfortunate human beings whose haunting fear of poverty and the poor-house is a pitiful sight to behold. Where, Mr. Emminger, is your sense of right and justice, and where is the milk of human kindness that should flow in your veins?"[48] Emminger's own crews finished the road-work after the WPA contract ended.

Despite the frustrations and the constant financial drain, Segerstrom's need for ore kept his hopes alive, and Silver Dyke crews working, for another two years. But a much more promising prospect was the Oreana, a unique scheelite deposit in a pegmatite dike located about eighteen miles south of Mill City on the western slope of the Humboldt Range. After its discovery in the early 1930s Heizer had leased it in 1934 for the newly formed Rare Metals subsidiary. A small crew developed a series of thin but rich stringer veins that in some places tested up to 5 percent scheelite. After Rare Metals reopened the 120-ton Toulon mill in 1936, Oreana mining began in earnest. Truckers brought raw ore to the rail yard on the nearby Southern Pacific tracks for shipment to the mill twenty-five miles south. After milling, concentrates from various mines were stored until a 30-ton railcar could be loaded for shipment to eastern customers. In the fall of 1937 Oreana was yielding up to 20 tons of ore per day that averaged 3 percent scheelite, but the deposit was too small to last long. Total production for 1938 was only 305 tons of concentrates, or 25,000 units. By 1939 it was nearly mined out.[49]

For all of this expanded effort, Nevada-Massachusetts confronted fundamental economic realities it could not overcome. One was the inevitable consequence of nonselective mass mining. Stoping all the ore in sight rather than taking only the best increased the volume of ore milled but not the amount of tungsten produced. Total ore production rose from 64,000 metric tons mined and milled in 1934 to 71,000 metric tons by 1939. During the same period the output of concentrates dropped from 643 tons to 465 tons, and average WO_3 production declined from 0.7 units per ton to 0.5.[50] Accelerated development also shortened the life of the Mill City mines without substantially increasing production, although for a few years company profits rose significantly. By 1939 Heizer and his underground crews, after hoisting almost all of the good ore and much of the lesser grades, had nearly reached the bottom of the main Mill City orebodies. With war just around the corner the company faced a dim future unless it could find and produce more tungsten, still considered one of the "key metals" of democracy.

NEW TECHNOLOGY: PROMISE AND PITFALLS

Although ore dressing methods within the industry had improved over the years, Segerstrom and Heizer saw little need for a major upgrade so long as the principal ore was high-grade scheelite that contained relatively few impurities. In the early 1930s they had improved milling efficiency by adding roasting ovens and magnetic separators to the milling sequence. Roasting dried and calcined the concentrate, driving off sulfur in the form of SO_2. It also oxydized pyrite and magnetized it along with garnet, by far the largest bulk of unusable concentrate. Magnetic separators caught tramp iron as well as the magnetized pyrite and garnet particles. In 1933, at the suggestion of engineers from the Bureau of Mines, they had produced an even higher grade of concentrate by adding a two-cell Kraut flotation unit to remove molybdenum sulfide in the middlings.[51]

A renewed emphasis on milling efficiency and product quality grew out of rising production demands. One source of tungsten the company had not yet exploited was its own mill tailings. Between World War I and 1939, Nevada-Massachusetts and its predecessor had milled a million tons of ore. The processed sands and slimes, after passing through traditional gravity-flow crushing, grinding, washing, and concentrating machinery, were slurried out the back of the mill and allowed to flow uncontained over a gently sloping alluvial fan toward the Humboldt River. A company assay indicated that 100,000 units of WO_3 still remained in the tailings. Every ton of ore milled by gravity concentration dumped another half-unit on the pile that had spread like hot cream cheese for more than a quarter of a mile downstream. The tailings were an alluring prospect for further treatment if an effective process could be developed. Determining the best process, however, took much longer than expected. In 1934 the Dorr Company, a Denver firm specializing in milling equipment, sent one of its engineers, George Crerar, to Mill City on a sixty-day "loan" to see if the milling circuit could be improved. Crerar's visit was too short to do much more than test a few samples, but his confidence was infectious. After the loan period ended, Segerstrom put him on the Nevada-Massachusetts payroll to design a separate tailings plant. Early experiments seemed to indicate that the tailings could be worked successfully by differential flotation, but the financial and technical problems of installing a cost-efficient flotation unit took years to overcome.[52]

By the 1930s, the process of concentrating minerals by floating them in an oily froth that left gangue and waste behind had evolved into one of the key milling technologies of the modern age. Unlike older gravity-flow milling methods for gold and silver, flotation quickly passed from a pioneer period of desultory experiments to a modern era of industrial hegemony ruled by a multinational corporation, the Minerals Separation Company. Incorporated in London in 1903, Minerals Separation began as a production company, floating ore by the Cattermole process at Broken Hill in New South Wales, Australia. It grew rich and powerful by buying up the key patents to more successful flotation methods and equipment, and then tightly controlling the licensing of their applications around the world. Challengers faced this industrial monolith at their peril, for Minerals Separation rarely lost a patent fight in the courts.[53]

Segerstrom's personal experience with the Minerals Separation way of doing business dated from 1916, when William J. Loring, as Segerstrom's manager, installed an expensive flotation pilot plant at the Dutch-Sweeney gold mine in Tuolumne County, California. Loring had touted it as the harbinger of a new milling era on the Mother Lode, but the mine's financial support collapsed while flotation of gold ores was still in the experimental stage. Now as president of Nevada-Massachusetts, Segerstrom wanted to avoid both Loring's managerial extravagance and Mineral Separation's high-priced services. Unfortunately, the small flotation circuit installed at Mill City in 1933 could not handle low-grade mill tailings contaminated with phosphorus and other impurities. Upgrading to differential flotation, however, required the use of reagents and processes controlled by Minerals Separation patents. The Sonora banker bristled at the thought of having to pay a royalty to a "foreign" firm of 10 percent on net profits. He tried to negotiate a flat fee instead. When that failed he put off any further negotiations, hoping a better and less costly alternative could be found.[54]

In the meantime, Crerar's experiments with Mill City tailings continued. He was visited periodically by chemical engineers and metallurgists from the U.S. Bureau of Mines in Reno, whose advice to Segerstrom earlier had led to the installation of the sulfide flotation cells. Over the next three years they tested a variety of procedures for extracting the remaining values in the sand tailings and slimes.

The geology of the Mill City deposit greatly complicated the milling problem. Typically, skarn ores in the Great Basin contained garnet, epidote, quartz, pyrite, and other minerals formed with carbonate ores in contact metamorphic zones. Most of these were harder than scheelite, making fine grinding necessary to free it from the surrounding gangue and expose more of the mineral surface to the action of reagents. In the process, however, much of the soft and friable scheelite was "slimed," or pulverized to a pulp containing microparticles too fine to capture in traditional gravity separation milling circuits. Flotation had solved sliming problems with most sulfide ores, but the molecular structure of scheelite made it harder to float because the reagent used could not distinguish the valuable calcium tungstate from worthless calcite, a common carbonate mineral. Moreover, fine grinding even further slimed the scheelite, leaving microscopic scheelite particles in a colloidal solution, hard to capture by reagents in a bubbly froth.[55]

Early in 1937, acting on the recommendations of Crerar and the Bureau of Mines engineers, the Nevada-Massachusetts Company built a seventy-five-ton pilot plant next to the gravity concentrator to treat tailings pumped directly from the slime thickener of the older plant. To recover as much of the remaining scheelite as possible, two separate flotation units were used. The first replaced the older flotation cells with a newer four-cell machine that lifted pyrite and other sulfides and pumped them out on the waste pile. The second eight-cell unit floated the scheelite and left the waste behind. As a last stage to make the concentrate easier to market, it was pumped to a special slime table, which raised the WO_3 content of the final product to 50 percent or higher.[56]

The pilot plant had demonstrated the technical feasibility of improving tungsten recovery by flotation. Segerstrom now had to determine whether it was cost-effective, or even financially possible. Building a full-scale commercial plant to treat old and new Mill City tailings during a recession would place a heavy financial burden on the Nevada-Massachusetts Company. The high price of Minerals Separation Company's patented reagents alone brought the flotation costs close to the margin of profitability, not counting purification and other processing charges.[57]

Segerstrom pondered the issue for two years without coming to a decision. Meanwhile, a Minerals Separation agent paid a surprise call on Crerar at the Mill City pilot plant. After showing his unwelcome visitor everything a "visiting engineer would see," Crerar denied using any patented reagents, and claimed that American Cyanamid had supplied all the chemicals now in use. Privately, however, he wrote Segerstrom that "we had better dig into this matter a little further, aside from a possible defense in litigation with the Minerals Separation Co., as we may be able to locate another source of the product that would save us some money. . . . There is no definite threat of litigation or trouble from the Minerals Separation Co. but they will undoubtedly contact you in the matter."[58]

GOLCONDA: EXPANSION AND CHEMICAL PROCESSING

While his engineers were still testing Mill City tailings, Segerstrom took control of another tungsten deposit that required new methods of treatment. In 1935 Heizer had examined a "peculiar" tungsten-bearing sedimentary iron and manganese deposit in the Golconda mining district at the foot of the Edna Mountains in southeastern Humboldt County. The mineral complex contained micron-size tungsten particles locked in a partially soluble colloidal suspension of psilomelane (MnO_2), not amenable to concentrating "by the usual gravity methods," as Segerstrom told Haesler. But Crerar, ever the optimist, thought he could unlock it through chemistry. The orebody had been worked for manganese during World War I, but the mine shut down soon afterward, and the current owners were ready to deal.[59]

Acquiring Golconda was a risky business, but the timing seemed right for risk taking. As will be seen in the next chapter, that same spring, after years of deliberation by various government and private agencies, the House of Representatives approved modest funding to purchase tungsten and other strategic metals as part of a new package of naval appropriations. However, the Senate killed the stockpile money despite heavy lobbying by the domestic metals industry. Another try seemed inevitable, and Segerstrom wanted to be ready for the next metals boom.[60]

He was also encouraged by new tungsten demand. The latest commercial sensation was tungsten carbide, an alloy first developed by Osram Gesellschaft in Germany at the beginning of World War I to improve the performance of wire-drawing dies. The process involved carburizing metallic tungsten particles to increase hardness and durability, mixing with powdered cobalt as a binder, pressing and shaping the matrix into billets or blooms, then molding and fusing the alloyed metal by sintering at carefully controlled temperatures.[61]

As the product of a "breakthrough technology" that prefaced a major shift in industrial development and consumer use, tungsten carbide was rapidly diffused and commodified despite efforts of the global steel cartel to control its price and distribution. Soon after initial development, the Krupp company absorbed the controlling patents and eventually marketed an improved carbide product it called Widia. In the meantime, General Electric metallurgists in New York, working with samples of the early Osram experiments, developed and patented an American version of tungsten carbide. In 1928 GE organized a subsidiary, Carboloy, to "exploit the product commercially." Later that same year Krupp and GE concluded a secret agreement to pool patent information, divide marketing territories, and keep the price comfortably above the profit margin. The deal kept U.S. tungsten carbide production and price under the cartel's control right up to America's entry into World War II.[62]

Despite its high cost and limited production, tungsten carbide was a phenomenal marketing success by the late 1930s. In the machine tool industry, this superhard alloy rapidly replaced more expensive industrial diamonds as a coating to increase the hardness and durability of cutting-tool tips. Arms makers around the world also recognized its value in the development of a newer and more deadly variety of armor-piercing shells. By World War II half the world's tungsten consumption was in the form of tungsten carbide.[63]

These encouraging market signals spurred Segerstrom's interest in acquiring Golconda. With most of the company's profits already going into extensive new development at Mill City, however, he could not very well recommend a new venture to his conservative eastern board members. Under the circumstances, turning to the private family business was understandable. Without telling his Nevada-Massachusetts directors, Segerstrom optioned 320 acres for the Rare Metals company.[64]

The Segerstrom family papers do not reveal whether the patriarch was conflicted by divided loyalties because of his multiple business ventures, but a couple of years after the Golconda deal a teletyped conversation with J. J. Haesler offers significant clues. Both men had sons about the same age. Segerstrom's eldest, Charles Jr. (Tod), had periodically visited the Haesler residence near Clifton, New Jersey, while completing a business degree at Harvard. Praising the foresight of his western colleague, the senior Haesler wrote that "you surely must look upon Galconda [sic] as one of the finest things that you could possible [sic] get hold of to remain in your control for the rest of your days and then to go to Charles and the family." Segerstrom's reply hints at family ambitions looming large in financial decisions: "Yes I feel exactly as you do. In fact I do not want to make even the same contract as Nev Mass. I would like to keep it clear to sell to any one we pleased and not even guarantee to supply anyone and later go into some of the refining and etc that the others do."[65]

The Golconda venture, though an exciting prospect for future development, was secondary to more immediate production problems. Segerstrom could not risk alienating important eastern customers by failing to fulfill signed contracts for regular quarterly delivery of Mill City scheelite. Fortunately, Segerstrom's other operations made up much of the Mill City shortage. Supplemented by family earnings on Nevada-Massachusetts' profits, in May 1937 the Rare Metals company completed its Toulon mill rehabilitation and began concentrating ore from Ragged Top and other company properties, in addition to scattered lots purchased from small regional producers. The working agreement with Nevada-Massachusetts made Toulon essentially an integral part of the larger operation and a central point for ore processing and shipment. Periodic equipment upgrades and additions gradually brought the mill up to a capacity of one hundred tons per day by World War II.[66]

The technical and financial problems at Golconda were harder to overcome. With Crerar still preoccupied with the Mill City tailings puzzle, Segerstrom turned to Molycorp for help. Late in 1936 he proposed a deal by which the eastern refinery would purchase Golconda concentrates at a discount and develop an effective extraction process. Hirsch needed the tungsten, and his chief chemical engineer, Emil A. Lucas, was eager to test it. But Hirsch dickered too long and wanted too much for Molycorp's services.[67] They never got the contract.

The following March, after completing the Mill City pilot plant and turning to the Golconda milling problem, Crerar devised an extraction method he said could recover all of the available tungsten values and produce artificial calcium tungstate ($CaWO_4$). "Synthetic scheelite," as it was known in the trade, could be sold to direct smelting customers at a premium. It was easily reduced either to acid or powder and marketed as a secondary product for making filament wire, dies, inks, and other industrial and consumer products. Considering the industry shift toward tungsten carbide, Golconda synthetic scheelite would be the purist on the market, giving Nevada-Massachusetts an edge over any competition. In a burst of proprietary enthusiasm, Segerstrom said Crerar's process was "entirely revolutionary" and worth patenting, but similar treatments had been patented long before. Whether a patent application was attempted is not clear from the available records, but no patents were ever issued in either Crerar's or Segerstrom's name.[68]

The process Crerar developed was adapted from metallurgical techniques dating from at least the mid-nineteenth century. After crushing to less than one-half inch and mixing with salt, coke, and soda ash, the ore was sintered in a reducing furnace, and then finely ground and sorted. The fines went through a series of washes and thickeners, with sodium salts and solvents mixed in to form soluble sodium tungstate. After settling, the clear filtrate was pumped through two precipitation tanks containing calcium chloride and other reagents. The first tank dropped out the sulfates, phosphates, molybdates, and other impurities, and the second precipitated bright-yellow crystals of calcium tungstate. After filtering and drying, the final product was sacked and weighed for shipment.[69]

Crerar's flowchart went through several refinements before it was perfected. While Crerar worked at Mill City, Segerstrom commissioned a separate study by Professor Colin Fink of Columbia University. Fink's one-page "Final Report on the Recovery of WO_3 from Low-Grade Ore" arrived in the fall of 1936, nearly a year after Crerar's initial work. It served as independent confirmation that Golconda ores were amenable to treatment by heat and chemical reduction to recover calcium tungstate. How much Crerar relied on Fink's brief analysis is hard to tell. Although Crerar did most of the work, Fink's name and fame drew more attention. By World War II the recovery method was known as the "Fink Process."[70]

ECONOMIC UNCERTAINTIES, 1937–39

No matter who received credit, the Golconda process remained theoretical until Segerstrom decided when and where to build a chemical plant, or even if it should be built at all. J. J. Haesler had warned his California friend to avoid the mistakes of other companies by building a bigger plant than the market— or the orebody—could bear. Neither Segerstrom nor his Boston backers wanted to leverage the company back into debt, which the company president had just eliminated by a prudent management of profits. Nor did they wish to be in direct competition with the company's biggest customer, Molybdenum Corporation of America, which used Nevada-Mass concentrates to make and market its own brand of synthetic scheelite. Given the uncertain economy, the unknown capital expenses involved, and Molycorp's looming presence in the tungsten powder business, it was not an easy decision.[71]

The Mill City tailings puzzle also remained unsolved. Crerar's pilot plant, after months of testing and tinkering, demonstrated that a full-scale flotation plant to treat ponded tailings as well as those from active operations could boost WO_3 production by 10 to 20 percent. But would this increased production offset the added costs of patented reagents? The margin was too close to call.[72]

Segerstrom was still mulling things over when the "Roosevelt Depression" began late in 1937. The steel industry took much of the blame. To pay for higher wages following the settlement of a protracted labor strike, Bethlehem and U.S. Steel had raised prices on steel products. It was too much for the fragile economy. Orders fell for rails and autos during the summer and dipped sharply in the last quarter. The wholesale price index for metals and metal products, which had climbed close to its pre-1929 high by November 1936, fell ten points in 1937. Food and fiber prices followed the same downward trend.[73]

Though causal factors seemed "somewhat mysterious" to some analysts, the recession's origin was no mystery to Leon Henderson. Months earlier he had told Roosevelt that Big Steel's price hikes would cause recession. When the downturn came, Roosevelt was furious. He blamed the steel cartel for "ruining" his recovery program. His letter to the Justice Department in the fall of 1937 asking the antitrust division to look into "inadequacies and defects in our anti-monopoly laws" signaled a shift in administration business policy. In the spring he sent an address to Congress warning of the cartelization of American industry and the dangers it posed to free enterprise. Taking cues from Henderson and Thurman Arnold, a progressive lawyer and author of *The Folklore of Capitalism,* one of the era's most influential books, the president proposed a series of new laws designed to "stop the progress of collectivism in business."[74]

To back his words while Congress deliberated, Roosevelt put Arnold in charge of the Justice Department's antitrust division. For the next three years federal attorneys probed and prosecuted. More than two hundred investigations and ninety-three suits were conducted during Arnold's tenure. The administration shift pleased antimonopolists but gave conservatives further cause for alarm. Every government challenge to old ways of doing business seemed fraught with totalitarian implications. The arrest in 1940 by U.S. marshals of several western lumber company officials for price-fixing evoked a sardonic comment from Segerstrom. "I'm afraid that they will try Ges[ta]po methods in this country before long," he told Haesler.[75]

That same year the Justice Department filed an antitrust suit against General Electric and two subsidiaries, Carboloy and International GE, charging them with conspiring over a twelve-year period with the Krupp company of Germany to "dominate the entire interstate and foreign commerce of the United States in the field of hard metal compositions."[76] Segerstrom's reaction to the government's case is unknown, but he must have been conflicted by economic and political incongruities. His business instincts favored cartelization if the result was domestic price and market stability, but he also considered himself a patriot and distrusted foreign conspiracies in restraint of American trade. Moreover, his company had more than passing interest in the suit, for GE was a customer and potential investor—indeed, it later purchased the Nevada-Massachusetts Company from the Segerstrom family. But he never learned the outcome of the case. Wartime priorities delayed the prosecution, and he died before it finally went to trial.

As in previous economic downturns, the recession during Roosevelt's second term volatilized already unstable tungsten prices. The metals market broke in mid-November 1937, then rallied in December before falling again in January 1938. The slump extended into 1939 as speculators and small producers dumped ore on the market to raise cash. Metal prices rallied with the opening guns of World War II, but dropped again during the "phony war" and did not begin resurging until Hitler's barrage on Western Europe in the spring of 1940.[77]

Whatever its cause, the recession slowed operations at Nevada-Massachusetts and its affiliates. To appease nervous Boston bankers, Segerstrom had to retrench. He finally gave up on Silver Dyke, a money loser from the start. In the spring of 1938 he pulled out all the men and machinery and wrote it off as a total loss. "Our experience at the Silver Dyke mine teaches us that one cannot be too careful in taking up tungsten prospects," he ruefully told another investor. But he argued against shutting down the Mill City operations, resisting suggestions to close the pilot tailings plant and to cut development work. Despite running at only 50 percent capacity, company production costs as a whole still remained within reasonable margins of potential profitability. With the world on a war footing, tungsten demand was bound to rise. The only remaining questions were when and who would be ready to take advantage of it.[78]

The economic slowdown gave Segerstrom more time to work on improving milling efficiency and plan a Golconda campaign. During the winter of 1937 he built a new testing facility with family funds just below his palatial home on Knowles Hill in Sonora. Early the next year he hired Donald Read, a young chemical engineer completing graduate studies at Columbia University. Crerar soon joined him, along with all his equipment and data from Nevada. Their joint experiments over the next few months on Mill City tailings and Golconda ores confirmed what Crerar had already discovered, that both could be beneficiated by gravity separation and flotation, and then refined chemically to produce a highly purified calcium tungstate. By the summer of 1939 the most pressing technical problems had been solved. All that remained was to put technology to work by upgrading the milling facilities at Golconda and Mill City.[79]

It could not be said of Charles Segerstrom, as it was said of Loring and other extravagant mine managers, that the "management was too highgrade for the lowgrade ore." The boss of Nevada-Massachusetts had learned by long experience to avoid unnecessary risks. Even though he still faced serious ore shortages, building a chemical plant was a "tremendous investment" that he was reluctant to make during a "spell of uncertainty." Serious work on both the enlarged tailings plant at Mill City and the chemical plant at Golconda did not commence until a rearmament panic swept over the industrial world after Hitler's invasion of Western Europe.[80]

For all his complaints against the "seven kinds of New Deal itch" that he thought were continually aggravating businessmen, Charles H. Segerstrom had prospered, personally and professionally, during the Roosevelt years. The corporate history of Nevada-Massachusetts in the later 1930s illustrates what economic historians have concluded long ago, that the antibusiness rhetoric of the Second New Deal was more smoke than fire. In 1938 Segerstrom admitted as much in a cautiously optimistic letter to his Boston banker George I. Emery: "I feel we are all to be congratulated on . . . [retiring the outstanding debt] during the boom period we had with the New Deal, and while I do not see any great future, I do believe if we struggle along, we will make some real money during the coming years."[81]

From 1934 to the beginnings of World War II, the main problems of the Nevada-Massachusetts Company were geological and technical rather than geopolitical. They concerned internal issues of deposition, development, and production rather than external threats such as free trade or government tax and labor policies. External matters were easy targets for attack, but the real problems were those that mine managers everywhere had to confront on a day-to-day basis. Segerstrom not only managed to keep his business interests out of trouble during troubled times but also gave his corporate partners as well as his family reason to expect better days ahead despite the darkening international horizon.

Shaft house at the Cold Springs Mine, Nederland, Colorado. Before World War I, Colorado was the nation's largest producer of tungsten, with ferberite the principal ore. Tungsten production in the Nederland District continued intermittently until the late 1950s. Much of it came from the Cold Springs Mine, operated for many years by the Wolf Tongue Mining Company. Covering the headframe with wooden housing protected men and equipment from inclement weather at 8,500 feet. Artist Richard Turner's rendition shows the shaft house in its declining years, just before it was torn down and the shaft covered as part of the state's mine reclamation program. Courtesy of C. R. Turner, Nederland, Colorado.

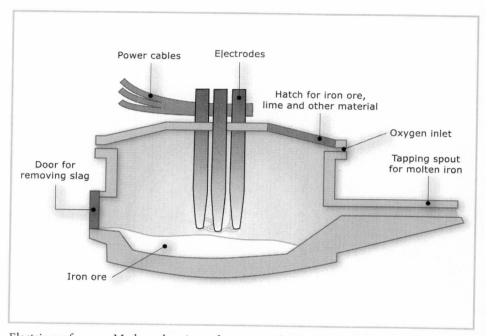

Electric arc furnace. Modern electric arc furnaces evolved out of designs developed before World War I by a French scientist, Paul Héroult. By the 1920s, electric furnaces had largely replaced the crucible for making small batches of high-quality specialty steels. Bulk steels were still made in Bessemer and open-hearth furnaces until the development of the basic oxygen furnace after World War II. The illustration above shows one design of a modern three-phase, three-electrode, alternating-current electric furnace, used primarily to remelt steel scrap in a batch process. Courtesy of *Te Ara: The Encyclopedia of New Zealand*, © Crown Copyright 2006–2009 Ministry for Culture and Heritage, New Zealand.

Overview of Tungsten, the mining camp on the eastern slopes of the Eugene Mountains, seven miles northwest of Mill City in Pershing County, Nevada, about 1935. This view looks northwest from the flanks of Stank Hill toward Springer Hill to the left. Established early in 1918 and originally called Sutton after one of the original locators, the camp was a service center primarily made up of worker cabins, barracks and warehouses. It grew rapidly for a few months after consolidation of the major claims, but then shut down when tungsten prices collapsed at the end of World War I. In 1925 it revived under a new owner, the Nevada-Massachusetts Company. Most of the buildings shown here, plus many more constructed with funding from the Federal Housing Authority during World War II, were sold and removed after the Nevada-Massachusetts Company closed in 1958. Later the General Electric Company demolished what remained and built a new surface plant. Note the trestle at lower left, part of the 3.5 mile tramway that carried ore from the Springer, Stank, and Sutton mines to the Pacific mill downslope out of the picture to the far right. From the Segerstrom Collection, Holt-Atherton Special Collections, University of the Pacific Library. Courtesy of the Segerstrom family.

The Golconda chemical plant was completed in 1941 by the Nevada-Massachusetts Company in anticipation of wartime tungsten demand. Through a process of milling, magnetic separation, sintering and chemical treatment, this 100 ton–per-day facility treated complex manganese-tungsten ore from the Golconda Mine as well as low-grade concentrate from the flotation plant near Mill City. At its height in 1944, the plant employed 35 men in 3 shifts on a 24-7 schedule. By June 1945, however, a combination of declining federal subsidies, high chemical expenses, and falling ore grade forced the plant to close. Courtesy of the W. M. Keck Earth Science and Mineral Engineering Museum, Mackay School of Earth Science and Engineering, University of Nevada, Reno.

Nightingale mine and mill. Located on the eastern flank of the Nightingale Mountains near the western border of Pershing County, tungsten deposits were discovered here in 1917 and mined briefly until prices collapsed at the end of World War I. A Colorado firm headed by J. G. Clark bought the property in 1929 and built a 100 ton mill, but chronic indebtedness and erratic prices during the Depression years kept the mine closed except for a few brief periods of development. Forced into foreclosure by his creditors, Clark lost the property in a sheriff's sale in 1943 and died soon afterward, never learning that the buyer was an old competitor, C. H. Segerstrom. After the senior Segerstrom died in 1946, his heirs organized the Wolfram Company and operated the Star-Nightingale mines until the mid-1950s. Courtesy of the W. M. Keck Earth Science and Mineral Engineering Museum, Mackay School of Earth Science and Engineering, University of Nevada, Reno.

Pacific mill and tailings plant at Tungsten, Nevada, about 1939. Opened as a small gravity-flow crushing and concentrating plant near the end of World War I, the Pacific Mill was enlarged and upgraded by adding flotation during the Nevada-Massachusetts Company's first major development campaign when tungsten prices began to rise after 1933. In 1936 the company built a 250-ton pilot plant using differential flotation to test the recovery of WO_3 from old mine tailings that the gravity plant had failed to capture. The success of the pilot led to the construction of a 1,000 ton tailings-flotation plant that was completed in 1941. During World War II much of the company's tungsten production came from this highly efficient reprocessing operation. From the Segerstrom Collection, Holt-Atherton Special Collections, University of the Pacific Library. Courtesy of the Segerstrom family.

Pacific mill from tailings pond. Uncontained by impound dams or other regulatory measures later imposed on mining and milling operations, millions of tons of tailings from the Pacific Mill flowed freely for more than thirty years over the alluvial fan below. The old gravity mill burned to the ground in a raging fire of unknown origins in November 1943, but the tailings-flotation plant escaped damage and continued to operate for another nine months until all the values were recovered. Courtesy of the W.M. Keck Earth Science and Mineral Engineering Museum, Mackay School of Earth Science and Engineering, University of Nevada, Reno.

Tramming ore to the Humboldt Mill. Built by the Nevada Humboldt Company on the southern flank of Humboldt Hill, 200 feet above the mining camp, the Humboldt mill ran briefly in 1918 but closed in less than a year after tungsten prices plummeted. In the early 1930s the Nevada-Massachusetts Company absorbed the Humboldt property and refurbished the mill. Gas-powered locomotives trammed ore in side-dump steel cars from the mine portal down a 4 percent grade to an ore bin in front of the mill. Underground, muckers trammed cars by hand from the stopes to loading stations until late in World War II, when small electric locomotives became available. Courtesy of the Nevada Historical Society.

Charles H. Segerstrom posed for this jaunty portrait in the
mid-1930s, at the peak of his career as president of the Nevada-
Massachusetts Company. From the Segerstrom Collection, Holt-
Atherton Special Collections, University of the Pacific Library.
Courtesy of the Segerstrom family.

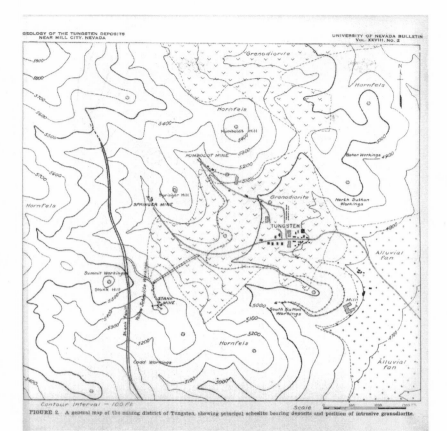

GEOLOGY OF THE TUNGSTEN DEPOSITS
NEAR MILL CITY, NEVADA

UNIVERSITY OF NEVADA BULLETIN
VOL. XXVIII, NO. 2

FIGURE 2. A general map of the mining district of Tungsten, showing principal scheelite bearing deposits and position of intrusive granodiorite.

Map of Nevada-Massachusetts' mines and mills near Mill City, 1934. This 1,200-acre contour map shows the four consolidated mine groups and two mills that made up the company holdings. W. J. Loring, as president and general manager of the Pacific Tungsten Company, brought together the Stank, Springer, and Sutton workings in 1918. The Humboldt remained under separate ownership until 1928, when it was absorbed by Pacific Tungsten's successor, the Nevada-Massachusetts Company. From Paul F. Kerr, "Geology of the Tungsten Deposits Near Mill City, Nevada," *Nevada Bureau of Mines and Geology Bulletin* 28, no. 2 (1934). Courtesy of the Nevada Bureau of Mines and Geology.

Toulon Mill. Built by the Billsby Company in 1916 on Southern Pacific's main line 12 miles north of Lovelock, Nevada, the Toulon mill during World War I served as a custom mill for the Nightingale, Humboldt, and Pacific Tungsten Co. mines until they built their own mills. In 1929 Nevada-Massachusetts acquired the Toulon plant, but it remained idle until 1936 when it was taken over by Rare Metals, a Segerstrom family corporation in partnership with Nevada-Massachusetts. After rehabilitation, enlargement, and upgrading with flotation cells, sintering furnaces, magnetic separators, and other equipment, the mill played an important role during the war years by processing and purifying off-grade tungsten concentrates from mines in both Nevada and California. Though slowly deteriorating, the structure still can be seen today along the railroad tracks, a reminder of Nevada's rich tungsten mining heritage. Courtesy of the W. M. Keck Earth Science and Mineral Engineering Museum, Mackay School of Earth Science and Engineering, University of Nevada, Reno.

Magnetic Separator. This Wetherall Magnetic Separator, used in the Pacific Tungsten plant near Mill City, Nevada, improved ore grade by capturing magnetized pyrite, garnet, and tramp iron from finely ground and sintered concentrate. Purification was a necessary step in the milling process, for off-grade tungsten ores could not be used for direct smelting to make high-quality steel alloys. Cleaning ore at the mill site added expense and time to processing, but purified concentrates brought premium prices and steady sales among the nation's steel alloy makers. Courtesy of Special Collections, University of Nevada-Reno Library. UNRS-P2003-15/17.

Rare Metals label. Processed tungsten concentrates from the Toulon mill were sorted and stored in bins to await shipment by rail to eastern buyers. Depending on the size of the order, concentrates were packaged either in burlap sacks or steel drums. Colorful labels like these identified the source and promoted the company to both shippers and customers. Courtesy of the Nevada Historical Society.

6 STRATEGIC METALS AT THE START OF WORLD WAR II

"The situation is very unsettled and no one knows what is going to happen in the next few weeks. The lid may be blown off the whole thing in Europe if Germany starts in with Poland and then we don't know what will happen."[1]

The Blitzkrieg that launched World War II was still six months away when Joseph J. Haesler, sitting before the teletype in his New York brokerage, pecked out that message to his friend and client Charles Segerstrom in Sonora, California. Both were seasoned professionals. Regardless of what course a European war might take, Haesler knew it would unsettle international metals markets. Segerstrom needed little reminding of the consequences. Remembering the lessons from the first global conflict, he recognized the symptoms of instability early enough to plan a flexible mining and marketing strategy that could readily adjust to changing conditions.

As we have seen, during the midthirties the boss of the Nevada-Massachusetts Company had implemented an ambitious tungsten exploration, development, and acquisition program. He actively acquired properties and leases but waited until the demand rose to invest the capital necessary for full-scale mining and milling. As international tension increased he anticipated a metals boom, but the conflicting domestic and foreign market signals after 1936 upset his plans. During the most turbulent times he moved cautiously, keeping costs down and avoiding speculative long-term capital investments. But the yellow light turned green as Germany blasted its way across Poland, and then turned to take on the West. Conflict in Europe did what the Sino-Japanese war could not do—weaken isolationist sentiment in America and strengthen the voices calling for mobilization. Like a chilly autumn breeze, rearmament was in the air, and that was good for miners producing strategic minerals.[2]

THE METALS INDUSTRY ON THE EVE OF WAR

The metals industry in the interwar years provides a good behavioral model to compare business words with deeds. Before the 1960s, studies of business attitudes, based largely on trade publications, concluded that business opinion mirrored views of the larger society. Despite a generation of accusations after World War I, no causal connection was ever proven between profiteering and influencing war policy. Most business leaders opposed war as an unpredictable disruption of the normal business cycle. Following from the discredited Nye Committee hearings of 1934–36, late in 1939 McGraw-Hill executives asserted the business viewpoint in an indignant denial: "To say that Industry and Business want war or will encourage, directly or indirectly, our participation in the present war, is a vicious and deliberate lie."[3]

Opposition to war did not mean opposition to the commerce of war or its preparation. In a study of the business press during the 1930s, Gabriel Kolko concluded that American businessmen hated totalitarianism but drew a disconnect between its moral and economic implications. Patent and licensing agreements drawn during the prewar years remained in force even after America entered the war. These collaborative arrangements, along with export quotas, territorial divisions, and other arrangements, document the close relationship between German cartels and leading American corporations. We have already noted the 1928 agreement between Krupp and General Electric that fixed tungsten carbide prices at five times higher than the cost of production. Antitrust prosecution, started in 1940 but set aside during the war, ended with a district court decision in 1948 finding GE and its collaborators guilty on all counts.[4]

Arms merchants may not have promoted wars, but wars promoted arms sales. Italy's war on Ethiopia in 1935 drew arms dealers from Belgium, Germany, Czechoslovakia, Switzerland, and Japan. During the Spanish civil war, despite a nonintervention agreement among the advanced European powers, the Soviet government, Germany, Italy, and Portugal, all rushed in with arms and aid. In 1937, with superior aircraft that easily penetrated weaker Nationalist defenses, Japan in five months of savage warfare controlled Shanghai and much of Northeast China, and threatened the mines and markets of the Southeast. American missionaries in China, witnessing Japanese atrocities firsthand, condemned Japan but also blamed American business interests for supplying the aggressors with military equipment and supplies.[5]

The missionary appeal to Christian morality helped change both American public opinion and foreign policy. Neutrality gradually gave way to a more Wilsonian view that rewarded good international behavior and punished bad. Japan's invasion of the Asian mainland and the Nazi threat in Europe by 1939 had convinced all but the most intransigent isolationists that war was preferable to fascist domination. The change was reflected in the title of a 1941 book by a retired U.S. diplomat: *You Can't Do Business With Hitler.* American prosperity, he concluded, depended on victory over totalitarianism—"not to save the world, but to save ourselves."[6]

Regardless of policy shifts and attitude changes, foreign conflict directly affected the prices of strategic metals. As in the First World War, prices of all metals rose during the fascist wars of the 1930s. In contrast to the postwar slump of 1919–21, however, they remained steady during the recession of 1937–39, reflecting tensions on the eve of another global conflict.

Tungsten's volatility was due mainly to China's internal instability and its conflict with Japan. World tungsten prices shot upward on reports that the Chinese government would stop exporting ore during the Sino-Japanese war. While it lasted, the high tungsten market stimulated American domestic production but also attracted unscrupulous buyers and sellers. With spot prices at twenty-five dollars per unit, Haesler's Metal and Ore company had

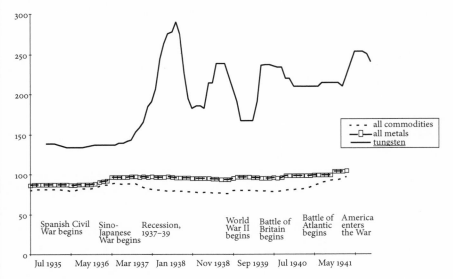

U.S. metal price index and international events (1927 = 100)

to sue a small western producer who refused to deliver on a contract for fourteen-dollar ore. In December, thieves broke into a warehouse at Mill City and made off with twenty-eight hundred pounds of concentrates, which they sold to an innocent Los Angeles metals dealer. A suspicious assayer recognized the ore and notified Segerstrom. Company officials promptly upgraded warehouse facilities, but a slump in sales during the recession probably discouraged more thieves than new security measures.[7]

THE NATIONAL STRATEGIC STOCKPILE

The volatility of tungsten prices at the onset of World War II had strategic as well as economic overtones. Both military and industrial leaders wanted to avoid the mistakes of the previous war, when material shortages led to chaotic market conditions, frustrating mobilization delays, and frantic production efforts. As we have seen, even before the first war ended, Bernard Baruch, as head of the War Industries Board, supported legislation to build a stockpile of tungsten and other war materials. With congressional support and the president's signature, the nation launched its first stockpiling plan. It was strictly a wartime measure, however, and died when the war ended despite Baruch's postwar "crusade" to keep the nation prepared for future wars.[8]

After 1920 the War Department, with congressional authorization, assumed primary responsibility for military procurement and planning. Whether business executives and military officials could work together remained to be seen. For years the "brass hats" that dominated the Army and Navy Munitions Board, charged with drawing up a mobilization plan, viewed stockpiling and other preparedness issues primarily as logistical problems. They resisted Baruch's insistence that war planning must consider the nation's economic limits and civilian needs as well as its military requirements.[9]

Politics also complicated the planning process. Antiwar activists and isolationists in the 1920s and '30s led efforts to reduce military budgets and sponsor pacifist panaceas like the Kellogg-Briand Pact outlawing war "as an instrument of national policy." Their influence peaked during the Nye Committee hearings of the mid-1930s, which cast aspersions at the military and industrial elite. At the other extreme, the American Legion, backed by the military establishment and some congressional war hawks, called for a "universal draft" and more active preparedness efforts. Torn between these polarities, the political leadership faltered. War planning proceeded on various fronts, but before the mid-1930s there was wide disagreement over the details as well as the implications.[10]

Despite the political posturing and the internal disputes, in 1930 the Munitions Board completed its first Industrial Mobilization Plan (IMP). Intended as a comprehensive manual for identifying and addressing the nation's resource needs in wartime, it went through three revisions over the next nine years, but could never catch up with the changing times. When war finally arrived nobody in government or industry paid much attention to it.[11]

For the metals industry, however, the IMP provided an opening for domestic tungsten producers to promote the stockpile concept. Tungsten was one of twenty-nine "strategic and critical" materials the Munitions Board identified as either in short supply or entirely lacking among the nation's natural resources. The 1933 revised IMP was not published until 1935, but the timing was propitious, with a new administration anxious to promote economic recovery and Congress in a spending mood. The National Industrial Recovery Act's $3.4 billion budget included $238 million for ship construction and another $77 million for modernization. As the first phase of a massive public works program aimed at both creating jobs and upgrading the navy, implementation required annual funding supplements from Congress as well as coordinated planning by government, industry, labor, and military personnel. The Vinson-Trammell Act in 1934 made congressional intent more explicit, authorizing the navy to continue shipbuilding until it reached the full strength established by the 1922 and 1930 arms limitation treaties. Peacetime naval spending increased yearly thereafter. It more than doubled between 1935 and 1940, a budgetary bonanza for contractors and suppliers.[12]

While industry representatives and military officials were still working on details, an advance copy of the 1933 IMP came into the hands of Edward R. Hagenah, a metals lobbyist with more connections in Washington than money. He saw in it great commercial possibilities, both for industry and for himself. Anxious to attract new clients, in the fall of 1934 he sent a newsletter to Charles H. Segerstrom and other tungsten dealers. In capital letters he emphasized the recommendation that the government build "stockpiles of manganese ores, tin, nickel, mercury, rubber, chromium, mica, cobalt, radium, coconut shells and TUNGSTEN." The report's "greatest point," he concluded, was "in selling the services the idea of providing a reserve stock of tungsten FROM AMERICAN MINES." The news made an obvious impression on the president of the Nevada-Massachusetts Company. His decision to acquire Golconda in 1935 was based at least in part on the optimistic views of this Washington "insider." Not long afterward, Segerstrom hired Hagenah to represent company interests on Capitol Hill. The Sonora executive needed a voice in Washington after the death of Hagenah's predecessor, Nelson Franklin.[13]

Military-industrial collaboration increased after the upsurge of totalitarian aggression in North Africa and Europe. Fear of war abroad made war planners more conciliatory at home. After military leaders acknowledged their dependence on the "civilian economy," revised mobilization plans gave more credence to industrial needs.[14]

War worries also increased the calls for a strategic stockpile. In Congress the leading advocate was a Nevada Democrat with close ties to the mining industry. James G. Scrugham, a professional engineer and former dean at the University of Nevada, had risen rapidly in state politics after serving as an assistant chief of army ordinance in World War I. Elected governor in 1922, he promoted public works and tourism to spur new growth, but four years later lost to a Republican candidate backed by the powerful Wingfield political machine. For the next six years Scrugham led the "loyal opposition" as State Democratic Central Committee chairman, and kept his name in the public eye as editor and publisher of the *Nevada State Journal*. He won Nevada's only seat in the House when the Democrats swept the 1932 election.[15]

The mining West welcomed Scrugham's addition to Congress, and for good reason. On issues affecting mining and metals he joined a bipartisan western congressional lobby on behalf of regional and local interests. Despite the State Department's calls for tariff reduction and reciprocity in trade agreements, Scrugham stood tall with mining and farming protectionists in the West and South. During the international debates over monetary policy he sided with colleagues in Nevada, Colorado, Montana, Idaho, and Arizona in supporting gold and silver purchase legislation, measures designed to cheapen the dollar and raise prices at home.[16]

Scrugham's stockpile sympathies came naturally from his military background as well as his regional focus. As an ordnance officer in World War I he was frustrated by the "shortage of strategic materials," and vowed to correct the situation when he got to Congress. In March 1934 the "Colonel" joined Arizona congresswoman Isabella Greenway in calling for a federal program to purchase copper, tungsten, manganese, and other metals "to be used only in the event of war." They said it would stimulate jobs, and benefit both industry and the military.[17]

How much industry benefited Congressman Scrugham at the same time cannot be fully documented, but certainly he was under obligation to key members of the Nevada mining establishment, including the Nevada-Massachusetts Company. Charles H. Segerstrom, despite his Republican partisanship and his intense opposition to New Deal inflationary policies, cultivated Scrugham's support with special favors, as this delicately parsed letter to E. R. Hagenah in the fall of 1934 suggests:

Regarding Mr. Scrugham, I wish you would kindly see him and have a little chat with him and find out where he left our Chevrolet coupe which we loaned to him at the time Dr. Lee was in Nevada. I do not want you to put it to him in such a way that he would be hurt, but as a matter of fact we loaned him the car over a month ago and he has not advised us where he left the car or what he did with it, and if you could just find out in a very diplomatic way and wire us so we can reclaim the car wherever he might have left it, I would thank you very much.[18]

A month later he learned that the Nevada lawmaker had crashed while driving in Arizona. Scrugham escaped with only cuts and bruises, but the coupe was not so lucky. Though not covered by insurance, the company auditor thought it still had some trade-in value if "the [auto] Agent is long on [new] cars."[19]

Fully recovered in 1935 and serving on the House Military Affairs Committee, Colonel Scrugham took the first financial steps to build a federal stockpile. Responding to Admiral C. J. Peoples' testimony that the navy would need $20 million added to its budget "to provide adequate stock piles of strategic and critical raw materials," Scrugham proposed a $7.5 million addition to the 1936 naval appropriation bill for the purchase of "emergency war minerals, including manganese and tungsten." Our "dependence upon foreign sources of supply for many of the strategic minerals necessary for military and construction purposes," he told colleagues, was the "weakest point in the naval program and to a lesser degree the War Department program." That was a jaundiced view, but even modest budget increases had trouble because of executive policy disputes over the amount and nature of federal spending during the critical years of the First New Deal. With White House advisers torn between providing adequate relief and keeping a balanced budget, they downplayed the navy's recommendation for millions in new stockpile funding. Despite the lack of administration support, however, the amended bill passed the House with the stockpile money intact.[20]

In the Senate, Scrugham's proposal faced new difficulties. Antiwar activists opposed any increases in military spending. "Why all this madness?" Senator Nye asked time and again in a series of bitter rhetorical jabs at the arms industry. "There is profit in the game of preparing for war, and those who profit from it have made national defense a racket—a racket that is international in its scope." Nye's minority coalition of peacemakers and isolationists managed to delay action, but could not stop Senate war hawks from actually adding more funding for ship construction than spelled out in the House version. In the Senate Appropriations Committee, however, the Pennsylvania representatives of the steel industry had more influence than the North Dakota spokesman for pacifism. Big Steel opposed any measure that might raise prices for raw materials. Despite lobby pressure on behalf of domestic metal producers, the Senate stripped out the stockpile money before passing a $460 million naval supply bill. The final version was the largest shipbuilding appropriation since the five-power naval limitation treaty of 1922.[21]

The 1935 congressional debates underscored the growing importance of the stockpile concept, but they also highlighted long-standing internal divisions among leaders of government, industry, and the military over how to go about building a federal reserve. Scrugham's proposal was patterned on Bernard Baruch's old plan for an industry-dominated central agency to stimulate domestic development of raw materials, using tariffs and other incentives to encourage private enterprise. Like many domestic metal promoters, Baruch had assumed that America's resource base was larger than previously recognized. Rather than fixing prices, Baruch's market-driven model favored collaboration among producers and negotiation between buyers and sellers to establish reasonable price ceilings, even if that meant relaxing antitrust laws during national emergencies. Any sign of "profiteering" could be controlled by an excess profits tax.[22]

Baruch's entrepreneurial approach to resource planning influenced a generation of policy makers in government and industry between the two world wars. But his star began to fade as capitalist culture came under attack in the Depression decade.[23]

That resource planning and development took a different turn after 1933 was not surprising, given the liberal and progressive views of some of FDR's closest advisers. Conservationists in Interior and Agriculture emphasized comprehensive planning and "wise use" of natural resources. Instead of assuming that the nation's resource base was plentiful and that industry knew best how to exploit it, policy makers turned to scholars and scientists for information and guidance. The president's choice to represent the minerals industry on the National Resources Board was Charles K. Leith, professor of geology at the University of Wisconsin. He had impeccable credentials and considerable influence in both government and industry. After serving in World War I as minerals consultant for the War Industries Board, he chaired the Mineral Advisory Committee of the Army and Navy Munitions Board.[24]

Leith's recommendations began with a global perspective. The uneven distribution of natural resources divided the nations of the world into "haves" and "have-nots." The totalitarian bluster was understandable, considering that England and the United States together controlled three-fourths of the world's mineral resources. Each nation had different wants and needs, however. Studies by the U.S. Geological Survey, the Bureau of Mines, and the Munitions Board since World War I had all shown that America was rich in some minerals but lacked significant deposits of nickel, manganese, and chromium, with tungsten slightly more extensive but still in short supply. Finding a lasting solution to this global imbalance through trade or other means would be difficult. Leith cautioned against shortsighted appeasement policies, but acknowledged that war might come from trying to restrain the aggressors through some form of "minerals sanction."[25]

Leith's views reinforced those of Cordell Hull. The secretary of state considered reciprocal trade the best mechanism for correcting global resource imbalances. Internationalists in the State Department objected on principle to any form of economic inequality in foreign relations. However, by the mid-1930s even Hull had to admit that reciprocal trade in the short run could not protect the Western democracies from the impending fascist threat. He agreed with Leith that the "defense of democracy and the defense of the military position more or less coincide." Thus, planning for conservation and "orderly development" of minerals had to be consistent with the immediate needs of national security.[26]

So far this was nothing new. For years Baruch had been using the same national security rationale to justify his own recommendations for a mineral reserve. Similar ends could be reached by different means, however. Leith's plan reflected his progressive heritage and his academic rationalism. If the nation needed minerals, it did not matter where they came from. The primary criteria should be based on science and technology, not politics or economics. National security was not the same as nationalism.

Leith had no objections to building a reserve with domestic minerals, provided it was technically feasible and cost-effective. But he questioned the efficacy of private enterprise in developing the nation's vital resources. Instead of relying on market incentives, he placed the burden of implementation and enforcement directly on the shoulders of the federal government. The Baruch proposal had identified certain minerals as strategic but had left it up to private industry to discover new deposits and develop the technology needed to exploit them. Those jobs Leith thought should be primarily the responsibility of government geologists and engineers. Whereas Baruch and Scrugham were staunchly protectionist, Leith, like Hull, said tariffs do more harm than good by raising consumer prices and stifling trade. Finally, Leith's plan called for a tighter regulatory regime and vigorous enforcement of antitrust laws. While accepting the need for industrial stability and efficiency, the Wisconsin geologist warned that the adverse effects of cartelization on consumers and taxpayers outweighed any practical advantages.[27]

Leith's prominence and close ties to the Roosevelt administration gave his views wide circulation. They helped build the momentum for a political shift away from the protective policies of the past. Reducing the price of raw materials loomed large in procurement and stockpile studies by the War and Navy departments. Efficient resource planning and preservation on an international scale also blended well with the State Department's reciprocal trade program. At Cordell Hull's urging, the president established a Standing Liaison Committee to coordinate policy and ideas between the Departments of State, Navy, and War.[28]

Congressional legislation soon reflected this rising internationalist influence. In 1936 an expanded stockpile proposal put the mining lobby on the defensive. The sponsor was Charles I. Faddis, a Pennsylvania Democrat with ties to the steel industry. He had served with distinction in World War I, and would serve again in the next conflict. His bill, inspired by Senator James F. Byrnes and favored by the State and War departments, authorized bartering to trade American surplus wheat and cotton for tin and rubber from the British Empire. Faddis also thought Britain could be persuaded to pay down its old war debt with strategic minerals to meet current U.S. needs. Only after barter efforts failed could the War Department purchase strategic materials outright from the cheapest sources of supply. The bill never made it out of committee that year, but Faddis gathered allies and planned a new campaign after the fall elections.[29]

Coming at the same time as a new round of reciprocal trade talks with Argentina and Bolivia, protectionists saw the Faddis bill as part of a combined assault on tariffs. Through their trade associations and their political allies they fought back. Congressman Scrugham and his colleague from Arizona John R. Murdock said the Faddis proposal did not recognize the danger that foreign supplies might be cut off in wartime. The Nevada delegate tried for the second time to fund a modest naval reserve with domestic resources, but once again his amendment died in committee. If stockpile ideas were ever to be implemented, clearly it was time for clarification and compromise.[30]

Prior to adjournment in the fall of 1936, Congress asked the administration for a thorough review of the nation's stockpile needs. The request prompted new studies by the Standing Liaison Committee as well as the Munitions Board, of which Leith was a civilian consultant. A special report, delivered to Congress just before Christmas, concluded that the strategic needs of the army and navy could not be separated, and "must be considered as a national problem rather than as a purely naval problem." The report did not deter Colonel Scrugham's successful third effort to build a separate naval reserve, however. The five million dollars he secured in the annual navy supply bill was only a sideshow to the major stockpile campaigns fought over the next three years, but his amendment did contain an important domestic incentive that authorized a 25 percent premium for American ores over foreign products that met the same specifications.[31]

The main stockpile campaign of 1937 opened in the House of Representatives with a revised proposal by Charles Faddis to build a national stockpile from foreign sources. It won the War Department's endorsement as "realistic" because of the numerous studies indicating a shortage of domestic stocks of manganese, chromium, tungsten, and other strategic minerals. The mining lobby, however, challenged both the evidence and the economic implications. Colonel Scrugham cited the government studies as reason to encourage domestic development and start stockpiling minerals as soon as possible before foreign sources were threatened or cut off by war. Arizona congressman Murdock, in hearings before the Committee on Military Affairs, played on Depression concerns. He claimed that there was an "ample supply" of manganese in Arizona that could be tapped and new jobs created if domestic mining were encouraged. J. Carson Adkerson, the intrepid manganese king, raised the intriguing possibility that American mines could produce more ore in an emergency than previously thought possible. With an exclusionary duty on foreign ore, he asserted, Americans could mine and mill ore as cheaply as the current cost of the imported product, and recover enough manganese to "make this country self-sufficient on a wartime basis." Pressed for clarification by Faddis himself, Adkerson insisted that there were no "physical limits" on how much manganese American mines could produce. The only limiting condition was the market price of ore.[32]

Coming from a mining engineer, this was sheer hyperbole that ignored the practical limits of economic geology. Adkerson well knew, if the politicians did not, that ore price alone does not determine the success of a mining enterprise. His claims sounded plausible to nationalists in the tense environment of the late 1930s, however. By linking America's manganese potential to national security, he reinforced the arguments for a mineral reserve and made all the more obvious the need for additional research and development.[33]

Tungsten producers were only indirectly represented during the 1937 strategic-metal hearings, but they used Adkerson's arguments to promote their own industry. Domestic ore the previous year accounted for only 10 percent of world tungsten production, but American industry consumed nearly a quarter of the world's supply. Would America's reliance on foreign imports make it vulnerable in wartime? Adkerson had replied with a resounding YES, and tungsten producers followed his lead. As chairman of the American Tungsten Association, Charles H. Segerstrom assumed the role of national spokesman. Despite the production problems his own company faced in this same period, he gathered nationwide data and published the association's first educational booklet to demonstrate the current extent and future potential of America's tungsten resources. J. J. Haesler also thought the association should "take the lead in protecting" consumers by agreeing to a reasonable price "maximum" on domestic ores regardless of price swings on the world market.[34]

The Sino-Japanese conflict provided an arena in which to test the strength and stability of the domestic tungsten market. As we have seen, fighting in 1937 on the Asian mainland threatened Chinese tungsten exports. American consumers, fearing supply shortages, triggered a "spurt in demand" at home and a consequent increase in domestic production. But the war soon bogged down, and Chinese exports resumed, ending the "spurt" and triggering a downward spiral of prices and production. Producers said the lesson was not that America lacked sufficient ore deposits but that domestic mining could not prosper under such unpredictable international market conditions. The solution they proposed was industry-wide production and price agreements, along with federal subsidies in the form of protective tariffs and domestic ore purchases. Arizona's voluble Senator Henry F. Ashurst later summed up the mining strategy in a speech on the Senate floor: "If we will develop our own internal resources, protect our shores, [and] encourage the small mine owners in the United States, they will produce not only all the ferrous metals we need but also what we call the non-ferrous metals, and rare metals and precious metals [as well]."[35]

Despite growing support for a significant stockpile program, the administration turned cautious and even contradictory during the recession that began late in 1937. The president accepted the need for deficit spending to stimulate the economy, yet simultaneously ordered cutbacks in programs that strained the federal budget. In David Kennedy's words, Roosevelt was a "decidedly reluctant and an exceedingly moderate Keynesian."[36]

White House spending limits posed a sizable stumbling block to funding a major strategic reserve. Before building a budget, however, stockpile advocates had to reconcile the deep divisions that remained between nationalists who wanted to promote domestic metal development and internationalists who remained skeptical of domestic claims and looked to foreign sources of emergency materials. Although the nationalists had strong allies in Congress, internationalists gradually gained the upper hand in government and military circles. Early in 1938 they united behind a House resolution introduced by Samuel D. McReynolds, a Tennessee Democrat, which would require government procurement officers to buy strategic materials from foreign suppliers if domestic stocks were deemed inadequate for future emergencies. As Haesler warned Segerstrom, by emphasizing the need for "cooperation" between the United States and its trading partners, the bill indirectly threatened the tariff on tungsten and other metals. The McReynolds resolution failed to pass in 1938, but the idea was incorporated into subsequent stockpile legislation.[37]

The internationalist coalition also showed its strength in opposing a bill sponsored by Senator John E. Miller of Arkansas to buy and store a million tons of low-grade domestic manganese ore. Along with Miller, delegates from Virginia, Montana, and South Dakota—all with potential deposits to be tapped— rallied to its defense, but opponents buried the bill under a barrage of criticism from senior government and military officials. As a letter from Secretary of Interior Harold Ickes bluntly concluded, the proposal was a subsidy that could not be justified by any standard—financial, technical, social, economic, or strategic.[38]

In the face of growing opposition, the nationalists tried to salvage core objectives through compromise. In May 1938, Utah senator Elbert D. Thomas introduced a comprehensive stockpile bill that included features both sides wanted. To please the domestic mining lobby, Thomas paid lip service to the Buy American Act as a procurement guideline. As noted earlier, that Depression-era law required federal purchasing officers to favor American-made goods over foreign whenever possible, but loopholes in the measure made it impossible to enforce. Any department head could demur if he decided that the purchase was too expensive or "inconsistent with the public interest." Sensibly, the Thomas bill cited the need for domestic mineral development, but left economic and technical decisions regarding content, source, size, and quality of the emergency reserve up to the secretary of war, working through the Army and Navy Munitions Board. Meanwhile in the House, Congressman Scrugham, with a bow to Leith and the Bureau of Mines, introduced a companion bill that authorized $500,000 a year to pay for government exploration and technological development of domestic metal resources. The combined proposals won sympathetic letters from the heads of the State, War, Commerce, and Interior departments, and even outright endorsement from the chief of naval operations. But the $100 million price tag, spread over four years, was too high for the White House. Official replies from key cabinet officers were uniformly negative: "The Bureau of the Budget advises that the proposed legislation is not in accord with the program of the President."[39]

Budgetary constraints were soon overshadowed by the rush of events abroad. Early in January 1939, following the Munich "Diktat" and the Nazi march into Czechoslovakia, military planning gained new urgency. The president, taking "immediate steps for the protection of our liberties," sent Congress an emergency budget that called for $552 million in new defense spending for planes, bases, and training. The budget did not include money for a metals reserve, but stockpile prospects rose appreciably with every foreign crisis.[40]

A week before the president's message, Charles I. Faddis staked out the internationalist position by reintroducing his bill in the House to acquire four strategic metals abroad, either by purchase or by barter. The domestic mining lobby, represented by a coalition of Democrats from the mining West and South, took a different tack. They lined up behind a proposal by Scrugham that was absorbed into a $100 million collective bill by Congressman Andrew

Jackson May of Kentucky, chairman of the Committee on Military Affairs. The measure required the secretary of war to give "priority and preference" to domestic minerals "wherever possible." The phrase expressed more hope than reality, for a new government analysis underscored the lack of adequate American sources of chrome, cobalt, manganese, nickel, tin, tungsten, and vanadium. "I have seen this report," a worried Hagenah wrote Segerstrom, "and it paints a dismal picture of the seven minerals studied."[41]

Internationalists and metals consumers used the new government findings to their advantage. Rumors surfaced during the legislative session that a "combination of mineral importers and mining men" led by C. K. Leith and the Bureau of Mines wanted to kill the May bill because of its domestic bias. In February, T. M. Girdler, the Republic Steel chairman, warned that wartime supplies might be cut off for "long periods" unless the government began buying at least a two-year supply of the essential metals from "foreign sources." The government study also added leverage to the arguments of army and navy procurement officers. In hearings on the stockpile legislation they emphasized cost efficiency and urged buying foreign materials in peacetime while transport was still available and prices low. Nationalists found themselves on the defensive, but they did manage to add an amendment by Scrugham to the May bill. It called for a $500,000 annual appropriation over four years to the Bureau of Mines and the U.S. Geological Survey for a domestic survey and for beneficiation studies. Endorsed by the Bureau of Mines, the provision was "considered vital to the domestic industries," as Hagenah explained in a private letter to Segerstrom.[42]

Five additional stockpile bills, similar but slightly different to either the Faddis or the May approach, were introduced over the next six weeks as House members jockeyed for leadership and responded to constituents. Pragmatism won in the end, with both sides massaging the language of the various proposals down to a workable compromise.

At the other end of the Capitol, Senator Thomas waited until the House deliberations were nearly over before reintroducing a near duplicate of his 1938 compromise. Since it incorporated all the essential features of the 1939 House version, Congressman May used procedural motions to win approval of the Thomas bill in the House, and then to substitute it for his own proposal. During the floor debate in the Senate, Senator Byrnes asked if the president's budget guidelines still applied. Thomas responded that the executive branch no longer opposed the stockpile on budgetary grounds, yet the president "suggests" that first-year spending be held to $10 million out of the $100 million proposed over four years. A Byrnes amendment incorporated that suggestion into the final version. After a few more changes to bring all the major players aboard, early in June both houses passed the Thomas-May legislation. Roosevelt considered it a diversion from more pressing security measures, yet the $104 million price tag over four years represented less than 7 percent of the combined spending of the War and Navy departments for 1939 alone. He signed it on June 9. Two months later Congress approved the first appropriations under the legislation, but because of volatile metal prices in the lead up to war the War and Navy departments delayed the bidding process. In theory at least, if not in actual implementation, the Thomas bill, as it came to be called, was a significant step toward building the nation's first comprehensive strategic reserve.[43]

MOBILIZATION AND THE METALS INDUSTRY

The shift from a peacetime economy to one geared for war meant increasing productivity for the tungsten industry, but also increasing uncertainty, at least in the short term. Tungsten prices crept upward after 1938 as war talk grew in Europe, but large ore reserves in the United States and the lingering effects of recession on steel plants kept domestic demand low. Even the outbreak of war in September 1939 did not immediately stimulate domestic production, in part because of price instability in the tungsten market. After the Munich crisis the pound sterling had dropped to its lowest level since Britain went off the gold standard in 1931. Over the next year it rose slightly, but then fell sharply when the war started, in effect lowering the price by 20 percent of Chinese tungsten imported through Hong Kong. With the American economy still recuperating and isolationist sentiment still strong, mine owners feared that foreign producers might dump tungsten on American markets, further eroding prices and hurting domestic producers.[44]

The fall of France was a dividing line marking the end of uncertainty and the beginning of all-out production. Within six months after the Nazis marched into Paris, American opinion changed dramatically from aiding Britain short of war to "aiding Britain at the risk of war." In Congress, the shifting sentiment boosted support for Lend-Lease, the administration's response to Churchill's plea to "give us the tools." Hailed as a patriotic victory, its approval in the spring of 1941 drowned out the lingering voices of isolation but committed American industry to unprecedented production levels.[45]

Ensuring an adequate and steady supply of strategic raw materials was the first priority on the wartime production agenda. Despite the proximity of Mexico and other mineral-rich hemispheric nations, America faced glaring shortages of strategic materials after 1940. Defense industry consumption in the war years mushroomed two or three times over the 1919–40 base period. During the critical years 1942–44, 90 percent of chromium supplies came from abroad, 86 percent of manganese, 100 percent of nickel, and 61 percent of tungsten. At least a third of all copper, lead, and zinc—metals abundant in the United States—also came from foreign suppliers in these years, although the reasons had more to do with "a lack of organization and manpower," as one congressman put it, than a lack of domestic deposits.[46]

Stockpiling momentum started slowly but accelerated as national defense took on increasing urgency. Of the $100 million authorized in the Thomas bill, only $70 million was appropriated and only about $10 million actually spent before the summer of 1940, when Congress revised the law, resulting in a much larger and more comprehensive program. The new measure gave the Reconstruction Finance Corporation broad authority to purchase and control stockpiled materials. President Roosevelt followed with an Executive Order establishing two RFC subsidiaries, the Metals Reserve Company and the Defense Contracts Company. First used to build stockpiles of rubber and tin, these lending and spending firms rapidly expanded both funding and function.[47]

As rearmament mushroomed, the Roosevelt administration devised new structures and strategies to meet wartime production needs. Industrial mobilization had been the chief preoccupation of federal war-planning agencies since the 1920s, but advisers had differed sharply over administrative organization, resource allocation, production priorities, inflation controls, and other details. Late in 1939 FDR had established a new civilian advisory body, the War Resources Board (WRB), to assist the Army and Navy Munitions Board in drafting revisions to the Industrial Mobilization Plan. However, its preliminary report leaned too heavily toward the old Baruch model of an independent, industry-led "superagency" in charge of industrial mobilization. Wary of the business influence, Roosevelt dissolved the WRB early in 1940. In its place he restored an older body that reported directly to the president, the National Defense Advisory Council under the Office of Emergency Management. The council, in turn, organized an unwieldy seven-member National Defense Advisory Commission (NDAC), which could recommend and approve contracts for military orders, but lacked the power to compel industry to boost the production of military equipment and supplies.[48]

Production uncertainties in the face of accelerated military demand led to another administrative change. Early in 1941, responding to criticism that the NDAC lacked "central direction and control," Roosevelt replaced it with the Office of Production Management (OPM). "There will be no 'bottlenecks' in our determination to aid Great Britain," he told the American people late in 1940, yet the president's leadership style complicated decision making. To contemporary critics the seemingly perpetual revolving door of offices and staff produced conditions in Washington approaching "near-chaos." More recently, historian Matthew J. Dickinson characterized Roosevelt's managerial conduct as a willful system of "competitive adhocracy." Establishing overlapping agencies with redundant staff functions, issuing vague orders, encouraging informal input from a variety of sources—by these and other means FDR retained executive authority and control. Whenever he felt that a bureaucratic shake-up was necessary he used the presidential prerogative rather than seeking congressional authority to reorganize administrative units and change personnel. In the ensuing turmoil the man at the top carefully protected his personal power and prestige, both from political enemies and from conservative businessmen. As Henry Wallace once quipped, "FDR could keep all the balls in the air without losing his own."[49]

The OPM was a classic example of "competitive adhocracy" at work. Under the dual leadership of General Motors president William S. Knudson and Amalgamated Clothing Workers Union boss Sidney Hillman, ostensibly it represented the wartime spirit of labor and capital working together for the common good. However, it had no authority over military procurement, and after a separate Office of Price Administration and Civilian Supply was established in April 1941 (later reorganized as the OPA), it had no power to control wages or prices. In practice, the OPM was little more than a discordant collection of line agencies operating without central focus or command. It lasted only long enough to demonstrate the need for a stronger and more centralized emergency management effort.[50]

Slow to recover after years of depression, American industry was swamped after 1940 by the flood of military orders from home and abroad. The "quick change to a war economy," White House attorney James Rowe later recalled, increased production pressures and tested the government's incipient economic controls. The steel industry, underutilized just a few months earlier, at first resisted any new capital expenditures, remembering all too well the glut at the end of 1918 and the recession that followed. When steel prices began to rise industry leaders resisted calls for stabilization, claiming, like farmers, that higher prices were necessary with costs rising after a decade of depression. Labor made the same argument for higher wages, adding to the incongruous clamor that engulfed wartime Washington.[51]

Fearing shortages of raw material for arms and armaments, federal officials used a combination of threats and promises to stimulate military production. Military leaders, dedicated to winning at all costs, demanded more from the civilian economy and pressed business to shift resources by canceling civilian contracts. Thurman Arnold, the antitrust chief in the Justice Department, warned of a steel industry conspiracy to limit production and thus increase profits at the expense of national defense. OPA head Leon Henderson threatened to seek stronger price control legislation if voluntary commodity price agreements with industry did not keep inflation in check.[52]

The threats proved less effective than the promises. Winning the war was more important than keeping faith with older socioeconomic reform programs or keeping big business in check. Congress offered numerous incentives. One of the most important was legislation authorizing military procurement officers to switch from competitive open bidding to negotiated contracts, by which individual sellers and buyers by mutual agreement determined the baseline cost of the product, plus a "reasonable" profit. By 1941 "cost-plus" was standard operating procedure in the defense industry. Another boon to business was legislation approved late in 1940 allowing war industries to depreciate capital investments over five years instead of twenty. The strategic mining industry got a special break in 1940, when Congress lowered the excess profits tax on producers of tungsten, manganese, chromite, tin, and other strategic minerals. The following year, beset by cries of unfairness, it repealed the exemption, but restored it in 1942 after intensive industry lobbying and a special appeal from the War Production Board.[53]

However welcomed by industry, these incentives did not increase steel capacity, as some intended. In 1943, with the Allied offensive overseas consuming vast quantities of war goods, the government took a more direct approach. It negotiated a $2.9 billion joint venture with twenty private steel corporations to build twenty-nine new plants. After the war private industry purchased the public's wartime assets for a fraction of the cost.[54]

Financial aid, tax breaks, accelerated depreciation, "cost-plus," and other subsidies approved by the Roosevelt administration signaled a pragmatic shift away from the antibusiness rhetoric of the Second New Deal. With taxpayers funding two-thirds of the capital costs and government agencies guaranteeing a 7 percent minimum return on private investments, war industries earned handsome profits during the war years. This "carrot and stick" approach, in the words of historian Robert Higgs, amounted to shifting the "risks from the military suppliers to the taxpayers."[55]

Whether the domestic mining industry stood to gain much from a more friendly government remained to be seen. Many mine owners saw inherent contradictions in shifting federal attitudes. New Deal labor laws, especially the imposition of collective bargaining and the minimum wage, seemed incompatible with new business subsidy proposals. Segerstrom's lament to Hagenah during the recession of 1937–39 had expressed a common managerial view: "I have talked with many people, and they all seem to be frightened to even investing in the primest of enterprises. They fear for labor; they fear for taxes; they fear for their very existence, and I believe that until the administration makes up its mind to stop trying to remake the world, and go back to old time business economics, that we are going to be in for a very hard time."[56]

Roosevelt's reelection in 1940 had disappointed many industrial leaders, and the imposition of wage and price controls as the economy heated up received mixed reviews from businessmen. Mine managers welcomed efforts to hold wages in line, but complained loudly about price ceilings for minerals they produced. From their perspective commodity prices still reflected Depression conditions while production costs soared in the red-hot wartime economy.[57]

TUNGSTEN STRATEGY: CONSERVATION AND IMPORTATION

Building a strategic stockpile during a military crisis, with defense industries competing with domestic manufacturers for limited supplies, was next to impossible without both prioritization and conservation. Through the Office of Production Management, the government in 1941 began a systematic effort to conserve strategic materials, with consumer products taking the biggest hit. To save tungsten, the OPM first urged, and then ordered, the nation's domestic steel producers outside the defense industry to find substitutes for the gray metal. Molybdenum was the recommended substitute, based on studies begun by the Army Ordinance Department in the early 1930s suggesting that tungsten's "twin" could be used for most applications. At first, steel companies had been reluctant to explore molybdenum's possibilities. They grew more enthusiastic after 1937, as Climax production expanded and the market price for molybdenum rose at a slower rate than prices for tungsten.[58]

In the summer of 1941, an OPM order to cut the tungsten content in "molybdenum type" tool steel by 25 percent brought protests from the tungsten lobby. Warning that the order would force Nevada-Massachusetts to cut production by 75 percent and "throw the market into chaos," Segerstrom "alerted" Senators Hiram Johnson and Pat McCarran, "the two strongest anti-administration men that I could find." In response to their inquiry, OPM deputy director A. I. Henderson said the new directive would still leave about 5 percent tungsten in the formula. Rationing of tungsten, still the nation's "most important" strategic material, began three months before the United States entered the war.[59]

Strategic planning also meant denying vital materials to enemies, real or potential. In wartime that strategy might be pursued aggressively through blockades and bombing raids, but commerce and diplomacy played equally important roles, both before and during the war. Soon after Germany invaded Poland, the State Department imposed a "moral embargo" on molybdenum and other domestic metals scheduled for delivery to Axis powers. When some mining officials protested, Cordell Hull suggested they mollify their stockholders with a newsletter explaining that the embargo applied to "those who had bombed open cities killing women and children."[60]

Outside the United States, Hull's moral suasion had little impact. Strategic ores produced elsewhere went to the highest bidder. During the late 1930s, as Hitler prepared for war, the German arms industry consumed over four thousand metric tons of tungsten annually—more than twice the rate of U.S. consumption.[61]

With virtually no domestic tungsten supplies and no sizable stockpiles—despite rumors to the contrary—Germany was the eight hundred-pound gorilla on the international tungsten market. Before 1940 the Germans imported most concentrates from China, partly through a barter agreement that resulted in exchanging Chinese tungsten for German arms, and partly by quietly purchasing Chinese ore from other dealers on the international market. Sino-German trade policy abruptly changed after Japan invaded China in 1937. Hitler had welcomed Japan as an anticommunist ally after it signed the Anti-Comintern pact in 1936, and as an Axis partner following the 1940 Berlin pact. Japanese militarists, in turn, pressed the Nazi government to end arms shipments to China, although German industry needed Chinese raw materials. After the war began the British blockade cut the last remnants of Sino-German trade by sea, but some Chinese ore continued to reach German plants via the Trans-Siberian Railroad until the Nazi dictator launched the Soviet invasion in June 1941.[62]

By that time the American government had intervened in an effort to keep Chinese ore out of Nazi hands and supply the stockpile program at the same time. With RFC money, the government-backed Metals Reserve Company had begun purchasing Chinese concentrates late in 1939 at a fixed price slightly higher than the domestic price. The ore arrived by ship from Hong Kong until the Japanese blocked that route in 1941. Despite enormous transportation costs, shipments then came overland via the Burma Road until that route was also cut off. For a few desperate months in 1942 and 1943 some American airmen risked their lives flying tungsten over the Himalayas to freighters waiting in the Indian Ocean. Some eight thousand tons of Chinese concentrates went into the stockpile during this period.[63]

Bereft of its Asian ore sources, German industry could still import low-grade wolfram concentrates from Portugal and Spain, but they came at a high price. Most European tungsten is produced from the vein and placer deposits of the Iberian Peninsula. Wolfram is the main ore, with the largest deposit in the Trasos-Montes region north of the Tagus River in northeast Portugal. Unfortunately for the Germans, the British were heavy investors in Iberian mines and controlled much of the tungsten production. However, rising prices after 1939 stimulated new output, particularly among independents who sold wolfram on the open market.[64]

For both military and economic reasons, keeping the Iberian Peninsula neutral was important to belligerents on both sides. Though dictators ruled both Spain and Portugal, British control of the Mediterranean made any thought of joining the Axis problematical. On the other hand, British economic interests kept the Western allies from putting too much pressure on the Salazar regime in Portugal or the Franco regime in Spain. To keep wolfram out of German hands the British government resorted to "preclusive purchasing" of all available supplies, forcing German dealers into a bidding war. The market price of tungsten shot upward from eleven hundred dollars per metric ton in 1940 to twenty thousand dollars in 1941. Franco refused to intervene, letting small Spanish tungsten producers enjoy their windfall. Early in 1942, however, Salazar declared "strict neutrality" and gave both sides a share of the independent wolfram output, but fixed the price at twice the prewar market quotation. Since Portugal was by far the largest producer, his intervention kept Germany supplied with at least some ore at inflated prices until the inland routes were finally closed with the Allied invasion in 1944. Nevertheless, the constant threat of an oil embargo by their Western suppliers kept both dictators from drifting too far outside the orbit of Allied strategic interests.[65]

For two years after Pearl Harbor American procurement officials followed Britain's "preclusive buying" strategy, applying it wherever necessary to prevent Axis powers from obtaining critical materials even if not needed for the stockpile. Prices paid for Spanish and Portuguese wolfram were three or four times higher than those obtained by American tungsten producers under contracts with the Metals Reserve Company. In the Western Hemisphere, Roosevelt's "Good Neighbor" policy promoted mutual cooperation and war planning. At periodic meetings beginning in 1938, U.S. and Latino foreign ministers set policies for bilateral and multilateral trade, leaving specific terms and conditions up to private or government negotiators. The result was a free flow of American dollars and manufactured goods in exchange for substantial amounts of "strategic and critical" raw materials from Latin America. Even corn and other farm products the United States had in abundance were purchased by government agents as a gesture of American goodwill and support for economic development south of the border. During the war, Mexico, America's largest Latino trading partner, shipped the bulk of its strategic exports to the United States.[66]

Farther south, Bolivian tin and tungsten deposits in the La Paz, Cochabamba, Oruro, and Potosí districts ranked high on the list of American strategic priorities, but they were also coveted by the Japanese. When Japan offered to buy Bolivian tungsten in 1941, Bolivian officials rejected the bid for "economic and political reasons," it was reported. Doubtless they felt the Americans breathing down their necks, but higher prices paid by U.S. Metals Reserve provided the proper business incentives. Most of Bolivia's tin and tungsten came to the United States during the war. Keeping strategic metals from enemy hands was expensive, but national security justified the cost. Overall, the RFC bid from ten to twenty times the normal market price for the most critical metals.[67]

One American bidder overseas had a personal connection with the Nevada-Massachusetts Company, although Charles Segerstrom never knew to what extent until after the war. In April 1943 he received a telegram from an official at the Board of Economic Warfare, inquiring about the "ability and loyalty" of Joseph J. Haesler, who was under consideration as "Foreign Service

Head Minerals Purchase Negotiator." Endorsed by Segerstrom and others, Haesler spent a year in Brazil and another year in Turkey buying ore for Metals Reserve, outbidding—and outfoxing—German and Japanese competitors for strategic metals.[68]

THE RESPONSE OF DOMESTIC PRODUCERS

Within the mining industry, prominent spokesmen welcomed the government's emphasis on building strategic reserves, but not at the expense of domestic producers. As we have seen, the early talk of scarcity alarmed military planners, and preliminary studies of domestic mineral supplies were not encouraging. Many American ores seemed too marginal to provide a steady and reliable source of strategic minerals, but domestic mining men thought differently. At the annual meeting of the American Mining Congress in September 1940, they sought to reassure Americans there would be no shortages of lead, mercury, copper, and other metals in case of war. As spokesman for the tungsten industry, Charles H. Segerstrom insisted that domestic mines could supply enough high-quality ore for the indefinite future.[69]

Not on the agenda of the AMC meeting, but clearly on the minds of mining representatives, were the adverse consequences of foreign ore purchases on domestic production and profits. These were old economic fears, reflected in both testimony and legislative language dating back to the tariff debates of the late twenties. Old wine could be packaged in new bottles, however. National security gave new meaning to the mining industry's call for military and industry procurement agents to "Buy American," but the loopholes in that legislation, as noted earlier, made the tariff the last bulwark against foreign competition. Through the Depression years free-trade pressure had come not only from the State Department but also as a result of internal divisions between producer and consumer that weakened the voices of protectionism. During hard times some sharp metal traders had purchased domestic ore from hard-pressed mine owners at cut-rate prices, or bought and stockpiled low-priced foreign ore until prices began to rise, then dumped it on the domestic market.[70]

A more direct threat to tungsten producers was the 1938 reciprocal trade talks with Britain. Any deal with the British would automatically extend to any country with "most favored nation" status, including China. Tungsten powder and alloys were on the State Department's preliminary report of dutiable items considered for reduction. Cutting the tungsten duty by 50 percent, Segerstrom feared, "would in all probability make unprofitable every tungsten mine in America." As president of his trade association he filed a brief with the Tariff Commission opposing any lowering of the duty. When Haesler suggested the brief emphasize the danger that Chinese ore imports might be soon be cut off entirely because of the Japanese invasion, Segerstrom responded that he "did not go into these matters at all and only tried to show that we had the ore, that we were in a position to get it out, that Chinese or coolie wages could not be competed with in the UAS [sic] and as to the effects I thought better to lay off and not say anything that would get them up on their ear." After a quick trip to Washington in March to meet with the tungsten lobby, Segerstrom returned to Sonora, confident that the "tungsten situation [was] in very excellent shape."[71]

His confidence was not misplaced, but national security weighed more heavily with the Tariff Commission than old protectionist arguments. The western mining boss learned a lesson in lobbying from Van Rensselaer Lansingh, Molycorp's vice president, whose fieldwork in 1936 Segerstrom had privately ridiculed. In a trade letter distributed early in 1938, Lansingh made the case both for protection and for stockpile legislation, the latter still pending in Congress. Without a national reserve, he argued, the United States

is absolutely dependent on a continuous flow of tungsten from foreign sources especially China. . . . If this supply is put off by Japan or other influences, this country would be compelled to fall back on its own mines and refineries, and if these were not in operation it would take from one to two years to get them going again. From a standpoint of national policy, therefore, it would seem a most foolish move to attempt to wreck the tungsten industry in this country by the proposed changes in the present tariff.[72]

Open hearings on the trade talks with Britain concluded in March 1938. Through the summer the tungsten lobby worked furiously to influence the outcome. Under intense pressure the State Department lowered import fees by 30 to 50 percent on hundreds of commodities, but left the duties on chromium and tungsten intact. "This is the very best of news!" Segerstrom exclaimed in reply to a telegram from Washington informing him of the decision.[73]

Free traders made little headway during the war years despite metal shortages and rising prices after 1940. The economic burden fell entirely on consumers. Strategic materials imported under government contract were not subject to duty. "Obviously nothing is obtained by having the Government pay tariff duties to itself," said one official. The State Department still opposed any "special interest" legislation that tended to undermine the expansion of trade, but the duty of fifty cents per pound of tungsten contained in imported ore or concentrates stood firm until 1948.[74]

As tariffs continued to be enforced for many industrial commodities, foreign exporters naturally were puzzled by the contradictions between American policy and practice. Mexican officials, for example, said if the United States wanted more critical supplies it should lower the tariff on eleven strategic minerals sold "exclusively" to American firms. Protectionists ignored such counterintuitive advice. Indeed, as foreign trade routes reopened later in the war, at least one congressman called for increasing the tariff to protect home industry from cheaper minerals pouring into the country.[75]

Regardless of classic protectionist arguments, tariffs did little to stimulate new mining activity. Since 1939 the Bureau of Mines had turned up dozens of low-grade tungsten, antimony, manganese, and chrome prospects throughout the mining West, but large operating companies usually avoided the financial risks involved in opening unproven claims. They remembered all too well the lessons of 1918. The accelerating pace of rearmament, however, caused strategic shortages and alarmed officials from both industry and government. With Nazi wolf packs lurking in the Atlantic and millions of tons of Allied cargoes at risk or already destroyed, even low-grade domestic ores seemed worth a second look.[76]

MOBILIZING NEVADA-MASSACHUSETTS

How the nation's largest prewar tungsten production company responded to these wartime conditions is fully revealed in the extensive corporate records on deposit at the University of the Pacific. Charles H. Segerstrom was politically active and well connected, but his conservative views on domestic economic and social policy had little national impact while Roosevelt was in office. During the war years, however, despite the production troubles of Nevada-Massachusetts, his position as leader of the tungsten lobby made him an active player in the political machinations behind the strategic metals program. But first he had to attend to the complex problems facing his own company—always his first priority.

As noted earlier, declining ore sales during the "Roosevelt Depression" had shut down many small domestic producers and forced even larger companies to retrench. Early in 1939, with cheap foreign ore still a glut on the market and steel mills "waiting for times to change," Segerstrom decided to bid on a navy contract under the special strategic reserve program approved two years earlier. On behalf of Nevada-Massachusetts, Hagenah and Haesler worked for months in Washington to complete the paperwork. On September 19, three weeks after the war started, naval officials agreed to purchase 17,250 units at twenty-five dollars a unit, more than 30 percent above the market price just a month before. War inflation figured only slightly in the price rise, however. With a nod to the "Buy American" terms that Congressman Scrugham had included in the law, the navy had previously announced a 25 percent premium for domestic ores, provided they met the specifications.[77]

One specification that almost killed the deal was the Walsh-Healey Act, a legacy of New Deal labor policy. It required government contractors to abide by federal guidelines for wages, hours, and working conditions. A Brookings Institution study recommended suspending restrictive working hours to meet growing manpower shortages, but they remained in force until 1943. Segerstrom complained that the Walsh-Healey provision would "shut down this plant or ruin us as the terms are beyond us." Haesler and Hagenah found a loophole, however. After talking with Labor Department officials, they wired Segerstrom that mine operators could avoid legal problems by selling the ore directly to private brokers, and then let the brokers deal with the government. That would break the direct link between the government and the mines.[78]

In the meantime Segerstrom confronted the realities of a superheated labor market. Responding to signs of discontent, in October he offered his men a forty-hour workweek and time and a half for overtime. That brought the company into compliance with federal guidelines and secured the navy contract, but it soured Segerstrom on doing any further government business. As he explained to a board member a few months later:

It was on account of this Walsh-Healey act that I have deferred taking any orders from the government at all, in fact, I would rather sell to a private consumer at $2.00 per unit less, and I am sure we would be money ahead against selling to the government direct, because they at once place inspectors in all parts of the plant, and begin an agitation for unionization of the men, and I am afraid if we had a union which would control . . . actions at Mill City, it would make it very difficult to make a profit.[79]

Despite the complications, anticipating war's impact on the mining business put the boss of Nevada-Massachusetts in a good mood. In November 1939 — ten days after Roosevelt signed a revised Neutrality Act lifting the arms embargo on belligerents abroad — Segerstrom penned a happy note to his

Boston colleagues: "As to the near future of tungsten I am very very bullish. I believe we should do everything in our power to produce all the ore possible during the next couple of years. I am very sanguine that the war will continue and outside of that I feel the U.S. Army and Navy will purchase at least 5 million dollars worth of tungsten ore, and there is no reason why we could not supply the major part of it, providing we have the supplies available." The trick was to keep production high and costs low, taking advantage of every opportunity to peddle tungsten while the demand lasted.[80]

To avoid an ore shortage just as business was picking up, Segerstrom expanded operations beyond Mill City. He and his associates took to the road again, investigating nearly every likely tungsten prospect in the Far West, and even some unlikely ones. When George I. Emery, a Boston board member, learned of the renewed quest for ore, he wondered if the company had overlooked the property near Bishop, California, that Union Carbide had acquired two years before. The inquiry prompted a reply that revealed not only Segerstrom's single-minded focus on tungsten but also his limitations as a nontechnically trained mine operator.

He replied that he knew all about the Pine Creek operation run by Union Carbide's mining division, U.S. Vanadium Corporation, and wanted nothing to do with it. After a brief run in World War I it had been rehabilitated by the Watterson brothers of Bishop, bankers and major investors in the local economy until they were convicted of embezzlement during the Owens Valley water wars of the 1920s. Early in 1934, while the mine was in bankruptcy and the Wattersons in San Quentin prison, the receiver had offered it to Nevada-Massachusetts for twenty-five thousand dollars. Segerstrom and Heizer, along with a team of engineers, had "carefully" examined the property, perched high on the spectacular eastern front of the southern Sierra Nevada. The ore was a complex mixture of powellite, molybdenite, and copper minerals, difficult and expensive to separate and process with the milling technology available at the time. They concluded the mine was too remote and the deposit too low grade to make money unless tungsten prices rose above twenty dollars a unit, double the then current market price.[81]

From Segerstrom's perspective in 1939, the Pine Creek mine was a costly failure. Union Carbide had taken it over two years before, trying to boost ore production and expand its customer base at the expense of Molycorp, its chief rival in the ferrotungsten market. Thus far the Connecticut company had spent half a million dollars on the California property, but it was still "not yet ready to produce ore," as Segerstrom explained to Emery. He had it on good authority from J. R. Van Fleet, the Pine Creek manager, that the mill concentrates were running as high as 3 percent molybdenum even after flotation and thus "absolutely valueless as far as making ferrotungsten." The Nevada-Massachusetts president doubted if Pine Creek would "ever be a factor in the tungsten game." Compared with the competition, he concluded, "we are without a doubt in the best possible position that any tungsten property is in the United States, and this includes the Union Carbide, who have any amount of money to invest in any venture they may see fit."[82]

Considering the contrast between one of the multinational giants in the corporate world and the regional firm that had struggled for years on the verge of insolvency, the trace of hauteur in Segerstrom's remark is understandable. Through a combination of prudent management and frugal financing during the worst depression in American history, he had built a successful family business. His objective after the war started was not to conquer or control the tungsten industry but to keep his firm "on an even keel" and make a reasonable profit while the wartime boom lasted.[83]

Union Carbide had big ambitions for Pine Creek, and its persistence paid off. Just before the United States entered the war, a team of metallurgical engineers lead by Blair Burwell perfected a successful new process to produce synthetic scheelite with molybdenum and copper as by-products. With a new mill and a high-priority authorization from the Metals Reserve Company, Pine Creek soon became America's single largest tungsten producer. Its wartime output was 432,091 units of WO_3, compared to 357,001 for the Mill City plant. However, the combined wartime production for the Nevada-Massachusetts Company, including Rare Metals, Golconda, and other subsidiaries, was 713,839 units, or 40 percent more than Pine Creek's during the same period.[84]

Segerstrom had a hard enough time keeping his own mills running without worrying about rivals. In 1939, on behalf of the Rare Metals Corporation, he purchased a half-interest in the Nevada Tungsten Company's dormant property near Gardnerville, one of several small mines that had started up in 1937 when metal prices were high but soon shut down after running into financial trouble. "We do not know if this property will ever amount to anything or not," Segerstrom frankly told Haesler, "but it is there ready for operation when prices or demand increases sufficiently to warrant some work being done on it." Rare Metals later explored the shallow skarn deposit that the previous owners had mostly mined out, but found nothing of commercial value at depth. Ruefully, Segerstrom admitted his mistake, telling his manager, Ott Heizer, that the mine was worth less than the debt owed on it.[85]

The Golconda orebody that Rare Metals had optioned in 1935 was more complex, and seemingly more valuable. To keep it away from competitors, early in 1939 Segerstrom organized another family syndicate, the Mantunoco Corporation, and leased the remaining claims not already under company control. After Segerstrom, in turn, leased the entire Golconda operation to the Nevada-Massachusetts Company, development plans proceeded under an integrated management. Construction began in 1940 on the thousand-ton tailings plant at Mill City and the chemical plant at Golconda. It took more than a year for both mills to reach full production.[86]

Plant expansion and mine acquisition helped the Nevada-Massachusetts Company stay in the front ranks of national tungsten producers, but it was difficult to balance supply and demand. Segerstrom had waited until demand rose to make essential capital expenditures, but he waited almost too long. By the fall of 1940, mushrooming rearmament contracts absorbed all available tungsten stocks. As his new mills were still under construction, production estimates fell short, and he had to scramble to find other suppliers to fill standing orders. Keeping track of constantly changing priority ratings and other federal guidelines caused many delays in the delivery of essential equipment and supplies. Jammed traffic on railroad lines cost more valuable time. The backlog grew worse during the winter months when bad weather closed many mines. Segerstrom was besieged with telegrams and letters from desperate customers. One steel company president, demanding a share of any tungsten produced, insisted, "We are working on important orders going into the National Defense Program and I daresay that our requirements are more important and more critical than those of almost any other producer."[87]

As metal shortages mounted, the company president worked out a complicated deal with the navy to "borrow" back enough tungsten concentrate from navy stocks supplied earlier to meet the needs of frantic steel companies. The procedure was highly irregular, but the navy cooperated by first "rejecting" the tonnage already shipped, then extending the time schedule for replacement. This solved the immediate problem. Soon carload lots of "rejected" ore began arriving at plants throughout the eastern Steel Belt. Not until September 1941 were the milling facilities at Mill City and Golconda fully operational and production back on a predictable schedule. By that time, however, imported ore from China and Bolivia had reduced the navy's tungsten needs, and the navy contract with Nevada-Massachusetts was canceled by mutual agreement.[88]

At the start of World War II tungsten still ranked near the top of the list of strategic metals, and the Nevada-Massachusetts Company still remained the nation's largest tungsten producer. The status of both was challenged but not changed as the nation mobilized in the two years prior to Pearl Harbor. New metallurgy boosted molybdenum's use as a tungsten substitute, and extensive development of molybdenum and tungsten deposits in Colorado and California narrowed the Nevada company's margin of leadership. With government and industry buying and consuming more tungsten than ever before, Charles Segerstrom had to work harder than ever to keep his mines productive and his mills operating. At the same time, as the recognized leader of the tungsten industry, his political agenda grew ever more crowded. In September 1939 he had wanted to "just sit still for a minute and see which way the world is going to jump," but actively managing both family and corporate interests allowed no time for quiet contemplation. Time for anything but work seemed a luxury after America entered the war.[89]

7

AMERICAN WARTIME METAL
POLICY AND PRACTICE

America's belated entry into World War II changed the focus if not the scope of national security. Before Pearl Harbor, the chief aims of policy makers had been to mobilize the military to keep American shores safe and spur industry to keep Britain in the war. During the two-year conversion to a wartime economy, government planners had imposed a series of "mechanisms" to encourage greater production of war goods, prevent or offset labor and material shortages, and counter inflationary pressures. Mobilization brought the nation out of a prolonged economic slump, but trying to balance competing industrial and military needs required coordinated planning and an increased federal presence.[1]

After Pearl Harbor the pace of military and industrial mobilization accelerated, and so did the size and power of the federal bureaucracy. FDR's "trial and error" approach to emergency management complicated wartime planning and gave his political enemies much to criticize. Internal differences continued to surface between liberals in the administration and conservatives from business and industry called upon to help make and implement wartime economic policies. Frequent changes in policy and personnel added to the impression that Washington was a bureaucratic battleground.

To quell the confusion and ensure the success of the wartime economy, federal planners had to confront four tough but familiar issues: how to maximize production without running short of raw materials; how to distribute the workforce to serve both military and industrial needs, as well as increase its number and productivity, while at the same time remain responsive to basic labor demands; how to keep production high without inflating prices, and productivity high without inflating wages; and finally, how to plan ahead, both for war and for peace. Every American at every level of organization and industry was affected, directly or indirectly, by the government's effort to deal with these questions of production, labor, inflation, and planning. To Charles

Segerstrom they were the bedrock issues of the tungsten business. His career as industry spokesman and president of Nevada-Massachusetts depended on his ability to run a successful private business within the constrictive limits of a wartime economy. It was time to apply the lessons of World War I.

THE EVOLVING SUPERSTRUCTURE OF WARTIME BUREAUCRACY

The Roosevelt administration also tried to avoid repeating mistakes of previous wars, but formulating effective action plans during wartime was difficult under any circumstances. As the shooting war began on the Atlantic the slow pace of conversion from a peacetime to a wartime economy remained the administration's chief concern. Instead of replicating the Baruch model of a one-man economic "czar" as in World War I, the president had experimented with multiple mobilization agencies. By mid-1941, the mixed results proved only that America's wartime industrial capacity still lagged far behind its potential.[2]

That fall Roosevelt moved closer to the Baruch formula. He replaced the cumbersome two-person leadership of the Office of Production Management with a single director, Donald M. Nelson, a former Sears executive who had served since 1940 as coordinator of national defense purchases. After Pearl Harbor, another reorganization replaced the OPM with a centralized command structure under a new agency, the War Production Board. To run it the president turned again to Nelson, whose efforts to help small businesses secure government contracts had won friends in Congress. With the president's support and congressional approval, Nelson became "the Boss" of war production, with powers greater than Baruch's in World War I. In the patriotic fervor of a nation under attack, businessmen were expected to fall in line. "More and more," wrote a journalist, "industry will be told what to do and asked about it later."[3]

Nelson's job performance did not match his public image as "dictator of war production." Critics said he lacked the courage to exercise power, that "too often his flexibility became vacillation." The administration's deference to military opinion, as well as Nelson's lack of control over labor and prices, left him more cautious than decisive. His undue circumspection dismayed impatient New Deal professionals. John Kenneth Galbraith, an assistant to Leon Henderson in the Office of Price Administration, thought little of Nelson and other hesitant "industrial bureaucrats."[4]

In retrospect, the personality clashes within the War Production Board reflected deep-seated differences between military and civilian views of the nation's wartime resource needs and priorities. Some military advisers wanted more "fat" trimmed from the civilian economy. They paid little attention to government economists who argued for a more coordinated and balanced distribution of resources. In the private sector, industry spokesmen were also divided. Small-business interests complained that they did not get their fair share of defense contracts. Big contractors complained that government efforts to help small business led to inefficiency and production delays.[5]

Continual changes in key personnel clarified leadership roles but did not entirely reduce the bureaucratic infighting known in the media as the "battle of Washington." In the fall of 1942 Roosevelt stepped in to impose political order with the Office of Economic Stabilization (OES), housed directly in the White House. To head it he turned to a personal friend and party faithful, former senator and Supreme Court justice James F. Byrnes. For two more years Nelson continued to chair the War Production Board as a subordinate under Byrnes. In the spring of 1943 the OES itself was transformed into a superagency, the Office of War Mobilization, with Byrnes as "super czar" or "assistant President." Until V-E day, as a contemporary critic observed, he had "virtually complete powers over all home front war activities."[6]

FEDERAL AID TO THE MINING INDUSTRY

Conflicts involving critical metals illustrate the disharmony that complicated strategic planning. By 1941 shortages loomed as demand for aluminum, copper, tungsten, zinc, and other metals far exceeded supply. Some strategic metals had significant domestic ore reserves that could be tapped with proper incentives. Many others had almost no domestic sources and had to be imported.

The War Production Board's cumbersome supply allocation and distribution system exacerbated the problems. After receiving and evaluating three-month production schedules, WPB officials issued "preference ratings" to qualified contractors on a case-by-case basis without considering the total economic impact or the consequences if resources were unavailable. The flaws of this fragmentary approach became obvious during the critical months of 1942. Severe shortages of men and material crippled the nation's productive potential. War plants could not meet projected goals, even after metallurgical innovations and product redesigns reduced the amount of tungsten, chromium, and nickel in ferroalloys. The nation's steel capacity seemed dangerously low, despite the two-year effort, previously described, to stimulate steel output through various government incentives. Even new plant construction, authorized and partly underwritten by the government, stalled after consuming enormous amounts of steel and creating priority "chaos."[7]

The resource shortage was the most urgent problem and the first to be tackled by a systematic effort. Ferdinand Eberstadt, a securities broker before his appointment late in 1941 as chairman of the Army and Navy Munitions Board, led a coordinated effort to improve the handling and distribution of scarce resources. The result was the Controlled Materials Plan, modeled in part after the British Steel allocation system, whose manager, Arnold Plant, had crossed the Atlantic to help develop a similar plan in the United States. Introduced early in November 1942, the CMP first applied to steel, copper, and aluminum. Implemented for the metals industry by Metals Reserve, the CMP established an orderly process for calculating the amount of resources available and balancing total needs with production priorities. Eventually, the system "provided an effective control over the entire industrial economy," in the words of historian Robert Cuff. Ironically, Eberstadt, the one person most responsible for the CMP, soon lost his job. Nelson fired him, allegedly for insubordination because he "represented the viewpoint of the armed services within his agency," as the press reported.[8]

The Controlled Materials Plan greatly improved the allocation and distribution of scarce wartime resources, but it did not stimulate new production. As discussed in the previous chapter, war planners had been calculating the nation's wartime material needs since the early 1920s. They had anticipated the shortages that would result from mobilization, and, after 1939, with White House encouragement and congressional authorization, they gradually had implemented plans to relieve those shortages. They had cut civilian consumption, and imposed rules for substitution and conservation of vital commodities. They had also favored large production companies with cost-plus contracts, guaranteed loans, tax breaks, and other subsidies. Increasing the wartime stock of raw materials, however, required enormous new outlays of taxpayer dollars. With expanded powers and an inflated budget, the Reconstruction Finance Corporation became the government's principal financial link with the mining world. The RFC, in the words of Robert Higgs, could do "practically anything the defense and war-making authorities thought best for the nation's safety and the prosecution of the war."[9]

Under congressional authorization to "expedite the National Defense Program," the RFC and its subsidiaries soon began pumping money into the mining industry. Between 1939 and 1944 Metals Reserve spent fifty-five of the seventy million dollars Congress had appropriated under the Thomas bill buying cadmium, chrome, manganese, mercury, tungsten, and other strategic materials for the national stockpile. With separate RFC funding, Metals Reserve, besides regulating commercial ore sales, also contracted for three billion dollars in foreign and domestic metals, including copper, iron, lead, manganese, mercury, nickel, chromium, tungsten, and zinc. Initially intended for the national stockpile, some government stocks of metals and concentrates were resold to private industry when critical shortages arose.[10]

As we have seen with tungsten in the previous chapter, wartime expediency dictated both the government's price and the source of supplies. Internationalists were now in control, having stolen the nationalists' thunder by taking over their appeal to "national security." The result was a rush to restock no matter where or how costly the mineral sources. Both foreign and domestic suppliers benefited from the new emphasis.

Most of the foreign supplies came from resource-rich countries in the Western Hemisphere, although Africa and the Far East contributed significantly so long as shipping lanes were open and cargo space was available. Metals Reserve, the RFC branch that handled foreign metal contracts, patronized foreign contractors with periodic price increases and other incentives, at times offering better inducements than American contractors received. On the other hand, Metals Reserve subsidized domestic industry by reselling stockpiled materials at ceiling prices that did not include transportation, insurance, and other marginal costs of acquisition. Those the government absorbed.[11]

While Metals Reserve purchased available metal stocks, another branch of the RFC spent millions of dollars subsidizing private plant construction at home and abroad. Most of the money went to the nation's largest mining and smelting companies. Capital support for the steel industry has already been noted. To boost aluminum production, the Defense Plant Corporation signed contracts with Alcoa and Reynolds to build and operate two new reduction plants, using low-grade Arkansas bauxite purchased by the Metals Reserve Company for the purpose. In Texas, Defense Plant money largely financed the construction of a large smelter to recover tin from alluvial concentrates shipped in from Bolivia and the Far East. RFC funding also helped build a chromite recovery plant in Montana, two manganese processing facilities in Montana and Nevada, new vanadium milling operations in Utah and Colorado, expanded copper facilities in Utah and Arizona, and tungsten retreatment plants in Salt Lake City and at Glen Cove, Long Island. Outside the

United States, the Defense Plant Corporation supported the expansion of a copper mill in Chile, a nickel processing plant in Cuba, and exploratory work in Canada for new sources of copper, zinc, and lead.[12]

The combined effect of these measures helped organize and sustain the output of war goods, but acute shortages of copper and other metals made it clear by mid-1941 that more had to be done to keep the factories supplied with raw materials. Expanding subsidies to the domestic mining industry was the logical next step.

Just as it had in production planning, however, politics complicated efforts to boost domestic mine production. Since the start of mobilization, small firms had complained that they were not getting their fair share of defense contracts or raw materials. In July 1941 a special Senate committee to investigate the National Defense Program, chaired by a little-known but no-nonsense Missouri Democrat, Harry S. Truman, concluded that military procurement officers, as well as the defense establishment in Washington, were largely justified in bypassing the "little fellow" for reasons of speed and efficiency. The largest firms were simply the most capable of handling huge defense contracts. But the complaints persisted, and Congress later passed the Small Business Act in an attempt to address perceived or real inequalities. Those remedies in the long run did little to change the situation, however. Between 1939 and 1944 the amount of total U.S. manufacturing going to businesses with less than five hundred employees actually declined by more than 60 percent.[13]

In the summer of 1941 Secretary of the Interior Harold Ickes, one of several progressive Republicans in Roosevelt's cabinet, gave small domestic mining a boost in testimony before a special Senate subcommittee investigating the problems of small business. Ickes recognized the limitations of smaller firms in the mining business, but insisted that they were the backbone of domestic mineral production. To aid them he urged an accelerated program to discover, develop, and process new domestic ores. Even low-grade deposits he thought should be considered as part of the national stockpile if not actually mined until they were needed. To ease the financial risks and help smaller operators, the government should make cheap credit available and keep prices for domestic ores high enough to stimulate production, but not to encourage "large profits and control by large established companies." In his opinion, "Our only safety is to have a vigorous producing industry in every line of endeavor."[14]

Federal agencies soon responded with policy changes favoring small business. Metals Reserve offered small mining operators an unlimited market for marginal ore at a fixed price and a guaranteed profit. It established small-lot buying stations close to mines producing low-grade chromium, manganese, and mercury. The government-financed retreatment plant in Salt Lake City processed low-grade tungsten ores purchased directly from low-budget operators. The RFC, for its part, offered smaller companies a noncollateral development loan of twenty thousand dollars, with another twenty thousand dollars available if the operation showed promising results. Even a speculative dewatering project to open old workings could receive a five thousand-dollar RFC loan with few questions asked.[15]

The manganese purchase plan set the pattern for government strategic ore purchases from domestic sources. It was based on a Bureau of Mines report showing that scattered low-grade manganese deposits could be effectively processed using hydrometallurgical and electrolytic technology. Metals Reserve established a sliding scale of prices, starting with a base price for average ore (defined as containing 48 percent manganese), with premiums and penalties for higher or lower grades, including impurities.[16] Similar purchasing rules and requirements were used to acquire stocks of tungsten, vanadium, chromite, molybdenum, cobalt, and other metals.

Federal subsidies did not end with financial support. Miners had always welcomed government help in locating ore bodies and figuring out the best mining and milling methods—provided, of course, that government did not become a competitive threat to private industry. The 1939 Thomas Bill reflected this tradition in Congressman Scrugham's amendment providing special funds for federal research and development of strategic deposits. Bolstered by periodic funding supplements, field crews from the Bureau of Mines and the Geological Survey crisscrossed the country in an exhaustive investigation. They built pilot plants, conducted beneficiation studies, discovered new milling processes, aided and advised thousands of private operators, and wrote hundreds of reports. Focusing initially on seven critical metals—antimony, chromium, manganese, mercury, nickel, tin, and tungsten—the bureau enlarged its mandate over the years to include twenty-six additional metals and minerals—an indication of the shifting defense needs and technologies as the war continued. Many new deposits proved vital to war industries, and private industry was quick to exploit them. For example, the Yellow Pine mine in Idaho, discovered by bureau teams in 1941, supplied up to 40 percent of the nation's tungsten needs after the Japanese cut off Chinese wolfram exports.[17]

In the long run, as the graph below shows, all this federal effort on behalf of small mining companies did not materially alter the dominance of large enterprises, nor did it significantly boost domestic production from marginal companies. A similar picture emerges after reviewing both ferrous and non-ferrous domestic mine production data during the war years. Nearly 70 percent of U.S. copper output in 1941, for instance, came from just 11 companies, or 31 percent of the total number of active copper companies. Those numbers did not change significantly over the next four years. Similarly, between 55 and 80 percent of the domestic lead produced came from only one-third of the active companies in the lead industry. More than 130 domestic firms produced zinc during the war years, but between 50 and 80 percent of that yield came from just 16 companies on average.[18]

Wartime ferrous metal production showed even larger discrepancies between small and large producers. In 1941, three-fourths of domestic tungsten came from just two companies, representing less than 6 percent of the total number of active producers. Three years later, America's three largest tungsten companies—U.S. Vanadium in California, Bradley Mines in Idaho, and Nevada-Massachusetts in Nevada—accounted for 99 percent of the total output. Although Molycorp and U.S. Vanadium both produced molybdenum during the war years, Colorado's Climax mine remained by far the largest producer. The manganese industry was less concentrated, due to the wide

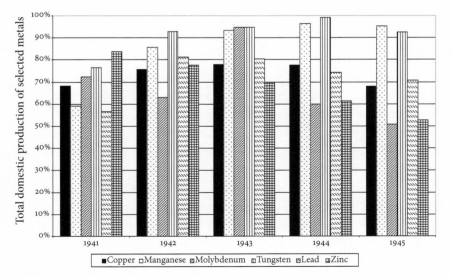

Wartime output of largest metal producers

distribution of low-grade deposits. In 1941 the Anaconda Corporation, with federal financial help, built a large plant in Montana to produce medium- to high-grade concentrates from low-grade ore. It remained the single largest manganese producer throughout the war years.[19]

INFLATION WORRIES AND PRICE CONTROLS

Developing an integrated production program that coordinated wage and price controls with industrial and military needs remained a central planning issue throughout the war. Before Byrnes took over as "super czar" late in 1942, the administration's approach to inflation had been as hesitant as its production planning, reflecting long-standing differences, both within and outside the government, over monetary policy and inflation controls. Policy makers were divided over how to expand the supply of vital war commodities and still keep prices under control. Economic conservatives argued strongly for a simple market solution: expand the profit margin by raising ceiling prices on affected materials. Most military procurement officers, and some corporate representatives in the Office of Production Management, favored this option. Inflation, after all, was secondary to marshaling all the resources necessary as quickly as possible to win the war.[20]

Leon Henderson thought differently. He had been the president's point man in the fight against inflation since the start of mobilization. Open market solutions to wartime emergencies made little sense to Henderson and his staff of professional economists and lawyers in government service. Since the mid-1930s they had fought to protect small business and consumers from the dominance of large corporations. But the war had disrupted the New Deal reform agenda. The marketplace was still burdened by the power of international cartels, high tariffs, "unwieldy capital structures," "inflexible wage and transportation rate structures," "trade taboos, chaotic labor conditions, . . . wasteful competition," and other limitations on both competition and production. With big business still untamed and market conditions skewed by war, they insisted that prices must be artificially controlled across the board.[21]

Early in 1941 FDR had reshuffled federal agencies and personnel, and later that year gave Henderson more authority over prices as head of a redefined Office of Price Administration. A tough bargainer and vocal critic of the military-industrial establishment, he had kept wholesale metal prices low in 1940 and 1941 by negotiating voluntary ceilings with major producers and suppliers. Voluntary controls lacked enforcement power, however, and as inflation pressures increased in the summer of 1941, the administration turned to Congress for statutory support.[22]

The price control debate lasted all summer and fall without resolution. Partisan rancor and ideological divisions complicated economic planning in Roosevelt's third term. Bernard Baruch discovered the congressional disconnect between production, prices, and wages when he testified on an administration proposal to strengthen Henderson's hand in fighting inflation. Wary of another wartime wage-price spiral, he wanted ceilings placed on all wages and prices, including farm goods. Anything less, he told Congress, would be wasteful and "piecemeal."[23]

Baruch's statist ideas, however, were decidedly premature before the critical months of 1942. Farmers complained that their incomes lacked "parity" with other segments of the economy. A powerful congressional farm bloc, led by southern Democrats, precluded any efforts to strengthen controls on farm prices. Small businessmen supported wage controls but not Baruch's "sweeping" vision of a command economy. Through local Chambers of Commerce across the country they voiced their objections. They were doubtless discomforted by the testimony of Treasury secretary Henry Morgenthau before the Senate Banking and Commerce Committee. Reflecting his New Deal reform roots, he opposed wage controls because "human beings are not property and should not be treated as such." Congress had previously imposed an excess profits tax without specifying any numbers, but Morgenthau also suggested taxing all corporate profits above 6 percent, claiming it was only equitable when millions of workers were being asked to fight for a dollar a day.[24]

Henderson backed away from the hard-liners and ideologues. A Keynesian economist, he preferred a flexible monetary policy to blanket wage and price controls. But the inflation rate climbed into double digits by the fall of 1941, heating up the debate and forcing the administration's hand. Henderson said stronger measures were necessary to prevent "chiseling," and threatened to use his "priority power" to curb inflation if new price control legislation failed to pass. He had the president's support, but Congress continued to muddle until the United States entered the war.[25]

Pearl Harbor ended the legislative impasse. In January 1942, with inflation above 11 percent, with severe shortages of labor and material threatening to stall the renewed rush to rearm, Congress gave the Office of Price Administration statutory authority to fix prices on strategic commodities. The following April the president announced a tough new anti-inflation program that included heavier taxes, rationing, and more stringent wage and price controls. For its part, the OPA issued the first "General Maximum Price Regulation," capping prices on most strategic commodities except farm goods and wages. Henderson opposed wage ceilings—which were outside his jurisdiction anyway—and remained flexible on commodity prices, despite public grumbling, consumer complaints, and attacks from the Far Right and Left against "communists" and "dictators" in the government. Specialized military equipment and supplies, including domestic ores and concentrates, were exempt from ceilings, and producers of "standard" commodities or other vital raw materials might be subsidized if special "hardship cases" arose or military necessity dictated.[26]

THE DEBATE OVER SUBSIDIES

The explosive demand for war material after Pearl Harbor triggered intense inflation pressures the government was determined to ease. During the spring and summer of 1941, the Office of Price Administration systematically had set ceilings for many strategic metal products and scrap, based on current market prices. That fall, however, metal shortages loomed as demand far exceeded supply, especially copper, lead, and zinc for armaments and ammunition. Fortunately, all three metals had significant domestic ore reserves that could be tapped with proper incentives—unlike manganese, nickel, chromium, and to a lesser extent tungsten, all of which were increasingly imported as the war progressed.

The government's answer to nonferrous metal shortages was not raising ceiling prices but subsidizing domestic metal producers to encourage greater output. This was the basis of the premium price quota plan, which the OPA, in conjunction with the War Production Board, introduced in mid-1941 for the copper industry but not fully implemented until 1942. Using data gathered by the Bureau of Mines, federal officials established annual production quotas for all major mine and smelter companies in the nation. After February 1942, Metals Reserve paid smelters a premium higher than the ceiling price for any copper, lead, and zinc produced above their quotas. The smelters passed the premium on to their suppliers. Any producer could qualify for the subsidy so long as he exceeded expected output, giving marginal operators and new companies a financial boost. By October the plan had sparked new efforts from 270 marginal copper mines, but they collectively contributed less than 2 percent of the nation's total copper output. The rest came from 15 large companies. They met their quotas but not much more, and thus were ineligible for the initial cop-per premium. In 1943, still facing supply shortages, Metals Reserve expanded the program so that all grades and almost all nonferrous metal companies that maximized production could receive premiums above the OPA's base prices.[27]

Ferrous metal mines also benefited from subsidized prices, especially small companies mining low-grade and marginal deposits previously considered uneconomical to develop. OPA ceilings did not apply to concentrates pur-chased for government stockpiles. During the war years, Metals Reserve paid premiums that varied from 25 to 50 percent above market prices for small lots of domestic manganese, chromite, molybdenum, tungsten, and other metals.[28]

Large or established ferrous metal producers did not receive the same benefits. In November 1942, for example, following a War Production Board report that listed tungsten among metals still critically short, Metals Reserve raised the price it offered for domestic tungsten concentrates by six dollars. However, only producers whose output for the previous year was less than one hundred short tons were eligible. When high-volume producers complained of discrimination and talked of raising ore prices on their own, the OPA stepped in. Citing a study showing that wartime tungsten prices were nearly 50 percent higher than in early 1939, it threatened to "issue a maximum price regulation to assure the flow of the commodity into needed war production." Later, Metals Reserve doubled the qualifying limit, but still left out the largest operators.[29]

The tungsten price increase also upset consumers, as Miles K. Smith, chief of the War Production Board's tungsten division, discovered after talking with Marx Hirsch, president of Molycorp. Smith had assumed that industry advis-ers had been consulted prior to the announced rate hike. He was shocked to learn that Hirsch, a member of the advisory board, knew nothing about it until he read the announcement in the newspapers. Smith immediately called a meeting with officials from the OPA, the WPB, and Metals Reserve, propos-ing to cancel the increase and restore the "whole tungsten picture . . . to the status quo." Hirsch was dubious of the outcome. He wrote Segerstrom that he "should be surprised if the $30 price is revoked only because of the reaction it will have on the public."[30]

Hirsch was right to be skeptical. The government's new price offer held for nearly a year, and was broadened to include a number of moderate producers not eligible under the original terms. It ended not because of adverse reac-tion but due to new assessments that revealed an embarrassing glut of many strategic metals. One official told a reporter late in 1943 that "some of the stuff is running out of our ears." Metals Reserve stopped paying the six-dollar tungsten premium after June 1944, and early in 1945 the War Manpower Commission (WMC), for labor purposes, downgraded the gray metal to "essential" rather than "critical," along with tin, aluminum, manganese, and chromium. Intended to stimulate new production, the thirty-dollar rate made little difference in the total volume of domestic output. As we have seen, during the period when the premium was in effect, more than 90 percent of domestic tungsten came from the three main producers not eligible for the higher price, Nevada-Massachusetts, U.S. Vanadium, and Bradley.[31]

Wartime government subsidies were not limited to metal producers. Insurance costs were subsidized if ceiling prices forced some small businesses to operate at a loss. Farmers got relief from transportation costs to help them survive without raising commodity prices. Congress rejected Henderson's efforts to broaden the subsidy program, however, thus confining price relief mostly to war contractors, farmers, and small businesses with severe financial hardships.[32]

Pragmatically applied and lifted when conditions warranted, subsidies helped assuage the most glaring economic inequities without seriously threatening the economic system or raiding the federal treasury. Many business conservatives, however, including the largest metal production companies, saw premiums and other subsidies as a slippery slope, a drag on free enterprise, and a prelude to massive social programs. They favored either a Baruchian formula—universal application of emergency ceilings—or abandoning government price intervention altogether. But most New Dealers and many pragmatic businessmen objected to "one size fits all" panaceas. So long as he had the president's backing, Henderson ignored the criticism and adjusted ceilings in special cases to "smooth out the rough spots." When some of the largest producers sought equal treatment, OPA officials said they could afford to absorb any loss, whereas the little firms could not. Everyone must sacrifice in wartime. Differential or "incentive" pricing proved erratic and sometimes inconsistent, but for most of the war it stimulated production of needed commodities while keeping inflation under control.[33]

The chief architect of price stabilization did not survive the political fallout. In the midterm elections of 1942 the public registered their displeasure with rationing and other impositions. Democratic incumbents lost heavily, and

Henderson was a political liability. A hero to New Deal reformers, he resigned late that year.[34]

LABOR DISLOCATIONS AND THE FEDERAL RESPONSE

In the perpetual commotion of wartime Washington, production planning was less complicated than mobilizing and apportioning the workforce to meet the needs of both military and industry. Until the summer of 1940, isolationist sentiment had been too strong to support any form of manpower mobilization. To a nation opposed to standing armies and wary of foreign wars, the idea of compulsory service in peacetime, though a key component of military planning, was universally unpopular. The looming presence of the Wehrmacht at the Channel crossings, however, weakened draft resistance and emboldened the political leadership. Led by Grenville Clark, a New York attorney already known as the "father of Selective Service" for his draft efforts in World War I, civilian advocates with good military and political connections helped write legislation and lined up congressional sponsorship. The endorsement of both presidential candidates in the 1940 campaign added a bipartisan boost that by mid-September propelled the nation's first peacetime draft into law.[35]

Col. (later Gen.) Lewis B. Hershey, director of the Selective Service from its inception through the Vietnam War, was a World War I veteran dedicated to the new law's "central ideal" of equality, or as he put it, "democracy at work." The belief that all citizens had an equal obligation to defend their country was fundamental to Western democratic values but difficult to implement. In the 1920s the American Legion had campaigned for universal conscription of material resources as well as manpower, but found little political support. The American Federation of Labor voiced the loudest opposition, suspicious that labor and not capital would carry the biggest burden.[36]

Compulsory service in Britain also rested on the premise of equality, although the centralized British Ministry of Labour and National Service issued blanket exemptions based on classification of employers by skill and occupational importance. Equality in the American system, however, meant a grassroots process. All citizens were obligated, but determining who would actually serve was a local decision made by civilian draft boards established in every county or parish in the nation. Occupational deferments were available only on an individual basis, and public opposition kept pressure on local boards to make few wartime exceptions.[37]

Labor dislocations festered for two years in the mobilization prelude to American entry, with nearly 80 percent of the workforce in jobs not vital to war production, and military officials demanding a full complement of men in uniform. Before Pearl Harbor the echoes of unemployment still reverberated across the country, providing a labor cushion to let manpower problems percolate before reaching a boil. In this period nearly a dozen federal agencies had some jurisdiction over workforce planning and operations, an indication of both the government's disorganization and labor's weakness in the face of a rising military-industrial establishment.[38]

The first serious effort to coordinate recruitment, training, and job placement for war industries came only after America entered the war. In April 1942 the president, in a typical crisis response, used his executive authority to create yet another emergency management agency, the War Manpower Commission, with Paul V. McNutt as chairman. A former Indiana governor and commander of the American Legion, McNutt was given sweeping authority over the civilian labor force, and could even set standards for Selective Service classifications. A second executive reorganization late that year grouped the WMC together with the Selective Service and the U.S. Employment Service (USES), the latter a New Deal agency with twelve regional directors, two hundred area directors, and fifteen hundred local offices. Consolidation gave the WMC vast operational as well as policy-making functions. It had broad visibility at the grassroots level and a direct impact on the training and redistribution of millions of wartime workers. Draft boards classified personnel according to military and industrial needs; the USES supervised the training, redistribution, and placement of industrial workers; and the WMC coordinated workforce rules and activities with other wartime agencies.[39]

If neatly packaged and functional on paper, the WMC fared poorly when labor scarcity reached crisis levels. Its policy-making responsibilities were vaguely defined and overlapped those of other agencies. Labor law and administration, including the rights of trade unions, for example, were largely the prerogatives of the U.S. Department of Labor, while wage rates could be adjusted, within limits, by the National War Labor Board (NWLB). The latter had been established to resolve labor-management disputes, following a no-strike, no-lockout pledge by both sides while the United States was at war. The WMC lacked jurisdiction in any case. Moreover, WMC operations at the regional and local levels were often inconsistent and uncoordinated. The army made separate contracts with industry, preferring large firms to small, thus impacting areas of full employment and adding to labor dislocations. McNutt was under constant pressure from the armed services to supply more soldiers and sailors, and from civilian interest groups to liberalize occupational deferrals. The War Department opposed any effort to interfere with military priorities, and draft boards generally complied with military demands.[40]

Wartime planning for men and materials was finally integrated and coordinated under a single federal agency in 1942–43, when the WMC came under control of the Economic Stabilization Board, later the Office of War Mobilization under James F. Byrnes. Labor's voice remained weak, however. The "super czar" relied on the same industrial and military advisers who had dominated earlier wartime agencies. As an adviser to Byrnes, McNutt had no power to enforce labor rights or WMC rules and regulations. Finding the proper balance to meet the incongruous demands of industry, agriculture, and the armed forces remained one of the chief issues of the war.[41]

THE LABOR CRISIS IN MINING

Wartime labor shortages in the metals industry exemplified the national problem. For domestic metal mines, the draft was a body blow to productivity. Miners in the 1940 census made up less than 2 percent of total employees, and only 15 percent of those worked in the metal mining industry. While the total number of miners rose in the first two years of the war, the metal mining workforce declined. Despite skyrocketing demand for base metals during the early years of the Selective Service, the draft drained men from the mines and smelters, especially single males under thirty-five. By 1942 the percentage of metal miners had dropped another 4 percent in comparison with the total mining workforce.[42]

Inflation added another dimension to the mine labor conundrum. Before the fall of 1942 the lack of a coordinated federal effort to control prices, wages, and production increased the risk of an inflation spiral. "They all hook together— inflation control, labor shortage, farm prices, war production and the draft," wrote Raymond Moley, Roosevelt's former brain truster turned newspaper columnist. Thus far, he said in September, the government merely "plays around the fringes of the problem."[43]

Moley had reason for concern. The consumer price index rose from less than 2 percent in May 1940 to more than 13 percent by May 1942. Skilled employees, the most mobile of the nation's workforce, responded by seeking either better wages or different jobs. Rising production costs caught many employers in a wage-price squeeze, exacerbated by OPA price ceilings on strategic commodities. The labor drain in the nonferrous metals industry was particularly worrisome. Wartime copper, lead, and zinc production had nearly doubled from prewar figures, yet supply still fell far short of demand despite premium price plans and other subsidies. In the Rocky Mountain West, base-metal producers were losing skilled underground workers to higher-paying and less hazardous defense jobs on the West Coast. Copper producers in July 1942 estimated that the draft and other "opportunities" had cost them five thousand workers, or 20 percent of the workforce. In lead and zinc mines and smelters the loss was about 12 percent. With draft boards under pressure to meet the military's insatiable demand for manpower, mine labor was squeezed from all sides.[44]

The labor shortage not only hurt ore production but also compromised underground safety as well. For seventy-five years the health and welfare of miners had been a top priority of organized labor in the American West. The union movement had helped build a mine regulatory regime at almost every level of government. Since its inception in 1910 the U.S. Bureau of Mines had emphasized mine safety, and most mining states followed the BOM's lead in upgrading regulatory codes for the protection of its mining workforce. But in national emergencies did safety or production count most? That question troubled Ott F. Heizer at Mill City in October 1942. He had to respond to a notice from the Nevada state mine inspector citing the Nevada-Massachusetts Company for violating the rule requiring at least two-man crews when any mining was done underground. A recent inspection found "only one man working alone" in the 400 stope of the Stank and the 700 level of the Humboldt. After explaining the failure to obtain additional men after an exhaustive effort "through private recruiting, private employment agencies and the U.S. Employment Service," Heizer humbly begged for consideration: "This letter is not written in a spirit of criticism of your rules and regulations, or a request that you disregard the safety measures demanded by the State Mining Code, but it is more to make a record of the situation existing here in the hope that some modification may be made in similar situations where the mine management feels that men may be safely worked alone."[45]

Federal officials stepped gingerly to meet the mining labor crisis. Caution was necessary, for political as well as economic reasons. Even after Byrnes took charge of the home front, policy makers were beset by too many divergent interests and opinions to develop a consistent and uniform labor strategy.

The result was similar to the method used to stimulate production: a desultory carrot-and-stick approach, with each step an experiment.

The response to the labor shortages in both the ferrous and nonferrous mines of the West illustrates how government strategy and tactics adapted to changing circumstances. In July 1942 the National War Labor Board mediated a dispute involving 157,000 steelworkers in plants not affiliated with U.S. Steel. Following administration anti-inflation guidelines, it rejected labor's argument for large pay hikes but recognized the need to compensate for inequities caused by inflation or "substandard" wages in any given industry. The board's business and government members, making up a majority, agreed that workers not already given cost-of-living adjustments were entitled to increases of up to 15 percent to match the 1941–42 inflation rate. The "Little Steel" formula brought steelworkers' average hourly wages in 1942 to $1.03, up 9 percent from the previous year. Labor representatives hotly protested what they considered a "token adjustment," but the majority decision established guidelines by which wage disputes were to be adjudicated throughout the war.[46]

The 15 percent solution was not enough to stop the drain of metal miners, however—as the NWLB concluded after investigating wage discrepancies between base-metal workers and those in other war industries. Two months after the "Little Steel" case, government members of the board joined the labor representatives in recommending a "fair wage adjustment" that would raise the average daily wage by $1 for 10,000 Utah, and Idaho copper, lead and zinc miners and metal workers. This time it was the industry members who howled. Despite Henderson's OPA opinion that the raise would not threaten ceiling prices or cause inflation, Roger D. Lapham, shipping executive and later mayor of San Francisco, said the decision would lead 76,000 other nonferrous miners to seek similar compensation. "A policy of granting wage increases to prevent or reduce the flow of manpower from one industry to another," he warned, "will only lead to chaos, economic and otherwise."[47]

The War Labor Board's divisive recommendations left director of economic stabilization James F. Byrnes in a quandary. For two months the issue lay on his desk while he fretted and copper production declined by 11,000 tons, mostly due to the lack of workers. McNutt's office estimated that meeting copper quotas would require 5,000 to 8,000 more miners. In September, to prevent any further job-hopping, McNutt imposed a "freeze" on workers already employed in the nonferrous mining and lumber industries. In the meantime, with the consumer price index still in double digits, Byrnes talked "tough" against inflation, warning that "every man and woman, dollar and thing, everything, must be utilized for the quickest winning of the war" even if it meant reducing living standards. He thought holding to the Little Steel formula of 15 percent would be eminently fair to both labor and management, but preventing the labor drain from the most critical base-metal industries was more important.[48]

Late in October the "super czar" announced his decision. He approved the NWLB majority recommendation to authorize a $1 per day pay increase for workers in the copper, lead, and zinc mines and smelters of Idaho and Utah, and a smaller increase for 4,000 "similar" workers in seven western states. The raise, Henderson had assured him, would still leave metalworkers' pay below the average strategic wage. The OPA director knew his econometrics. Though the average pay of metal miners rose to $0.88 per hour in 1942, up 20.3 percent from 1939, during the war years their wages never did catch up to those in steel and other vital war industries.[49]

The Rocky Mountain pay raise had little impact on copper production or labor recruitment. By the fall of 1942, bolder measures clearly were necessary. The Labor Department insisted that it was "absolutely essential" for the war effort to fill 6,000 vacant mine and 2,000 smelter jobs. With the head of the WPB's copper branch calling increased production prospects "black," General Hershey told local draft boards to reclassify any deferred worker "not supporting" or "adversely affecting" the war effort by shifting jobs for higher pay or better working conditions. Early in October the American and Canadian governments jointly issued mandates that closed all "nonessential" mining operations for the duration and pressured former gold miners to take base-metal jobs. War planners had grossly overestimated the number of available gold workers with mining skills, however, so the results were disappointing. Some copper mines began recruiting women for surface work in the repair shops and concentrators—adding a significant new dimension to "Rosie the Riveter"—but the low numbers involved did not materially ease the shortage. Skilled Mexican miners were also recruited, following from the success of the "bracero" program for farm laborers. The Manpower Commission wanted to import up to 10,000 experienced miners, but how many actually came and under what circumstances are difficult to determine. Fears of a labor drain by 1943 made the Mexican government reluctant even to support an increase in the number of unskilled braceros crossing the border.[50]

During the winter of 1942–43, the persistent calls for more copper miners reached one source of mining manpower not previously tapped—the U.S.

Army. Under growing pressure from industry, agriculture, and critics of the War Department's concept of total mobilization, army leaders agreed to furlough older men for temporary work on the home front. In 1943, 195,000 men over age thirty-eight were released under the program. For the armed services, the furloughs were part of a readiness assessment that recognized the physical and emotional problems dealing with older draftees with families or important jobs back home. The military view was reflected in a new law approved in November 1942 lowering the draft age to eighteen, and in new calls from the War Department for a cap on the upper draft age at twenty-six.[51]

Ensuring that furloughed men would be effectively used in the war effort required new rules back home. In February 1943 the Manpower Commission tightened the list of deferrable jobs, and the Selective Service made married men under thirty-eight draft eligible, despite congressional efforts to protect farmers and the family man. Paul McNutt advised men between eighteen and thirty-eight who were not in high-priority occupations to retrain (at government expense) if they wanted to avoid the draft. At the same time the Manpower Commission extended General Hershey's earlier order to "freeze" workers already in strategic industries. A few weeks earlier, the president had issued an emergency decree expanding the strategic workweek to forty-eight hours but leaving overtime wages intact. Director Byrnes, in the meantime, tried to head off any further wage increases beyond the 15 percent Little Steel formula, but had to back down under heavy criticism from labor and Congress.[52]

Some 4,000 men receiving furloughs in 1942–43 were skilled mine and smelter workers, happy to be home at their old jobs. Most returned to underground copper mines, but tungsten, lead, zinc, and molybdenum operations got furloughed men as well. Copper production rose by nearly 3,000 tons the first month after the furloughs went into effect, but the tonnage for the entire year of 1943 was only 1 percent above 1942 figures. For the same period, lead production rose less than one-half of 1 percent, molybdenum output gained nearly 9 percent, and zinc production increased less than 5 percent. Tungsten was the only metal with significant production increases after the furlough program began. The 1943 output was 28 percent ahead of the previous year.[53]

THE NATIONAL STOCKPILE AND SHIFTING PRIORITIES AFTER 1943

Labor shortages during the war's midyears were more visible and seemingly easier to solve than figuring out what to do with the national strategic stockpile once the war was over. The stockpile had grown slowly for the first two years after its creation by the landmark Thomas bill in 1939. Funding had been delayed while the domestic mining lobby fought to promote the "Buy American" provisions of the legislation and limit foreign purchases of strategic materials. Federal officials also moved cautiously, hoping to avoid mistakes that had caused inflation and overproduction during the last war. Though Metals Reserve coordinated the purchase and distribution of strategic metals, as a unit of the Reconstruction Finance Corporation it was bound by the fiscal policies of federal loan administrator Jesse Jones, who had tried to save taxpayer dollars by waiting for metal prices to drop before authorizing stockpile funding. In addition, rapid changes in wartime resource needs, and widespread disagreement over the nature and extent of domestic metal deposits, contributed to the slow pace of stockpile development. Perhaps the biggest reason for the slow growth, however, was the overwhelming production demands of mobilization. In wartime it was difficult to pursue a stockpile policy, since industry needed to use strategic materials "as fast as they were acquired."[54]

Building an emergency reserve took on greater urgency during the summer of 1941. War planners and pundits in the United States worried that American ship construction was too slow to offset the systematic destruction of British merchant shipping by Nazi wolf packs in the Atlantic. German ties to South America and warnings that Nazi agents were busy in Panama and other parts of the Americas added to the growing concern.[55]

Fear of the long-term consequences of unreliable foreign supplies and chronic metal shortages led to revisions in strategic thinking. Internationalists were gratified in August 1941 when Roosevelt and Churchill met together for the first time, pledging allegiance to democratic principles and promising to work for a peaceful and cooperative postwar world. In Article 4 of what became known as the Atlantic Charter, Britain and the United States recognized the right of all nations to equal trade and access to raw materials "needed for their economic prosperity." Honoring that pledge would require a more equitable distribution of resources between "have" and "have-not" nations. It would also test the willingness of imperial powers to compensate for centuries of colonial mineral exploitation, or what one historian called the "robber economy."[56]

The Atlantic Charter's focus on the future did not alter the immediate problem of defeating an implacable enemy. All notions of a short war ended with Germany's conquest of Europe. In May 1941 the Office of Production Management decided it needed to stockpile supplies for a three-year war, adding a year to the plan adopted in 1940. Even internationalists conceded the need for more rapid domestic mineral development. C. K. Leith, in testimony before Congress, urged support for a "greatly intensif[ied]" exploration and mining effort, even to the extent of "opening up of marginal and low-grade deposits." Pearl Harbor and the Japanese challenge to Pacific sea-lanes increased the sense of urgency. For the next two years, as we have seen, the government spared no expense to build strategic stockpiles from home and abroad both for current and for future wars.[57]

By 1943 fears of strategic shortages had given way to concerns over a growing mineral redundancy. While the subsidized domestic mining industry continued to turn out ores and concentrates as fast as possible, foreign metals once again flowed into American ports after Allied navies won control of both Atlantic and Pacific sea-lanes. Except for copper, industry metal inventories rose as well as warehouse stocks managed by Metals Reserve. To stem the tide and save their subsidies, domestic mining advocates sought to cap imports and raise tariffs. Internationalists, on the other hand, wanted to continue importing but slow the pace of domestic production. Cognizant of these "sharply contrasting views," WPB Chairman Donald Nelson recommended a compromise, and FDR apparently agreed. In an April 1943 memo to the president, Nelson advocated continuing a policy of unlimited acquisition of both foreign and domestic metals. Even if a surplus develops, he wrote, we have "lost nothing except the use of labor and equipment." Any excess would be available for "future war needs."[58]

Considered in the context of wartime strategy after the Casablanca conference and the Allied insistence on "unconditional surrender," Nelson's recommendations understandably encouraged unlimited resource development as a necessary cost of an all-out two-front global war. Wartime necessity also played into the hands of internationalists with an eye on the future. Partly to rebuild depleted stocks of resources, the State Department pressed ahead with a commodity-for-debt repayment component of its reciprocal trade program. Internationalists prevailed when the Eightieth Congress incorporated the State Department package as part of the "Economic Cooperation Act of 1948," popularly known as the Marshall Plan. Designed "to promote world peace and the general welfare, national interest, and foreign policy of the United States," the Marshall Plan was brilliant propaganda, but in practice it did not add significantly to stockpiled resources.[59]

Protectionists also had an eye on the future, but they looked backward as well. Domestic mine owners welcomed friendly assurances from government planners, but they remained wary of the potential market implications of inconsistent government policies. Certainly, domestic producers did not want the government selling metals directly to consumers after the war, as Nelson seemed to imply. Unrestricted "dumping" after World War I had contributed to an industry collapse that was too painful to forget.

Anticipating postwar economic problems led the domestic mine industry to press for modifications in stockpile policy long before the war ended. During the 1941 House debate on Lend-Lease, Nevada congressman James G. Scrugham wanted the American public to understand that after the war the administration planned to allow war-torn debtor countries to repay with commodity sales to the United States. Raising the familiar specter of "cheap foreign-labor conditions," he said the plan would undermine "whole segments of American industry," especially postwar mining and agriculture.[60]

The first signs of surplus appeared after Scrugham left the House of Representatives upon winning a Senate seat in the fall elections of 1942. Early the next year the War Production Board, acknowledging an excess of stockpiled lead, eased restrictions on domestic lead use. Yet other strategic minerals, especially copper, remained in short supply just when they were needed most. These glaring inconsistencies raised obvious long-term strategic and economic concerns. War planning was futile unless critical materials were available when actually needed. At the same time, mines could not be turned on and off like a spigot, and stockpiled surpluses could not be dumped without creating market chaos. Recognizing the need for a comprehensive revision of the stockpile program, government and industry members of the American Institute of Mining Engineers (AIME) came together in February 1943 to approve a resolution supporting the development of a postwar strategic stockpile to be used only in emergencies. The idea of a peacetime reserve was amorphous enough to attract both internationalists and protectionists, but the devil was in the details when Congress attempted to draft new stockpile legislation.[61]

Senator Scrugham, ever the champion of western mining interests, opened the first legislative drive. In June 1943 he introduced SB 1160, incorporating the main points of the AIME proposal but adding a controversial stipulation. It required the government, for a year following the end of war, to purchase material for the stockpile exclusively from "small or marginal mining enterprises." The hearings that followed exposed the ideological, economic, and geopolitical divisions that had dogged strategic resource policy making for fifty years.

Industry protectionists and their political allies lined up behind the

Scrugham bill, while administration spokesmen and academic professionals led the internationalist opposition. Most of the technical experts in government and academe agreed that foreign imports must continue to meet peacetime stockpile requirements, but they differed over the domestic impact. Some opposed any bill that encouraged the mining of low-grade deposits that could not be commercially mined in peacetime. A Columbia University geologist said it was better to leave marginal ores in the ground until technology improved to make beneficiation more cost efficient. R. R. Sayers, however, director of the Bureau of Mines, countered with an argument that made sense to most mining men: until you develop the deposit to see what exists, you can't develop the technology to mine it. Furthermore, added a wise Michigan engineer, marginal mines already operating should remain open, even if they required continued subsidies, until the orebody was exhausted, or else whatever values remained would be lost. But others thought depending on domestic mining was counterintuitive, given the larger picture of diminishing American resources. William Elliott of the War Production Board dismissed the bill as hopelessly biased. Hoisting ore from a mine and putting it in storage aboveground adds nothing to the stockpile, he argued. Even Secretary of the Interior Harold Ickes, a recognized friend of domestic mining, opposed the Scrugham bill as too restrictive and costly. He claimed—disingenuously given the loopholes—that preference for domestic sources was "adequately provided for" in the "Buy American" bill of 1933.[62]

Scrugham's first bill died in committee, but the Senate hearings had exposed the shortcomings of current stockpile policy, and the administration showed some willingness to negotiate. In August the War Department, reflecting industry "misgivings" about postwar dumping, offered reassurances it would support legislation freezing the stockpile for all but emergency use. It also revised emergency reserve estimates from three years to one year, thus reducing the immediate demand for copper and other strategic materials. In the meantime, military and civilian material supplies and estimates continued to change, reinforcing the need for better monitoring of current and future stockpile needs. A summer drop in lead production led some industry spokesmen to warn of future shortages. Even foreign imports were unreliable due to the enormous demands of military transport—a sign of new Allied offensives ahead—which tied up all available air and shipping cargo space.[63]

Senator Scrugham and the mining lobby tried again in December 1943 with a redrafted bill. It left in the postwar mandate requiring the government to buy American ore for a year but softened it to be inoperative if domestic resources were unavailable. Incorporating the "Buy American" idea from Secretary Ickes, Scrugham's revision required any government purchases of foreign ores to be 25 percent above the current market price, plus any existing duties. Stockpiled ore could not be released to industry without congressional approval. Another round of hearings followed in December, but by that time the government was swimming in surplus metals, with only copper still in short supply. That made planning less urgent, and gave opponents a chance to stall. How or even whether to protect domestic industry at the expense of consumers and taxpayers, what materials to stockpile in light of changing technology and changing national security needs, how to ease pent-up postwar demand in the face of a growing surplus for some commodities and possible shortages for others—these remained the toughest unresolved issues. After three more years of wrangling, a fragile consensus came together long enough to legislate the nation's first permanent peacetime stockpile bill. It would not be the last.[64]

It took two years of mobilization and another two years of war before American strategic metal policy and practice began to converge into an effective program. By emphasizing labor efficiency; by mobilizing women, students, seniors, prisoners of war, and other "labor reserves"; by shifting workers from low- to high-priority positions; and by working with the Mexican government to devise a "temporary" alien work program that actually lasted more than twenty years, administration officials eased the labor crisis in the mines, fields, and factories. Ironically, just as the mine production incentive program reached full-scale development it began to wind down. Soon after American military operations shifted from defense to offense, Washington policy makers shifted gears. Remembering the postwar economic consequences after the Great War, they warned of surpluses instead of shortages in strategic materials. By mid-1943 the same agencies that had stimulated domestic production now began canceling contracts, eliminating subsidies, lowering priorities, and forcing marginal producers to face real market conditions. Only copper remained on the critical list after 1943. With two years of intensive warfare still to come, American military and industrial planners were already anticipating the problems of postwar strategic stockpiling and economic reconversion.[65]

8

NEVADA-MASSACHUSETTS IN WORLD WAR II

Though distant from political and economic centers of wartime decision making, Charles H. Segerstrom kept careful track of events from his company headquarters on the western slope of the Sierra Nevada. Unless on a business trip he was never far from the teletype machines that connected his office in Sonora with the Nevada mines and his eastern brokerage in New York. An inveterate reader and writer, whenever he had a free minute he scanned and clipped the voluminous newspapers and trade journals that daily piled up on his desk, and pounded out correspondence on a manual typewriter. For fast contact with business and government associates outside the teletype network, he used telephone and telegram.

The frenetic pace of his wartime business affairs required all available forms of communication. At the time America entered the conflict he headed or controlled five different mining operations. Three were producing tungsten: the Nevada-Massachusetts Company at Mill City and Golconda, Nevada; Rare Metals Corporation at Oreana, Toulon, and Wadsworth, Nevada; and Charles H. Segerstrom's family mines and mills at Milford, Utah, and Topaz Lake, California. Two others were gold producers: the Boston California Mining Company at Coulterville, California, and Keystone Mines Syndicate at Amador City, California. The government's order in 1942 closing gold mining for the duration reduced the diversity of his operations, but by that time the insatiable demands of the tungsten business kept him totally preoccupied.[1]

PRAGMATIC PRODUCTION

Segerstrom's efforts to prepare for the war's impact on tungsten supply and demand have already been described. As president of the American Tungsten Association, he kept lobbyists busy in Washington defending the tungsten tariff and monitoring real or potential legislative threats. When the Senate Finance Committee tried to drop the excess profits tax exemption on strategic metal producers, he "immediately got busy," as he told a Boston board member. Alerted to constituent concerns, Nevada's Senator McCarran and Congressman Scrugham got the exemption restored in the 1941 Revenue Act. After the OPM pressured steel mills to use more molybdenum and less tungsten in tool steels, Segerstrom took umbrage. Assuming the government was more concerned with prices than scarcity, he wrote H. K. Masters, chief of the Tungsten Branch of the OPM—the same man whom he had escorted a month before through the western mining districts and thought was "one man in Washington who knows our problems." Tungsten producers, he told the OPM official, are "loyal Americans. . . . I can assure you if greater production in America is desired, it can only be obtained by giving the domestic producer more encouragement, either freezing the prices of labor or increasing the prices of tungsten."[2]

Domestic tungsten producers did get government help, as we have seen, but not exactly what Segerstrom prescribed. While openly advocating wage freezes and price increases, privately he recognized that for his own companies the most important government support was its guarantee to buy all the tungsten he could produce. Along with the boom in industrial demand, the new government stimulus ensured a virtually unlimited market for the gray metal, at least so long as wartime conditions prevailed. Segerstrom well remembered an important lesson of World War I: maximize production while demand was "hot," but look to the future and avoid overcapitalizing at all costs.[3]

Meeting the upsurge in demand required not only increasing production but also maintaining profitability, either by cutting costs or by raising commodity prices. The OPA had not yet set a ceiling price on tungsten when the government quietly began expanding its strategic metals purchase program in 1941. In October the Nevada-Massachusetts president signed a contract with Metals Reserve for 75,000 units to be delivered in quarterly installments over a twenty-month period beginning January 1, 1942, at the prevailing market price of $23 per unit. Segerstrom also agreed to sell Metals Reserve another 50,000 units from Golconda concentrates sent to Molybdenum Corporation for secondary processing. The government contracts provided a comfortable hedge against erratic industrial demand, but the company president thought the price too low for high-grade Mill City scheelite. In September 1941 he headed east on the train for preliminary talks with OPM officials. After "a long conference" in Washington, he came away with an agreement that Nevada-Massachusetts could supply the government with concentrates from any source at the base price, provided they met minimum specifications. Premium Mill City ore, however, could be reserved and sold to steel mills and other direct smelting customers for $2 above the base price. On the way home he took advantage of the deal to stop in Pittsburgh and sign new long-term contracts with tool steelmakers Braeburn, Universal, and Vanadium Alloys at premiums ranging from $26 to $27 per unit, based on the assumption that the OPA would eventually establish a ceiling of $24. He could hardly contain a tone of triumph in his report of the trip to the Boston board: "This . . . will increase our monthly income very materially . . . and will in a great measure take care of the increased costs for some time."[4]

The report was overly optimistic. Production costs, which Segerstrom had calculated at $11 or $12 per unit in September 1941, rose rapidly in the per-fervid economic climate after Pearl Harbor. In the first three months of 1942, Mill City costs averaged $20.91, in part due to higher prices for scarce materials. Tires, for example, were "almost impossible" to find, but Heizer sent out "scouts" in January to every service station in the area. They came back with "all the second-hand recapped tires we could find." Costs also rose above estimated expenses to fine-tune the new 1,000-ton tailings plant after its completion in March 1941. Moreover, new exploration and development to offset an alarming decline in ore grade from the Stank, Springer, Humboldt, and North Sutton mines in 1941 added significantly to material and labor costs.[5]

GOLCONDA, MILL CITY, AND RARE METALS

High costs also marred the rosy predictions about Golconda. "The situation [there] if anything is worse," Segerstrom lamented in March 1942 to a West Coast engineer working for Metals Reserve. Two years earlier the president of Nevada-Massachusetts had sold his directors on the idea of building a 50-ton chemical plant on the leased property to make "almost pure" artificial scheelite. Before it was finished the plant had been expanded to treat 100 tons of concentrates per day. It was expected to produce 15,000 units per month at a cost to the company of only $6 to $8 per unit. Based on the optimistic test reports of George Crerar and his technical staff, Segerstrom had touted the plant like a stock promoter. "If the present [tungsten] prices would stay at $20 per unit," he wrote in 1940 to a Boston banker, "the proposition almost reads like a Mississippi Bubble."[6]

Completed in March 1941, the Golconda facility never lived up to expectations, either in volume and quality of output or in costs of operation. Testing and tinkering took an inordinate amount of time, as Segerstrom complained in a letter to Boston: "There are so many different machines and tanks and mag- nestic [sic] separators, roasters, dryers, etc., that it is very difficult to estimate just when we are going to get . . . in good shape." In the fall of 1941 he added to the complications by deciding to double the plant capacity. A bigger chemi- cal facility was needed to treat off-grade ores, including manganese-bearing oxides stripped from the Golconda pit as well as gravity-mill concentrates from leased properties that were high in impurities.[7]

Inflation and delays made a mockery of Segerstrom's 1940 cost estimates. By early 1942, Golconda expenses had climbed to more than $28 per unit. The chemicals alone cost up to $15 per ton of ore treated. These excessive outlays looked bad compared with contracts agreed to earlier with Metals Reserve and Molycorp for $23 and $22 a unit, respectively. The Sonora boss tried to speed up expansion to maximize production efficiencies, but material shortages and slow deliveries gave him "nervous prostration."[8]

By the summer of 1942 most of the technical difficulties at Golconda had been resolved. Crerar's continued experiments in the Sonora lab improved the sintering process and greatly increased tungsten extraction. Although chemi- cals remained costly, Segerstrom's report to the Boston board for the first six months of 1942 contained only good news. Nevada-Massachusetts had pro- duced nearly 70,000 units in the period, with the Golconda plant contributing two-thirds of that amount. The gross profit of $700,000 from total operations was more than enough to pay back all capital outlays to bring Golconda on line. Profitability, he exclaimed, went hand in hand with patriotism: "This is the best record we have ever had for current funds. . . . I feel we are doing our full duty as a company engaged exclusively in war work. All of our material is going into war armaments and are [sic] going directly into the war zone."[9]

Despite Segerstrom's glowing early reports of "at least 100,000 units there in proven ore" and "no end of ore in sight," Golconda turned problematical as operations expanded. The 1941 decision to mine Golconda's low-grade ore complex by open-cut methods began well, but as the pit deepened the added amount of waste in relation to ore, or stripping ratio, increased costs to nearly prohibitive levels. Production there never raised much above 20 percent of the total output of company mines, and the unit costs of running the chemical plant remained up to 25 percent higher than income from company sales per unit throughout the war.[10]

The main source of company profitability during the war years was not

Golconda but the 1,000-ton tailings flotation plant at Mill City. Completed in March 1941 to reprocess tailings directly from the gravity mill as well as a million tons of old tailings from previous milling operations, the new mill proved its worth over a forty-two-month life span. More efficient and less costly to operate than the gravity mill, flotation captured most of the remaining tungsten values in the tailings pond. At its peak of production the mill turned out 4,000 to 5,000 units a month of concentrate that tested up to 70 percent WO$_3$.[11]

Milling old tailings was much more productive and profitable than finding and mining new orebodies. Within three months of opening, the new flotation plant was already turning out half of the company's total tungsten production. Despite Segerstrom's relentless—and costly—exploration and development efforts at the Mill City mines, ore grade remained low, at least until late 1943. Occasionally, Heizer's crews found pockets of good ore in the Stank, Humboldt, and North Sutton, but those were soon stoped out and mixed with lowgrade in the gravity mill to produce commercial-grade concentrate. As the accompanying chart shows, filling contracted production requirements meant grinding and floating much more ore in the company mills. In 1940, Nevada-Massachusetts mined 80,226 metric tons of ore, or 96 percent of the tonnage it milled in company plants. By 1943 it was mining less than 16 percent of its milled tonnage. Much of the rest came from its tailings pond.[12]

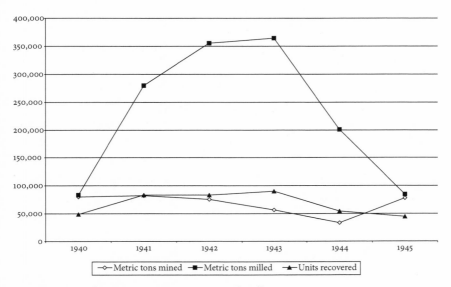

Nevada-Massachusetts: Wartime mining and milling

Segerstrom called the pond the company's "lifesaver," but he recognized its limitations. Of the nearly 1 million tons of tailings accumulated between 1918 and 1941, only 27 percent remained to be processed by 1944. That was gone in less than a year. Long before the tailings plant shut down, however, the man in charge had anticipated the end. "If it were not for the war and patriotic reasons," Segerstrom told an official of the War Production Board in 1942, "we would not attempt to take the ore out as rapidly as we are doing, but we realize the intense need for tungsten and have cooperated from the beginning with the authorities to get out every ounce of ore that could possibly be produced."[13]

The only affiliate of Nevada-Massachusetts that grew substantially during the war years was the Rare Metals Corporation. Financially independent but managed as an integral part of the larger company, Rare Metals processed ore hauled into the Toulon mill by truck or rail from company mines. It also reground tailings from its own millpond. As these resources diminished, Segerstrom expanded by leasing properties in Nevada, Utah, and California. In 1941 he leased two small scheelite operations near Milford in southwestern Utah, and a third at Topaz Lake, California. The next year he took control of the Long Lease mine southeast of Lovelock near the Humboldt River, the Alpine mine in the Nightingale district of Pershing County, and later the Palomar Scheelite mine in the Sonora desert forty miles east of Escondido, California. Toulon was also used as a custom mill for smaller independents, with Segerstrom or his brokers usually handling the concentrates on consignment. From a modest monthly output averaging 400 units in 1941, by war's end the Toulon mill was turning out nearly 3,000 units per month, or one-third of the total volume for Nevada-Massachusetts.[14]

As an independent company operating under Nevada-Massachusetts' unified management, Rare Metals had connections and advantages other small firms lacked. With the government's encouragement, Segerstrom planned to double the capacity of the Toulon mill and add a flotation plant to improve recovery. Early in 1943 he signed a two-year contract with Metals Reserve to produce 60,000 units. Financing mill construction with investment capital from a few eastern speculators spread the risk without losing control— another lesson learned from previous wartime experience. Labor and material shortages, however, slowed plant construction and forced Segerstrom to cut the contracted amount nearly in half. Only 12,833 units were actually delivered before Metals Reserve canceled the contract in 1944. In the meantime, the Toulon and Golconda mills continued to process off-grade material that

Molycorp absorbed at $22 per unit, a dollar less than Metals Reserve paid
Nevada-Massachusetts for its Mill City tungsten.[15]

SEARCHING FOR A SUBSIDY

As we saw in the previous chapter, to stimulate domestic production during a
severe shortage, Metals Reserve late in 1942 raised the price it paid high-cost
marginal tungsten producers by $6 a unit. Despite complaints of discrimina-
tion from larger firms, government inflation fighters rejected appeals to lift
ceiling prices across the board for the gray metal. Tungsten prices were higher
than anytime since 1918, and strategic metal producers were partially exempt
from the tax on excess profits. The OPA's answer to complaints from the iron
and steel industry applied as well to the big tungsten producers: profitability
was not the highest priority in wartime.[16]

Inflation also worried the president of Nevada-Massachusetts, but as costs
rose and profits fell his priorities changed. The first three quarters of 1942
were a financial bonanza for the tungsten company. With net profits averaging $30,000 or more per month, Segerstrom basked in the fiscal sunshine. He
warned his fellow producers that raising the tungsten ceiling would be "as a
ripple in the sea" and just add pressure for higher prices in steel and other
commodities. In the fourth quarter, however, his own companies felt the economic squeeze. Labor shortages slowed production, and higher costs shrank
profits. New wage pressures, additional overtime rules, and a proposed "victory tax" on corporations all foretold difficult times ahead. By year's end he
was lobbying Metals Reserve to raise the contract price it paid for ore from
Nevada-Massachusetts. Confidentially he told a Boston director that "I do not
know of any way we could get it because if we made a firm statement they
would probably claim we made too much money already." In a private note to
Haesler, he considered making a "horse trade" with the Washington crowd: "If
they dont give us any [price] raise and then put on the raise in wages and regulation of hours[,] . . . we will just shut down at Mill City and then only open
after we get some adjustment." While the price hike lasted and with tungsten
still critically short, talk of closing Nevada-Massachusetts was more of a bargaining chip than a serious threat. Segerstrom and his Boston bankers had too
much invested in the Mill City operation to shut down altogether. He wanted
Metals Reserve to qualify all his companies for the $6 raise, but failing that he
was prepared to accept less.[17]

His appeal did not fall on deaf ears. Segerstrom had good connections in

Washington. Many of the metals specialists and industrialists staffing wartime positions in Metals Reserve and the War Production Board knew the president of Nevada-Massachusetts, either personally or by reputation. In response to his request, Henry Carlisle, a consulting engineer in the ferroalloys branch of the WPB, offered some friendly advice. Under the qualifying rules, only small independent companies were eligible for the new price, not large firms "which have more than one property under the same name." Rare Metals and Milford should qualify, he thought, and perhaps others if they were not "considered as one" division of the same company. At any rate, the company president should "discuss the issue" with officials in Metals Reserve.[18]

Segerstrom took the hint. Rare Metals had never been financially tied to Nevada-Massachusetts and easily qualified for the new rate, but Golconda and Milford required some legal and financial rearranging. Segerstrom had person-ally borrowed money from Marx Hirsch to build the Milford plant in exchange for an agreement brokered by Haesler that Molycorp would take all the Mil-ford concentrates. With Hirsch's cooperation, and with advance approval from Metals Reserve, Segerstrom in March 1943 canceled the Molycorp contract. That "automatically put Milford in the position of a new producer supplying ore to Metals Reserve . . . at the maximum price limit of $30.00 without a con-tract," as Segerstrom described the arrangement to Haesler.[19]

The Golconda problem was more complicated. As a leased property it could not be easily disentangled from Nevada-Massachusetts. Metals Reserve rejected Segerstrom's first efforts to qualify Golconda for the $30 price, but the guide-lines allowed "special consideration" based on "demonstration of the necessity." That was not hard to prove during the first half of 1943. Unit costs had more than doubled over the past year, rising from $12.06 to $26.47. Much of that had been due to Golconda's "abnormally high" expenses, running at times above $30 per unit. Contract prices for company ore had climbed only about 1 percent since 1941, yet labor costs had risen 50 percent over the same period. At the Mill City mines, production had declined by 40 percent due to labor shortages, thus cutting revenues and preventing any significant development. For the first six months of 1943 Segerstrom reported a net loss of $111,448.69, with losses from the Mill City division running at $2,000 per month.[20]

Worried that inflation would slow tungsten production, officials at Met-als Reserve in the spring of 1943 loosened its pricing policy, allowing nine moderate-size firms to receive the $30 payment but leaving out the largest producers. For months Segerstrom lobbied for dropping all distinctions based on size and volume of production, emphasizing his own company plight in particular. "We have gone along with this operation without any complaint on our part," he wrote in May, "but we have lost money on practically every ton we have put through." A September letter, disingenuously vague but politely patriotic, was more urgent: "We are up against the worst problems that miners were ever plagued with, shortage of men; ore difficult to mine and mill; prices; deliveries, etc., but we are doing our level best and in wartime we must not complain." To reinforce the written appeals he spent weeks in Washington talking with officials on two separate trips.[21]

Persistence paid off at least for Golconda, which Metals Reserve eventually qualified for the higher rate, but not for the Mill City division. Unfortunately for major tungsten producers, the squeeze on company profits came just as the dynamics of America's war effort began to change. In the summer of 1943 home front production met or exceeded military demands abroad for the first time in the war. Henry Carlisle, sympathetic to Segerstrom's plight, frankly advised him to consider closing all but the tailings plant. "Back here [in Washington] the general feeling is that we have an awful lot of tungsten," he wrote in October.[22]

Metals were among the first commodities hit by production cutbacks. With the Axis powers in retreat, strategic stockpiles mushrooming, and Allied defense industries churning out more planes, tanks, ships, and ammunition than were needed on the battlefronts, war planners in Washington put on the brakes. In a series of orders beginning in September 1943, the War Production Board, in conjunction with the War Manpower Commission and Metals Reserve, ended military furloughs for miners, terminated most premium price plans, cut back imports of overstocked metals, and reduced incentives for maximizing production. Segerstrom, in Washington to assess the situation, described the impact in an excited teletype to his son Tod:

The whole metal situation is in a serious condition due to the fact that there is a large surplus of every kind of metal The copper and Zinc as well as moly are the worst off They cant sell current production and dont know where to turn The same is true of every one of the metals and in the arms factories and other plants men are being laid off right along In tool steel they are only running one shift where they ran three eight hours shifts a short time ago. In the plants making shells and loading them its almost at a stand still This is also true in many other smaller plants.[23]

This was at best a short-range view. Copper and zinc remained on the critical list, and nonferrous metal premiums were actually extended for the duration of the war and beyond. Large mining operations with standard government or industry contracts remained open, though many lowered production goals as demand dropped and prices declined. Those most seriously hurt by cutbacks were the "war baby" mines almost wholly dependent on government subsidies to stay in business, including marginal producers of bauxite, mercury, vanadium, chromite, molybdenum, and tungsten. Without a premium contract from Metals Reserve, they were unable to compete as markets deregulated and metal prices fell. Many closed when ore grade fell below economical recovery; others shut down or sold out under a mounting burden of debt.[24]

FORTUITOUS CIRCUMSTANCES

Segerstrom was still searching for subsidies when an unwelcome but all too frequent visitor to mining camps arrived at Mill City. Just a month after Carlisle's advice to consider a partial closure, the company was partially closed by disaster. On November 11, 1943, the old Pacific Tungsten gravity mill, built in 1918 and upgraded to 250 tons per day in the late 1930s, burned to the ground in a fire of unknown origins. The flotation tailings plant, about a thousand feet away, escaped undamaged. Mill City production immediately dropped by half. Segerstrom's wire to Washington announcing the bad news ended with a gratuitous cry for help: "The general public will probably never appreciate the great contributions that have been made as we have not capitalized [on our] . . . strategic position but have tried to get the greatest production at fair and reasonable prices and we hope something can be done to assist us now."[25]

Regardless of the momentary glut of metals, big producers like Nevada-Massachusetts were too important to ignore. Federal officials responded quickly, agreeing to remodeling plans, cutting red tape, expediting construction and replacement of lost equipment. At Segerstrom's request, Metals Reserve cut its contract requirements by 40 percent and permitted the company to ship low-grade Mill City ore to the Golconda chemical plant for treatment. An inveterate opportunist, the company boss took advantage of circumstances to upgrade and modernize. Instead of rebuilding the old gravity plant, he replaced only the crushing equipment, but located it near the collar of the South Sutton mine shaft, where development crews had recently found a significant new ore body. After primary and secondary crushing, the ore was trucked to a new ball mill in the flotation plant for fine grinding. Depending entirely thereafter on the flotation plant cut mill capacity by 20 percent, but also reduced labor and machinery costs just as tungsten prices were dropping.

With insurance money funding half the bill, Segerstrom expressed cautious optimism in his January 1944 report to the company board: "I do feel that if tungsten is going to be mined in America that we are going to have our share of it, because I do believe that under normal conditions we can produce tungsten about as cheap as any one I know of in the country, and while we may have many dark days ahead, I do not feel at all in despair, and with the funds we now have in the treasury, we are in a position to finance or handle the property as we see fit."[26]

"COMPETITIVE CHAOS"

Nevada-Massachusetts was large enough to escape much of the instability that affected marginal operations, but declining markets, labor shortages, and high costs after 1942 hurt profitability and jeopardized the company's financial future. During the winter of 1943–44, the competitive pressures stemming from changing market conditions added a new concern. Segerstrom was no stranger to competition. Despite periodic efforts over the years to organize domestic tungsten producers into a cooperative cartel, only the smaller firms took the first step by joining his tungsten trade association. They were fickle partners at best, jumping ship whenever it was to their financial advantage. Unlike corporate leaders in steel, copper, and vanadium, the largest tungsten producers were even less inclined to cooperate. Only the interlude of war regulations and price controls tempered their determination to enlarge, or at least protect, their resource base and market share from any real or perceived competitive threats at home or abroad.

In a 1944 New Year's letter Segerstrom confided to his old partner W. J. Loring that Nevada-Massachusetts was in a "very tight jam." With the government canceling premium contracts, he feared, soon "everything [will be] thrown back into a competitive chaos." The comment stemmed from a rumor in December 1943 that U.S. Vanadium, whose unrefined Pine Creek concentrates he had once written off as "valueless as far as making ferrotungsten," was about to "cut the price of ferrotungsten to a point where it would drive competition out of business, particularly, the Molybdenum Corporation of America and the Nevada-Massachusetts Company." Incensed, Segerstrom had written a hot response to J. R. Van Fleet, the company president, demanding clarification of the company's intent.[27]

He was still awaiting Van Fleet's reply when he learned of a more immediate threat from another major tungsten producer. The Bradley Mining Company, a long-established San Francisco mining and engineering firm, owned and operated the Yellow Pine mine at Stibnite, Idaho. The Bradley name had been associated with tungsten since the early 1900s, when Baruch money and Bradley engineering combined to develop the Atolia mine in California. For a few years during World War I Atolia was the nation's largest tungsten producer, but the mine was nearing the end of its productive life by the time Nevada-Massachusetts rose to national leadership in the early 1930s. Early in 1937 Segerstrom, hoping to expand production, arranged with the Bradleys to treat Atolia mine tailings at the Nevada company's new flotation plant in Mill City for a fifty-fifty share of proceeds. The deal fell through, however, when the recession hit and ore prices dropped. Atolia then installed its own tailings plant and came back into production during World War II.[28]

The Yellow Pine mine had started in 1932 as a gold producer. Nine years later, as part of the nationwide search for strategic metals, government geologists examined diamond drill cores from the property and identified significant tungsten deposits. In August 1941, using both open-cut and underground methods, the Bradley company began mining high-grade scheelite from nearly vertical mineralized fractures in a quartz shear zone a hundred feet wide in some places. Over the next two years, aided by high-priority government ratings and help from the U.S. Employment Service, the operators expanded the flotation mill and increased the mine crew from 100 to 650. Yellow Pine was the nation's second-largest tungsten producer in 1942 and the largest in 1943 and 1944.[29]

Before the metals glut became apparent in 1943, Metals Reserve was Yellow Pine's primary customer, holding substantial contracts for mill-run concentrates and low-grade mill tailings for the national stockpile. The Bradleys also marketed high-grade ore to industrial customers. Over the summer and fall, however, changing market conditions required new marketing strategies. The Bradley firm responded to announced government contract cuts by cutting its sales prices and expanding direct sales to industry. To enlarge its marketing outreach the company built a processing plant and warehouse in Pittsburgh and hired as eastern agent Miles K. Smith, a well-known former executive of Metals Reserve. All this made the Bradleys a serious competitive threat to Nevada-Massachusetts.

In January 1944 Segerstrom learned just how serious. A. C. Daft, Segerstrom's Pittsburgh agent, reported that Smith was aggressively pushing direct sales, intending to attract new business by selling higher-quality concentrates at a lower price than any competitor could offer. Even worse, Smith had approached Vanadium Alloys Steel, an important customer of Nevada-Massachusetts, offering high-grade scheelite units at $2.40 lower than the Nevada firm's price. "I know their tactics so well, know exactly what they are going to do," Segerstrom explained to one of his board members. "At any given price they try to obtain the business by giving 50c less. This gives all the steel mills a chance to begin chiseling, and no one knows where it will end."[30]

The Bradley gambit could not go unchallenged. Vanadium Alloys, one of several alloy steel specialists in the Pittsburgh area, had been buying tungsten from Nevada-Massachusetts and its predecessor since 1918. After 1930 Vanadium Alloys metallurgists charged their electric furnaces directly with Mill City scheelite, paying premium prices for the best ore grades but saving money and time by avoiding the normal refining steps required for lower-grade concentrates. Vanadium had remained a loyal customer even when business was slack or ore shipments late. Segerstrom had repaid that loyalty with high-priority service, even at the expense of other customers. As he once told his Pittsburgh agent, "I think we should hold [the] Vanadium [contract] at all costs so as to have an out at all times."[31]

Segerstrom traveled east on the streamliner as soon as he could get train reservations after receiving Daft's letter, prepared to play marketing hardball. He met with the Bradley sales agent at their company headquarters in New York. Right after the meeting he reviewed the results in a teletyped message to his son Tod: "Ive just had a hell of a time here . . . and have spent all morning with Miles K Smith. . . . I talked to him for four hours and cant go over it all But I scared him within an inch of his life I told him that we wouldnt bother meeting his price that we would get more than that from other sources and would hold our present price or not sell at all and that if they wanted a price war I was ready to sell at $20 a unit and asked him to figure up what Bradleys would lose at that."[32]

Segerstrom's repost was more bluff than bluster. He told Tod that he had "all the aces this time" and was "going to crowd them to the limit," but just a few weeks earlier he had been "quite fearful" for the company's future if demand slowed and prices dropped to $20. With ore grade at Mill City "lower than ever in our history," he worried that the company could not sustain the heavy costs of new development once the low-cost production from tailings ended. "When we reach a position where we are operating at a loss," he concluded, "we should look the situation squarely in the face and shut down."[33]

From a different perspective, the Bradley tactics had less to with beating the competition than reaching the limits of the Yellow Pine orebody. Smith admitted to pushing Yellow Pine production to maximize profits while opportunity lasted. That was common strategy in a hot wartime market. But Smith also told Segerstrom that "they would be out of ore in 9 months and [that the] Bradleys wanted to get the stuff out as quickly as possible." By August 1944 Yellow Pine production had dropped by two-thirds, and Segerstrom no longer felt threatened. He gave up the idea of a price war, conceding the market for high-grade tungsten to the Bradleys while they were still in business. They closed in 1945 after mining out the ore, with a total production of 825,970 units.[34]

Late in 1944, while Segerstrom was still dealing with the Bradleys, his dispute with John R. Van Fleet died when Pine Creek shut down. Although keeping its mill running for secondary processing of low-grade concentrates from Nevada, Idaho, and California, U.S. Vanadium closed the high-altitude mine while it drove a new adit sixty-five hundred feet below the old portal. Mining resumed in April 1946 on a part-time basis. Thus, at war's end, of the Big Three domestic tungsten companies, only Nevada-Massachusetts was still producing ore.[35]

THE NIGHTINGALE IMBROGLIO

Whether for his own independent family operations or for Nevada-Massachusetts, Segerstrom controlled as many sources of ore as possible during the war years. He also kept in close touch with metals dealers on both coasts and abroad, anticipating the need to buy ore on the spot market to keep his customers supplied if his own resources ran short. Early in 1941, for example, he contracted with a Japanese firm to ship five hundred tons of Chinese concentrates at $16 a ton on the first available vessel. The ore was to be shipped from Shanghai, at the time under Japanese control. With American-Japanese trade relations rapidly deteriorating, before he confirmed the purchase Segerstrom had his Washington attorney ask the U.S. State Department for some assurance that the deal would go through. Segerstrom was still waiting for an official reply when he received a wire from J. J. Haesler who had helped arrange the purchase. Haesler said the Japanese government would be happy to release the ore if it received a shipment of high-octane gasoline of equal value in return. A few days later Haesler reported that the State Department suggested they "drop [the] matter temporary and see what happens." In July, President Roosevelt issued a decree restricting the sale of aviation gasoline to the Western Hemisphere.[36]

Segerstrom also supplemented company ore supplies with domestic purchases. In the highly competitive prewar market, getting the best price for high-grade scheelite often required playing dealers and importers off against each other. Early in 1940, he quietly collaborated with Haesler to stabilize tungsten prices. The strategy involved first warning ore sellers that war would threaten markets and thus reduce demand, and then offering to buy up any small tungsten lots that became available. Through Haesler's efforts as well as its own aggressive purchases, Nevada-Massachusetts built up its company reserves, and at the same time kept other dealers from dumping cheap ore on the market.[37]

After the United States entered the war, Haesler encouraged Segerstrom to strengthen the domestic cartel while he had the chance:

I think it would be fine if you could get the producers large and small together and sell their production thru one outlet at an agreed price This would hold the situation and keep out every Tom Dick and Harry who would ruin the western producers if they are given too much liberty. Those in the tungsten business now have no place in it They are just trying to make a killing out of it and they should be killed off by strong action Every refugee and his brother is thinking of domestic tungsten as a good speculation and when the war is over you wont hear any more about them.

Segerstrom's reply provides some insight into his business ethics. Since Nevada-Massachusetts already dominated the tungsten market on the West Coast, he said, price-fixing would only hurt his long-term customers. They preferred high-grade scheelite from a reliable source. As he earlier told George Emery, "because we have treated them right they prefer to buy from us." He wanted to keep it that way, regardless of temporary swings in the market.[38]

One might argue that treating customers right was simply good business practice, a practical marketing ploy rather than a matter of ethics. Certainly, as we have observed in his dealings with Molycorp, Segerstrom weighed business decisions by their expected outcome, not by any ethical standard. He was not inclined to philosophize—nor did he have time for deep introspection. If judged by his actions as a businessman, he was the consummate utilitarian.

Segerstrom's pragmatic outlook is best illustrated in his relations with John George "Jack" Clark, a Rocky Mountain mine investor and president of the Gold, Silver, and Tungsten Production Company of Boulder, Colorado. By no means a "refugee" speculator in Haesler's terms, Clark was well known and well liked in Colorado mining circles. Like William Loach of the Wolf Tongue Mining Company, Clark had been involved in Colorado ferberite production since before World War I, although on a much smaller scale. In 1929, with tungsten prices rising, he ventured westward across two states to pick up a new tungsten property that had just become available in Nevada. Named for the mountains where it was located in 1917, the Nightingale mine lay in the southwestern corner of Pershing County. The purchase also included the adjacent Star property, part of which crossed over into Washoe County, about forty miles and two mountain ranges west of Segerstrom's Rare Metals mill at Toulon.[39]

How much Nightingale tungsten Clark produced has never been determined. In 1939 Clark himself said that in the decade he had owned the mine his mill had turned out $200,000 worth of concentrates, but that claim was never substantiated. The ore was granular and crystalline scheelite, some high-grade pockets assaying up to 3 percent per ton. A series of skarn deposits were mined by drifting under near-vertical lenses and dropping broken ore and rock through "glory holes" to cars waiting below. During World War I a few tons of ore had been hauled to the mill at Toulon for treatment. Clark built a new mill rated at one hundred tons per day, but shut down during the Depression and then reopened in the late 1930s, only to close again for lack of a good water supply.[40]

Segerstrom had known about the Nightingale property since the late 1920s, but his first contact with Clark came through J. J. Haesler. In October 1931 the Colorado investor approached the Metal and Ore Company in New York, offering to sell a carload of high-grade scheelite a month at $10 a unit. Haesler put him off and then wrote Segerstrom, asking whether the Nightingale was "capable of furnishing this tonnage." There was no immediate response. Segerstrom had his own Depression troubles to worry about. But he did respond in 1933, when Clark offered to sell Nightingale scheelite through the auspices of Nevada-Massachusetts and its eastern brokerages, paying commissions as well as freight. After sending Heizer to inspect the Nightingale property, Segerstrom told A. C. Daft not to proceed with a sales agreement. Heizer had reported that "they had some fairly good ore but not in sufficient amount to justify a forward contract."[41]

Whether Segerstrom's observation was fair to Clark or merely a shrewd business ploy is hard to determine. If nothing else it was premature, for within a month the Sonora boss offered to lease Nightingale outright so he could make up the tonnage he had promised Molycorp but was unable to produce from his own Mill City properties. Clark wanted a 25 percent royalty and a limit to the number of hours the mine was worked. After haggling for several months, Segerstrom eventually counteroffered at 15 percent, but Clark held firm. Segerstrom managed to buy a half-carload of Clark's Colorado ferberite to satisfy Molycorp's incessant demand, and later Clark consigned two tons of Nightingale concentrates, high in phosphate, which Segerstrom forwarded to Daft along with a Mill City shipment. Daft sent the Nightingale lot on to Molycorp for further processing, charging Clark a 2 percent commission for the service.[42]

Segerstrom's early dealings with Clark involved a mixture of business and politics. Both could be hard bargainers. As national spokesman for the tungsten industry during the First New Deal, Segerstrom needed Clark's support behind efforts to organize producers into an effective cartel under the auspices of the National Recovery Administration. Clark joined the Tungsten Association, but was critical of its position on tariff issues. Despite Segerstrom's role in defending the existing duties on imported tungsten, Clark and his fellow Colorado producers wanted even greater restrictions. Early in 1934, after learning of proposals in Congress to exchange Chinese tungsten for American wheat, he wrote a prickly letter to the tungsten boss in Sonora: "Do you not think it is time that something should be done by our tungsten organization; otherwise what would be the reason for such an organization?"[43]

Segerstrom did not bother to reply. He remained president of the producers, but the NRA's reversal of fortune after 1933 left the trade association without government sponsorship and little raison d'être other than fighting tariff battles. Managing his own company operations took most of his time over the next five years. Clark, in the meantime, struggled to make Nightingale productive and profitable. He contracted to sell ore to Molycorp, but never produced enough concentrates to pay for the financial advances he obtained from his eastern customer to complete the plant. By 1939 he was heavily in debt to Molycorp with payments in arrears, but determined to hold on until the market swung his way. With war festering abroad it seemed only a matter of time.

Clark was stubbornly independent, but how long could he hold out? In the summer of 1939 Segerstrom decided to find out. Aware of Clark's financial vulnerabilities, and worried about the continual decline of ore grade at his own company mines, he quietly reinvestigated the situation at Nightingale and Star. An unpublished geologist's update sent him by an informed insider was surprisingly impressive. It asserted "the certainty of a very large tonnage of ore established" by work completed during the 1930s, and estimated "probable and possible resources" at a gross value of nearly $11 million. Without telling Clark, Segerstrom sought Molycorp's help in approaching the hard-pressed Colora-dan with what the deal makers thought was a fair proposal. E. A. Lucas, Moly-corp's vice president and chief metallurgist, agreed to make a pitch for leasing at least part of Clark's holdings to Segerstrom in order to get out more ore and pay off the Molycorp loan. The war in Europe had already started when Clark wrote the Sonora executive: "As I understood him [Lucas], you would form a new company, supply funds to pay our indebtedness in Nevada, take over our contract and obligations with Molybdenum Corporation of America, etc. and [he] suggested our taking 25 percent of the stock of your new company for our interest. Frankly, the deal does not appeal to us, as we really have a lot of money tied up in Nightingale."[44]

Unsettled by the tone of Clark's letter, Segerstrom was quick to reply. He had no thought of taking over, he said, only "to be helpful both to you and the Molybdenum Corporation" by offering the services of the Toulon plant to process Nightingale ore so that Clark's "obligations might be liquidated, and after . . . [that] there might be a possibility of arranging some plans for future operations, providing the property would warrant it." When Clark again rejected the proposal but opened the possibility of a "straight lease" for a 12.5 percent royalty of gross sales, Segerstrom wired Marx Hirsch, Lucas's boss at Molycorp, for help in dealing with Clark.[45]

Added pressure only increased his resistance. The Colorado executive met in Denver with Lucas and Segerstrom in March 1940. He listened while Seger-strom offered to lease the Star property for a year and pay off Clark's debt to Molycorp with ore at prevailing market rates, with an option to extend the lease two additional years for a standard royalty to the owner. Then he walked out to think about it. Two months later, hearing nothing, Lucas pressed harder. "I believe that this is as good a deal as you can make and better than any I expected Mr. Segerstrom to agree to," he told Clark in a curt letter. "Since you are risking nothing and will receive your property back without any encum-brance I can't understand why you would hesitate."[46]

Segerstrom tried again in June 1940, meeting with Clark in Los Angeles during a West Coast mining conference. Van Rensselaer Lansingh joined them as Molycorp representative, along with one of Clark's company board members. To win Clark's favor both Segerstrom and Lansingh made conces-sions on terms of the proposed lease, but Clark was in no mood to bargain, and negotiations quickly broke down. Lansingh wrote to Lucas afterward that

"Clark didn't want to do anything at all and I felt that no matter what sort of a deal was made it was extremely unlikely that Clark would be willing to go through with it in his present state of mind. . . . [He] was really looking for an excuse to turn down the deal."[47]

Over the next two years, despite critical tungsten shortages, Nightingale lay idle. Determined not to be gobbled up by either Molycorp or Nevada-Massachusetts, Clark told creditors he was trying to interest other operators in his Nevada mines, but he never concluded any deals. Segerstrom, in the meantime, refused to lower milling costs at Toulon for processing Nightingale ore "unless we had the actual handling of the property," as he explained to Lansingh. When Clark's promissory notes to Molycorp came due early in 1942, the testy Coloradan vowed to fight any attempt to foreclose. Marx Hirsch, Molycorp's president, not easily intimidated, declared Clark in default and filed suit in Nevada. A district judge placed the property in the hands of trustees pending further action. Hirsch asked Segerstrom to consider operating the property when the dust settled, perhaps even taking control. The Sonora executive worried about the consequences: "I do not know what to think of the Clark outfit. . . . [T]hey claim if we are going to do anything, they will keep us in court for years."[48]

Molycorp's legal claim against Clark took months to adjudicate. While the case was still pending, Segerstrom, with his confidential banking and business connections, looked into the financial strength of Clark and his firm, Gold, Silver, and Tungsten, Inc. One Colorado credit bureau, reporting a good credit history, concluded that Clark's corporation "owns many patented mining claims of supposedly great value but is controlled entirely by J. G. Clark. The corporation has never paid a dividend and is apparently run for the sole benefit of J. G. Clark." Hirsch, in the meantime, had his own paid informants in the field. Their objectivity might be questioned, but after learning from them that Clark's Nevada properties had "no values," the Molycorp president decided "that there was no use in pursuing [the suit] further than the sale of Nightingale as nothing could be collected."[49]

Late in 1942 Molycorp won a judgment against Clark and proceeded to foreclose on the Nightingale property. The harassed Coloradan's attorneys countered by seeking a restraining order to prevent the sale. If Segerstrom hoped to avoid direct involvement, he was jolted by Clark's testimony in January 1943, during a hearing in Reno on the motion. William Kearney, Segerstrom's attorney, sent him the bad news: "Mr. Clark spent two days on the stand[;] he testified that every deal he tried to make in the last three years fell through because of your opposition. I have no idea yet what the court will do, although Judge Norcross has always been more than sympathetic with the pleas of debtors who want to hang on indefinitely in the hope of finding someone who will bail them out of debt." Segerstrom brooded for nearly a month before replying with a mixture of scorn and sanctimony: "There isn't an official in Washington who does not know the circumstances, and who doesn't think the mine is no good and in the hands of a promoter. This has been the main reason why Mr. Clark has absolutely failed to interest any one. . . . I trust we can clean up this Clark mess, so that at least the Judge will believe that we are honorable men."[50]

In the popular literature of mining, the term *honorable* is more likely used in mockery than in truth, as in "honor among thieves." Segerstrom loved the literature and lore of western colorists like Bret Harte and Mark Twain, but there is no hint of mockery in his self-serving letter to Kearney. Convinced that Clark was solely to blame for his own troubles, the boss of Nevada-Massachusetts determined to take over Nightingale if the opportunity arose. One might argue that Segerstrom, in quietly maneuvering to outfox his rival, crossed an ethical boundary in business relations. Yet Segerstrom was not the cause of Clark's financial troubles, nor did he do anything to jeopardize the Coloradan's effort to lease or sell his Nevada holdings. Once Molycorp's claim was upheld by the courts and the property was sold for payment of a legitimate debt, there was nothing dishonest in Segerstrom's effort to acquire and develop it. Considering the circumstances, however, he thought it only prudent keep Clark from finding out.

The Nightingale and Star properties sold for $10,000 at a sheriff's sale in Reno on July 15, 1943. There was only one bidder. Clark at the last minute had tried to postpone the event, but Hirsch told Segerstrom that the sale had proceeded on schedule. "I do not know who bought it," he said. Even before receiving Hirsch's news, however, Segerstrom had written to Kearney, his Reno attorney, enclosing a cashier's check for $10,000, with instructions to have the deed to the property made out to "James A. Adams, a single man." He lived in New York City, but his address was to be recorded in care of Kearney to avoid having the "Reno people know his whereabouts." Six weeks later Adams wrote to Metals Reserve, asking for certification as a qualified producer to operate Nightingale and have Rare Metals mill the ore at Toulon "as soon as the legal tangles involving the property can be arranged." In his own letter to Metals

Reserve at the same time and written on the same typewriter, Segerstrom confirmed that Adams and associates had "asked us to mill some of their ore." He had told Adams that "we might after their troubles with Mr. Clark were ironed out, but did not want to get mixed up in any fights with him."[51]

Mounting debts and illness had combined to take much of the fight out of Clark by the time Kearney tried to take possession of Nightingale in the fall of 1943 on behalf of its new owner. He made a hasty retreat, however, when Clark's nephew, who occupied the property, threatened to "resist with force" any efforts to take over. Experienced in such matters, and with Hirsch's cooperation, Segerstrom under the name of Adams bought Clark's promissory notes from Molycorp and sued in Nevada courts to collect. Armed with Molycorp's deficiency judgment from the earlier decision, Adams attached the Nightingale property and the local sheriff ejected Clark's representatives. Clark made one last stand in the courts to recover, but the judge slapped him with a permanent injunction. By August 1944 Clark was dead after a long illness, and Segerstrom was in full control.[52]

Winning the legal fight for Nightingale was a pyrrhic victory for the senior Segerstrom. The tungsten shortage was over by the time "James A. Adams" deeded the property over to the Rare Metals Corporation in the fall of 1944. With ore prices dropping and demand down, a crew watched over the property and camped on it during the winter of 1944–45 while working the adjacent Alpine mine for Rare Metals, but did only a "little work" on Nightingale itself. Not until after the senior Segerstrom died did his Nightingale gamble begin to pay. For twenty years his eldest son, Charles Jr., in partnership with Ott Heizer's son, John, operated the Nightingale and Star mines together with the Toulon Mill under a new corporate name, the Wolfram Company. In the mid-1950s it was one of Nevada's largest tungsten producers, second only to Nevada-Massachusetts itself.[53]

LABOR PROBLEMS

The labor shortage that had crippled the national copper industry in 1942 also threatened Nevada-Massachusetts and other metal producers. As a "qualified producer" of strategic metals, the Nevada company came under federal labor guidelines. Accepting the Walsh-Healey Act requirement for a forty-hour workweek and overtime pay was not enough to satisfy growing labor demands in a buyer's market. In the fall of 1941 Mill City employees had told Segerstrom they wanted another pay raise, plus sixty housing units built at the mining site, seven miles away. If the company did not want to build the units, it could provide free transport from Mill City boardinghouses to the mines.[54]

To the cost-conscious field manager, Ott Heizer, labor complaints were "small stuff." He reflected the hard-line attitude of most mine managers whose careers spanned the cheap-labor Depression era. Charles Segerstrom had a more pragmatic view. The company could not risk losing skilled workers during a mining boom. When Haesler expressed doubts about making any money after the government began setting ceilings on commodity prices, the Sonora boss told him not to worry. "They never will control prices. Labor is in the saddle and until you control labor you cant fix any price as nearly always over 60 percent of any article is labor. . . . Some of the boys just are getting out of their heads and thats the time to keep your head clear[.] I think we will get through this ok but dont play the game their way." At the time, labor costs at Nevada-Massachusetts were 30 percent of total operating costs.[55] The company could afford a more accommodating labor attitude.

Keeping a clear head also had economic advantages. Though ideologically opposed to any lingering New Deal social policies, Segerstrom welcomed the government's financial assistance for worker housing. The Federal Housing Authority (FHA), a Depression-era agency created originally for urban slum clearance and low-rent housing, after 1940 became the government's principal agency dealing with the wartime housing needs of defense workers. Administered through hundreds of local housing authorities across the country, the national defense housing program provided up to 90 percent of the funding for acquisition and construction of homes and rental units to meet the most critical shortages. Title VI of the U.S. Housing Act, added in March 1941, authorized the FHA to guarantee home loans with only 10 percent equity, and gave high-priority supply ratings to builders of such homes.[56]

Segerstrom contacted housing officials soon after learning of the labor unrest in his mines. Early in 1942, with tungsten demand critical, they agreed to construct forty units at Mill City and twenty more at Golconda if the company would provide electric power and water. When the first units opened in September 1942, Segerstrom wrote George F. Sawyer that they were "very fine buildings with hardwood floors, modern plumbing, two bedrooms, and good enough for any one who might live in any part of the United States." The remaining units were ready by October. At a total cost to taxpayers of

$439,000, these small cottages gave miners adequate shelter, but did little to resolve a mine labor shortage that continued to fester in the coming months.[57]

By compromising on some wage and hour demands, and with help from federal grants for workers' housing, Segerstrom managed to keep crews at work on a limited schedule. Labor costs at Nevada-Massachusetts rose appreciably as a result, climbing by 50 percent between 1941 and 1943. Yet inland mines continued to lose skilled labor to higher-paying West Coast positions. The volume of processed ore at Mill City dropped steadily, down by 30 percent in August 1942 and another 50 percent in September. With shipyards at Mare Island and South San Francisco paying up to fifty cents more per hour than inland mines, Segerstrom was desperate to find "any one who wants to go out on the desert." In September 1942 the War Labor Board stepped in, ordering defense workers to stay in their strategic jobs. By that time, however, Heizer's underground crew at Mill City had shrunk to just seventeen workers.[58]

The job freeze helped stop the labor drain but did nothing to replace workers who had forsaken their old jobs. Replacing men with machines had often been a management ploy in response to labor shortages in American industry. During the 1930s, however, cheap labor had delayed technological innovation in western mining, especially among smaller independent companies. The situation changed rapidly as skilled labor grew scarce during mobilization. By 1941 labor-saving tools and equipment were in great demand, but they were also in short supply, and after Pearl Harbor nearly impossible to obtain. Early in 1941 Heizer had started an open cut at Golconda, in part because "we can handle the material with a shovel far less than men can by hand underground," as Segerstrom explained to George Emery. Yet even with a high-priority rating from the War Production Board, it took the company nearly five months to acquire a modern power shovel for the Golconda pit. Heavy loaders, electric locomotives, mechanical muckers, and other labor-saving devices did not arrive at the company mines on a major scale until after the government began easing restrictions on industrial equipment and supplies late in the war.[59]

The labor crisis nearly closed Nevada-Massachusetts just as tungsten demand neared its wartime peak. Frantically searching for employees, Segerstrom kept up a torrid pace of travel and telegrams to federal officials in San Francisco and Washington, D.C. The workforce, he complained, was "leaving so fast that we cannot keep our ore in the mine broken fast enough to keep the mill running at full capacity." In November he managed to hire nine men released through the army furlough program, but one never showed up at the mine, and two others disappeared "immediately upon arrival." A month later he wrote the War Production Board that without at least fifty more men the mine would shut down in thirty days. To attract them he asked to raise wages by 15 percent to ninety-four cents an hour, still within the "Little Steel" formula but against the administration's anti-inflation guidelines. Metals Reserve took the request to the War Labor Board. While it was still pending, early in 1943 a few more employees trickled in, enough to stave off closure but not to materially increase productivity. The mine boss thought he knew the reason why. He told WPB officials that "miners seem to be scarce and those who do come are of such poor grade that it would take two or three men to do an ordinary man's work."[60]

With tungsten still on the government's high-priority list, in April 1943 Segerstrom's petition got a boost from Ira B. Joralemon, a San Francisco field representative for the Metals Reserve Company. After inspecting the Mill City district, he wrote a lengthy report praising the company's management and operations, but underscoring the importance of lifting the wage ceiling to attract better-quality personnel. "It is absolutely essential to the continued life of the greatest tungsten district in America that more men be secured," he concluded.[61]

That summer the War Labor Board granted Segerstrom's request for a wage increase, giving the Mill City division a recruitment advantage over other Nevada mining companies that paid less. They promptly filed a complaint. Coming at the same time as the threat of a coal strike by the United Mine Workers under John L. Lewis, the western agitation added inflationary pressures the administration tried to contain. Within the 15 percent guidelines of the Little Steel formula, however, the War Labor Board granted increases to adjust regional wage discrepancies if strategic industries were involved. Rural mine wages rose modestly as a result, but they never caught up with the urban pay scale in steel mills, shipyards, or other defense plants.[62]

The pay raise at Nevada-Massachusetts, combined with an ambitious recruiting campaign, brought in thirty more men. All were far above draft age, some in their seventies and eighties. Segerstrom was grateful for anyone he could get, even if "most of them would be good subjects for an old man's home," he wrote a WPB official. Putting up with decrepit workers seemed only a temporary inconvenience by the fall of 1943. With many government strategic stockpiles overflowing and many marginal mines closing, he thought the labor shortage was nearly over. "Remember this, that the heat will soon be off everything," he confidently teletyped his son Charles H. Jr. from Washington.[63]

This was a business view that could not anticipate the voracious manpower demands of the U.S. military as America and its allies opened new offensives in Europe and the Pacific. Indeed, the labor shortage in metal mines worsened instead of improved after 1943, and remained problematical even after the fighting stopped. Under pressure from the army as well as the public, the War Manpower Commission tightened eligibility rules and reclassified many technical and managerial staff formerly deferred. After February 1944 all eligible men between eighteen and twenty-six were to be drafted unless they obtained a certificate from the military attesting that their civilian job was vital to the war effort. That was still not enough to fill the ranks for the combined assaults on Germany and Japan. With the consent of General Hershey and the War Department, in December 1944 Byrnes issued controversial "work or fight" orders. Any American male under age thirty-eight who had left a vital job or had changed jobs without authorization, or who was not working in an essential industry, was subject to the draft. Only twelve thousand men were drafted under this order, however, and most of those were subprime candidates used in noncombat service.[64]

Public opinion welcomed the new restrictions on deferrals, but business leaders tried to protect key managerial and technical positions.[65] As a member of the Tuolumne County Draft Board, Charles Segerstrom saw firsthand the opposing polarities of egalitarian resistance to occupational deferrals and the elitist argument that exemptions were necessary to maintain high levels of industrial productivity and efficiency. As president of a major mining company he also faced conflicting personal and professional responsibilities.

Joseph Haesler's predicament exemplified the hard choices involved under the new Selective Service rules. After Congress lowered the draft age to eighteen, the Manpower Commission clamped down on student deferrals. Among the thousands of college students affected was Haesler's eldest son, Joseph J. Haesler Jr., whose status as a Notre Dame freshman and member of the Naval Reserve Officer Training Corps (NROTC) no longer exempted him. In January 1943, just before his own service as chief ore buyer for the Board of Economic Warfare, the elder Haesler teletyped his friend Segerstrom in California. He sought the latter's help in securing a Nevada congressional appointment to Annapolis for the young man. A few days later the Sonora executive replied that he had "taken the matter up" with Nevada's three congressional delegates, but they "all had long waiting lists of boys from Nevada who would rather go that way than enlist in the Navy." Young Joe remained in the NROTC at Notre Dame until 1945, then went on active duty at a weather and aircraft refueling station in the Far East during the last big push against Japan. Both he and his father served with honor, and managed to survive the war intact.[66]

Segerstrom had other deferment problems to worry about. His own son, Charles H. "Tod" Segerstrom Jr., was prime draft age. Tod had risen swiftly into higher management after receiving an MBA from Harvard in 1937. During the war his position as executive vice president with Nevada-Massachusetts was enough to justify his 2-B classification by the Tuolumne County Draft Board. Despite War Department objections, the War Manpower Commission resisted efforts to end individual deferrals based on occupational status. A journalist succinctly summed up the WMC rationale: "Raiding war plants to get soldiers would disrupt war production." By 1944, however, with military personnel needs rising and production goals easing for many metals, the draft was an ominous threat to midlevel managers like the younger Segerstrom. His father's inquiry to a friend in the War Production Board was little consolation. The Selective Service "cannot consider tungsten in a critical position," he reported, "and the best that could be expected would be a brief deferment for a critical position to allow time to obtain a replacement." Despite that opinion, the senior Segerstrom's periodic affidavits to the Selective Service over the next eighteen months attesting to his son's crucial importance to company operations were persuasive. Tod was 2-A when the war ended.[67]

Both Segerstroms labored hard to rescue Merrill Cronwall from the draft. He managed the chemical laboratory at Golconda. Temporarily classified 2-B in 1941 by his hometown draft board in Los Angeles at the urging of the senior Segerstrom, Cronwall faced an uncertain future. Every six months the L.A. board reviewed his status and required employment affidavits from his superiors, a duty that fell mostly on the younger Segerstrom. Tod's declarations of Cronwall's vital role kept him deferred until the great military buildup to D day. In March 1944 the L.A. board reclassified Cronwall 1-A, prompting the senior Segerstrom to step in with an appeal declaring that the company "will be forced to shut down our whole operation" unless it was "allowed to keep our key men." He insisted that it "would be impossible to replace Mr. Cronwall at this time. . . . Employment agencies, the U.S.E.S., and scientific schools and bureaus are unable to supply any one with Mr. Cronwall's experience, nor can they supply any chemist or assayer to train." The appeal won over the L.A.

board. It changed Cronwall's classification back to 2-B, which he retained until the end of the war despite tungsten's strategic downgrading from "critical" to "essential" early in 1945. In all of this effort on Cronwall's behalf, Selective Service personnel never learned that the young man was the brother-in-law of the elder Segerstrom's daughter, Martha.[68]

The draft had only an indirect impact on the retirement of Ott F. Heizer, Segerstrom's venerable general manager. In a period of labor scarcity, Heizer's stubborn resistance to unionism hurt recruitment and retention. After a site visit at Mill City early in 1943, the younger Segerstrom had teletyped his father that "the tales we heard out West are now echoes in Washington. . . . [W]e must find some way to . . . bring things up to date as far as men go. And that they too should have everything we can do within reason. I know its tough. But . . . [Heizer] just wont cooperate They bring up the cabins the beds etc water in the change rooms Better air in the mines And a lot of small stuff that wouldnt make much difference in cost But which of course he wouldnt do except over his dead body."[69]

A lack of confidence in Heizer's assessment of ore reserves at the Mill City mines added to the momentum leading to resignation. His increasingly negative field reports after 1939 had been partly responsible for the company's expansionist efforts at the beginning of the war. Charles H. Segerstrom had no reason to question his general manager's judgment until the spring of 1943, when he talked with Ira B. Joralemon after the latter's site inspection for Metals Reserve. Heizer had "cried for the past three years that the mine was all done," the elder Segerstrom wrote his son, but now "Jorolemen [*sic*] says he would bet a lot that between the Stank and Springer we have a lot of ore." As postwar developments proved, that turned out to be a good bet. But even without new discoveries, Joralemon's upbeat assessment increased leverage for a management change. To ease Heizer out, the Segerstroms delayed the move until the following spring, and then offered him a consulting job until he could find a suitable home in the East. The official announcement attributed his retirement to the "serious illness of Mrs. Heizer." His replacement was a Mackay School of Mines graduate, William G. Emminger, who had worked for Nevada-Massachusetts for a decade, first as the frustrated superintendent at Silver Dyke, then as plant manager at Mill City.[70]

Labor problems at Nevada-Massachusetts persisted through V-J day and beyond. Saving a few vital staff members did not solve the larger workforce shortages that kept many domestic mines and plants running at half speed or less in the last year of war. Early in 1945 the War Manpower Commission added to Segerstrom's production headaches by reducing tungsten's high labor priority. The reclassification came in response to the War Production Board's revisions to its list of designated strategic minerals, prepared after inventories showed stockpile surpluses of many metals. After January 1945 industrial workers ages twenty-six through twenty-nine were deferred only if they helped produce "critical" metals—iron, copper, lead, zinc, mercury, molybdenum, and vanadium. Those working in metals designated only as "essential"— tungsten, tin, aluminum, manganese, and chromium—were subject to reclassification as 1-A.[71]

Segerstrom and his allies fought the order with appeals to mining officials on the War Production Board, using old national security arguments reinforced by more immediate military demands. During the first six months of 1945 the War Department rushed to increase its inventory of tungsten carbide "pegs" for projectile cores in anticipation of the great battles looming on enemy soil. In April the chief spokesman of the tungsten industry received welcome news that the gray metal was returned to the "critical" list. "It is hoped," said the WPB announcement, "that the recommended improvement in rating will prove of some assistance to the tungsten mines and help them both to maintain and increase their level of employment." That proved to be wishful thinking. Four months later the war was over, and the "peg" program ended before it could have much economic impact on the domestic tungsten industry.[72]

Contrary to management expectations, war's end did not reduce labor scarcity at Nevada-Massachusetts and other inland mines. After the government rescinded its 1942 ban on gold mining, many former gold miners left the desert to return to their old gold prospects in California. Nor did western mines escape the postwar labor troubles that affected steel, auto, coal, and other major industries. Across the country the labor movement, freed from its unhappy wartime pledge of cooperation, rose in a crescendo of pent-up demand for better wages and hours.[73]

Segerstrom blamed the New Deal's progressive labor policies for much of the unrest—a reflection of the conservative backlash that soon coalesced into a political manifesto, the Taft-Hartley Act of 1947. The labor situation, he complained in September 1945, "instead of getting better, has steadily grown worse. . . . The men who have been let out at various plants, instead of going to work, are sitting around twiddling their thumbs and collecting unemployment insurance." He reiterated the theme early in 1946, the worst year for strikes in American history: "It is not only the demand for higher wages but the inefficiency of the . . . men who think that they should have double wages for doing a half days work."[74]

Through retrenchment, reorganization, and technological innovation using economies of scale, Nevada-Massachusetts survived the postwar labor troubles. Under the direction of Charles H. Segerstrom Jr. the company entered a period of consolidation, followed by massive new development and productivity that accelerated during the Korean War. By that time the senior Segerstrom no longer worried about labor unrest. During the Phelps-Dodge strike that threatened other mines a few months before his death in 1946, he was both stoically resigned and characteristically optimistic in one of his last letters to George Emery of the Boston board: "These things have always adjusted themselves in the past and they will adjust themselves in the present instance. . . . I am certain that the future holds something good in store for us in the tungsten business."[75]

World War II proved the efficacy of Charles H. Segerstrom's managerial skills. As tungsten spokesman and business leader, he benefited from the experience gained in war and peace over a forty-year career. Though hostile to big government in general and the New Deal in particular, he set aside personal predilections to cultivate a good working relationship with the "dollar-a-year" men and technicians in three federal agencies, the War Production Board, the War Manpower Commission, and the Office of Price Administration. At the same time, though beset by labor and material shortages, technical difficulties, high costs, declining ore grade, and other problems, he adroitly managed his own companies. At war's end Nevada-Massachusetts was the only major tungsten producer still operating, although milling less than 25 percent of the ore it had turned out two years earlier. Net profits in 1945 were only 20 percent of those at the height of profitability in 1942, yet the company was pared down and cost-efficient, ready to take on the postwar world in much better shape than its predecessor had been at the end of the Great War.

Late in 1943 Segerstrom had received a personal letter from Glenn J. Degner, a former OPA official who had resigned to join the navy. After thanking the tungsten leader for his wartime efforts, Degner summed up his work with industry over the past two years:

Unfortunately, because of [the] . . . technical nature [of strategic metals], the general public will probably never come to appreciate the great contribution which the ferro alloy industry has and is making to the winning of the war. I regard it as one of the highest points in my experience to have been privileged to work with you and other members of the industry. The record of the industry speaks for itself. There was never any question of "too little, or too late." At the same time, the industry did not capitalize on its strategic position. Instead, it made available to all of the United Nations an abundant production at most fair and reasonable prices. To have been, as I was, a close observer of this remarkable performance has been a real privilege.

Gratified by Degner's praise, the Nevada-Massachusetts executive responded that he believed "time has demonstrated that we all have worked for the best interests of our country."[76]

Degner's remarks may not be broadly representative of the government-business relationship after 1941, but at least they suggest that for some industry leaders, including Charles H. Segerstrom, patriotism and profitability worked well together during the worst years of war. Though the latter years were less profitable, Segerstrom remained the cautious optimist. At the cusp of the aerospace age, he predicted a rosy future for tungsten. "Tungsten has found a new place in the mettalurgical [sic] world," he wrote the Boston banker George Sawyer in mid-1945. "There are so many new uses that I feel we should be able to continue to produce as long as any mine in the United States."[77]

EPILOGUE

As a corporate entity, the Nevada-Massachusetts Company lasted more than fifty years. For the first thirty Charles H. Segerstrom Sr. was president and chief executive officer. He essentially raised the company out of the ashes of its predecessor, Pacific Tungsten, and he remained its primary source of inspiration and leadership through the critical thirties and early forties. Under his management the company successfully anticipated and adjusted to dynamic shifts in macroeconomic business trends. Learning important lessons from the chaotic aftermath of World War I, Segerstrom built a solid business foundation that held steady through the economic turbulence of depression and war. Though politically conservative and philosophically opposed to government intervention in private or corporate activities, he adapted pragmatically to New Deal policies and programs when they directly benefited him or his company. At the same time he learned enough about the political system to develop strategies for influencing decisions at the highest level. His efforts on behalf of tungsten producers made him a national spokesman for the industry and an effective lobbyist for domestic mining causes.

Working twelve to fourteen hours a day six days a week gradually took its toll. Exhausted and frequently brought down by illness during the war years, the elder Segerstrom died on August 2, 1946, leaving the family assets in the hands of his heirs. His wife and five children consolidated their business interests in the immediate postwar years, buying out nonfamily shareholders, closing marginal operations, and reorganizing the remaining properties under a family management led by the late president's eldest son, Charles "Tod" Jr.[1]

Unlike the tungsten market slump after World War I, world tungsten prices dipped slightly in 1946 but then rose again on the postwar industrial and military demand for high-speed tools and high-temperature superalloys in the auto, aerospace, and electronics industries. Prices skyrocketed when the U.S. government began an intensive effort to fill newly established stockpile quotas after China fell to the communists in 1949 and the Korean War opened a year later. With Chinese wolfram sales restricted, Nevada-Massachusetts and other domestic producers did a booming business. Between 1951 and 1957, while the government's intensive buying program was in effect, tungsten prices remained more than double that of the late 1940s.

Succeeding his father as president of Nevada-Massachusetts, the younger Segerstrom skillfully maneuvered family interests to meet these changing postwar social and economic conditions. Labor strife was a serious issue in western mining during the boom-and-bust cycles of the late 1940s and 1950s. Despite the efforts of W. G. Emminger, Merrill Cronwall, and Eldridge Nash, all managers more sympathetic to worker complaints than their predecessors, Nevada-Massachusetts faced several work stoppages until Tod bowed to union demands and recognized the United Mine Workers as the "sole and exclusive bargaining agent" for all but technical, clerical, and supervisory employees. Union success was short-lived, however. All the company mines shut down less than eighteen months after signing the new contract.[2]

While metal prices were still attractive, Tod diversified the family's mining interests. The Rare Metals Corporation remained on the books as a family holding company, but the mill and most of its operating mines were reorganized under new subsidiaries. As we have seen earlier, Tod and Ott Heizer's son John Madden Heizer, a graduate of the Mackay School of Mines, operated the Nightingale and Star mines as well as the Long Lease mine near Lovelock under a partnership known as the Wolfram Company. In the 1950s they shipped ore processed at the Toulon mill to eastern buyers, using the same Metal and Ore brokerage as Nevada-Massachusetts. A decade later, with prices depressed, the partnership dissolved after selling the mill and other assets. Tod and John Heizer also teamed up to lease an iron mine in the Buena Vista Hills of Pershing County, later subleased to the Dodge Construction Company. Most of the estimated half-million tons of ore produced in the decade after World War II went to Japan, helping spur that nation's postwar steel industry.[3]

Nevada-Massachusetts remained under control of the Segerstrom family through the postwar boom, but company prospects faded quickly in the late 1950s. Confronted by a combination of lower prices, slowing markets, declining ore grades, and changing federal policies, the Segerstroms shut down all tungsten operations in 1958. Merrill Cronwall, who had taken over as general manager after W. G. Emminger retired, advised the family to hold on to the property until the market improved, and then sell to the highest bidder. He thought it unlikely the company could ever resume operations and compete with such giants as Union Carbide and the Tungsten Mining Corporation, the latter a North Carolina producer about to merge with Howe Sound, a diversified holding company with mines and other properties in Canada and the United States.[4]

This was shrewd advice. In the uncertain strategic climate of the late Eisenhower years, the Segerstroms stopped mining but hedged their bets. After dissolving Nevada-Massachusetts and the Wolfram Company, Tod, along with his brothers William and Richard, organized Tungsten Properties, Ltd. The new partnership bought up the liquidated assets of Nevada-Massachusetts, kept the mines in good condition with maintenance crews, and began a new exploratory program. Whether the Segerstroms intended to reopen is not clear, but their efforts eventually attracted a buyer. In 1970 General Electric signed a contract to purchase any tungsten concentrates they produced, and provided the financing for a new tailings plant. By the mid-1970s, GE had taken an 80 percent ownership share in the property. Utah International, an operating company partnering and later merging with GE, acquired the remaining 20 percent. After sixty years of financial involvement, the Segerstrom family had quietly ended its stake in Nevada tungsten.[5]

STRATEGIC METAL POLICY AFTER WORLD WAR II

The senior Segerstrom was too ill to participate directly in the development of a postwar national strategic minerals policy. That idea evolved gradually, as we have seen, starting with a series of congressional hearings involving implementation of the nation's strategic mineral stockpile program inaugurated by the Thomas bill in 1939. It took a major step forward in 1946. Though he died the same year, Segerstrom's belief that the nation's strategic mineral production companies must be protected and sustained was reflected in the provisions of the Strategic and Critical Materials Stockpiling Act, which President Truman reluctantly signed on July 23, 1946. Responding to reports showing the war had depleted many vital resources, Truman said stockpiling was "in the national interest." Yet he opposed the "Buy American" clause and the restrictions on disposal of surplus materials without the prior consent of Congress. Most authorities have been equally critical of the 1946 legislation and its protectionist implications. In practice, however, the act's loopholes gave procurement officers plenty of room to acquire foreign materials when necessary.[6]

For nearly two decades after World War II, despite the objection of internationalists and their allies, domestic mining priorities weighed heavily in federal mineral policy choices. Cold war fears of Soviet strength and intentions added a sense of urgency to appeals in support of America's defensive capabilities. The high tungsten duty ended in 1948 at the Geneva trade talks, but domestic mineral producers continued to use national security as justification for federal economic intervention on their behalf. From the Berlin crisis through the Korean War, assumptions about stockpiling strategic materials and protecting the nation's mineral industry were important components of military planning. By reiterating national defense needs, by intensifying the exploration and development of domestic ores, and by continuing to pay premium prices for domestic copper, lead, zinc, tungsten, and other strategic metals, federal programs offset lower prices of foreign ores and helped subsidize domestic producers. That domestic mining became so dependent on federal support was an ironic outcome for classical capitalists like Charles H. Segerstrom, so openly hostile to the political and social implications of Big Government. But the disruptive decades of war and depression had forced him to forsake ideology for the practical needs of an industry in trouble.[7]

The active history of the Nevada-Massachusetts Company ended in the late 1950s, but issues involving the national stockpile still trouble us today. Between the Korean War and the events of 9/11, the stockpile was regarded as one of the biggest political boondoggles in Washington. Stockpile issues have provided political ammunition for party leaders on both sides of the aisle. Disputes over the nature, cost, extent, and even justification for stockpiling have plagued every administration from Truman to the second George Bush. Over the decades the list of stockpiled materials has changed as strategic concepts changed, but political and economic issues frequently got in the way of rational planning. In the 1960s one congressman from Long Island told a reporter, "The only strategic material in my district happens to be duck feathers, and you'd be amazed at the passion with which duck farmers tell me that we haven't got near enough duck feathers in the inventory." In the 1970s the stockpile, said one journalist, was the "mother lode for mining industry representatives and lobbyists."[8]

Disagreement over war strategy has also affected the nature and extent of the stockpile. Internal policy differences and budget cutbacks hindered stockpile development in the late 1940s, but communist fears and the Korean War in the next decade led to massive new and expensive government mineral purchases from both domestic and foreign suppliers. Both the Truman and the Eisenhower administrations recognized America's growing dependence on foreign strategic resources, but the domestic mining lobby remained powerful enough to seek protection and subsidies, especially in times of crisis. In the 1950s stockpile surpluses in copper, lead, and zinc arose in part because of efforts to aid domestic mining.[9]

After Sputnik in 1957 revealed the vulnerabilities of America's nuclear defense, the Pentagon won administration support to reduce war planning to a three-year scenario from previous "long war" estimates. Ike remained convinced that stockpile reserves were "good investments," but what to do with the surplus led successive administrations to manipulate the stockpile for economic and political purposes. During the inflationary Vietnam era, President Johnson sold surplus aluminum from the stockpile to keep market prices in check. President Nixon played the same game after pressuring the Pentagon to rethink war strategy and come up with a one-year war plan, thus creating even greater stockpile superfluities. Congress later settled on a three-year stockpile requirement regardless of what war plan was in vogue.[10]

Stockpile strategy changed again in the late 1970s and '80s in the wake of OPEC and the international oil crisis. Inflationary pressures and new fears of global resource shortages motivated the Ford and Carter administrations. President Ford restored the three-year war contingency plan in 1976, and his successor supported new stockpile funding. Washington war hawks in the Reagan years renewed stockpile purchases for cobalt, vanadium, oil, and other minerals. Industry spokesmen and nationalists in the Reagan administration increased speculative interest in "hot" strategic metals like germanium and titanium, and encouraged business-friendly neoconservatives to propose expanding mineral exploration and development on public lands previously declared off-limits. By the late 1980s, however, China's reentry into the international trade in tungsten, manganese, and other minerals drove down metal prices and reduced international demand. After 2001 metal prices again started upward, partly in response to the "global war on terror," but more significantly as a result of China's precipitous growth and insatiable demand for raw materials. By 2005 U.S. tungsten consumers, almost wholly dependent on foreign sources, faced both high prices and supply shortages after China again began restricting tungsten exports.[11]

These ever-changing geopolitical complexities and contradictions have often frustrated policy makers and planners in both government and industry.

One analyst in the late 1980s concluded that the "pursuit of a national non-fuel minerals policy is an exercise in futility." Yet strategic stockpiling remains an integral part of the U.S. Code. Though amended many times since its New Deal origins, the law still pays lip service to the mining industry by retaining language reminding us that strategic stockpiles are necessary to protect America from a "dangerous and costly dependence" on foreign sources in times of emergency. As of 2007, however, only twenty-eight strategic materials were still being stored in the national stockpile, twenty-five of which were available for sale on the open market, including eight million pounds of tungsten ore and concentrates, the largest single item in the inventory.[12]

Direct government subsidies and price supports for domestic mining ended after 1958, but industry allies in Congress have subsequently tried to protect domestic producers by extending and strengthening Buy American laws. The U.S. Code still requires government procurement agents to buy "only such unmanufactured articles, materials, and supplies as have been mined or produced in the United States." However, like earlier versions, exceptions in the law make the restrictions meaningless. Furthermore, since 1996, World Trade Organization agreements require signatories to waive discriminatory procurement policies. The State Department also has made good use of the "memorandum of understanding," a well-established but informal principle in international law. An MOU, now applied to at least twenty-one trading partners around the world, regards defense materials produced in those countries essentially as "American made."[13]

Whatever the long-range impact of twenty-first-century conflicts on the nation's defense strategy, it seems safe to suggest that stockpile policy will continue to reflect the internationalist influence that has prevailed since the 1950s. Most Americans now recognize that the United States is dependent on other nations to keep it supplied with many vital materials. Though the domestic mining industry is still active, most of our strategic metals now come from abroad, and that situation is not likely to change in the foreseeable future.

NOTES

KEY TO SOURCE ABBREVIATIONS

AMC	American Mining Congress
BOMMY	Bureau of Mines *Minerals Yearbook*
CR	*Congressional Record*
EMJ	*Engineering and Mining Journal*
HSUS	U.S. Bureau of the Census, *Historical Statistics of the United States*
M&SP	*Mining and Scientific Press*
MI	*Mineral Industry*
NMC	Nevada-Massachusetts Company
NYT	*New York Times*
PTC	Pacific Tungsten Company
R-M	*Review-Miner* [Lovelock, Nev.]
SAUS	U.S. Bureau of the Census, *Statistical Abstracts of the United States*
SC	Segerstrom Collection
TT	teletype
UOPWA	Holt-Atherton Library, University of the Pacific
USGSMR	*Mineral Resources of the United States*
WMRC	War Minerals Relief Commission
WSJ	*Wall Street Journal*

KEY TO NAME ABBREVIATIONS

ACD	A. C. Daft
CGF	Colin G. Fink
CHS	Charles H. Segerstrom
CHSJR	Charles H. Segerstrom Jr.
COH	Charles Oscar Hardy
EAC	Edward A. Clark
ERH	Edward R. Hagenah
GFS	George F. Sawyer
GIE	George I. Emery

HDS	H. DeWitt Smith
IBJ	Ira B. Joralemon
JCT	Josephus C. Trimble
JDC	Julian D. Conover
JFC	J. F. Callbreath
JFD	John F. Davis
JGC	John G. Clark
JJH	John Joseph Haesler
JKG	J. K. Gustafson
MH	Marx Hirsch
MKS	Miles K. Smith
OFH	Ott F. Heizer
PEC	Philip E. Coyle
PFK	Paul F. Kerr
RCM	Roy C. McKenna
RGE	Robert G. Emerson
RMM	Robert M. Morgan
WJL	William J. Loring

PREFACE

1. *Las Vegas Review-Journal,* 21 Aug. 2002; John D. Schell, "Tungsten Alloy and Cancer in Rats: Link to Childhood Leukemia?"; "Kennametal Inc.," *International Directory of Company Histories,* vol. 68 (1997).

2. *Chicago Tribune,* 27 Nov. 2006; S. 3627, introduced 29 June 2006, Library of Congress.

3. K. C. Li and Chung Yu Wang, *Tungsten: Its History, Geology, Ore-Dressing, Metallurgy, Chemistry, Analysis, Applications, and Economics,* 205–40.

4. W. C. Balke, "The Story of Tungsten"; Charles H. Segerstrom, "Operations in Milford District"; *Manual of Mineralogy,* 21st ed., 222–23, 430–31; L. P. Larson et al., "Availability of Tungsten at Various Prices From Resources in the United States," 3–7; Anthony P. D. Werner et al., "International Strategic Mineral Issues, Summary Report—Tungsten," 12–13; U.S. Geological Survey, "Tungsten Statistics and Information," in *Commodity Statistics and Information,* 180–81.

5. Werner et al., "International Strategic Mineral Issues"; "Incandescent Light Bulb," Wikipedia online.

6. U.S. Code, Title 50, Sec 98a, chap. 5; Defense National Stockpile Center.

7. Michael Shafer, "Mineral Myths"; Peter Trubowitz, *Defining the National Interest: Conflict and Change in American Foreign Policy.* On economic integration and multinationalism in steel before World War I, see Carl Strikwerda, "The Troubled Origins of European Economic Integration: International Iron and Steel and Labor Migration in the Era of World War I." For the impact of culture on technology, see Thomas J. Misa,

A Nation of Steel: The Making of Modern America, 1865–1925, 265–66. The conflicted resource question is explored in Harold J. Barnett, "The Changing Relation of Natural Resources to National Security"; Philippe Le Billon, "The Geopolitical Economy of 'Resource Wars'"; and John M. Dunn, "American Dependence on Materials Imports the World-Wide Resource Base." On determinism, see Stephen Frenkel, "Geography, Empire, and Environmental Determinism"; and Paul C. Ceruzzi, "Moore's Law and Technological Determinism: Reflections on the History of Technology."

1 || STEEL ALLOYS AND THE RISE OF MODERN INDUSTRY

1. Jared Diamond, *Guns, Germans, and Steel: The Fates of Human Societies*, 239–49; William O. Vanderburg, "Mining and Milling Tungsten Ores," 22–44; Werner et al., "International Strategic Mineral Issues," 6–13.

2. J. R. Partington, *A History of Chemistry*, 171. Oddly, Georgius Agricola's magnum opus, published in 1556 and translated in 1910 by Herbert Clark Hoover and Lou Henry Hoover, does not mention wolfram or its origins, and neither do Hoover and Hoover, despite their extensive footnotes on metallurgy of base and precious metals. See *De Re Metallica*.

3. Cyril Stanley Smith, "The Discovery of Carbon in Steel," 159–63; Will Durant and Ariel Durant, *The Age of Voltaire*, 524–25; Uno Boklund, "Scheele, Carl Wilhelm"; Li and Wang, *Tungsten*, 1–2; Partington, *A History of Chemistry*, 217.

4. Ramon Gago, "The New Chemistry in Spain," 172.

5. Smith, "Discovery of Carbon in Steel," 149–50.

6. Geoffrey Tweedale, "Metallurgy and Technological Change: A Case Study of Sheffield Steel and America, 1830–1930," 189–99.

7. Ibid., 191–98; J. Gordon Parr, "The Sinking of the *Ma Robert*," 211–13.

8. Parr, "Sinking of the *Ma Robert*," 213–17.

9. Langdon White, "The Iron and Steel Industry of the Pittsburgh District"; Gerald G. Eggert, *The Iron Industry in Pennsylvania*, 15–26, 79–80.

10. Tweedale, "Metallurgy," 199–210. See also Christopher Freeman and Francisco Louçã, *As Time Goes By: From the Industrial Revolutions to the Information Revolution*, 232–33; and Ronald H. Limbaugh, "Making Old Tools Work Better: Pragmatic Adaptation and Innovation in Gold-Rush Technology."

11. Eggert, *Iron Industry in Pennsylvania*, 2–5, 27–63; Joseph Frazier Wall, *Andrew Carnegie*, 227–306, 471–76; White, "Iron and Steel Industry," 116–19; E. Willard Miller, "The Industrial Development of the Allegheny Valley of Western Pennsylvania," 390–400; Peter J. Dunn, *The Story of Franklin and Sterling Hill*, 5–13.

12. Bela Gold et al., *Technological Progress and Industrial Leadership: The Growth of the U.S. Steel Industry, 1900–1970*, 533–35; Robert B. Gordon, "The 'Kelly' Converter"; Misa, *Nation of Steel*, 19–21.

13. Wall, *Andrew Carnegie*, 311–20; Freeman and Louçã, *As Time Goes By*, 233–34.

14. Arthur H. Hiorns, *Principles of Metallurgy*, 185; U.S. Geological Survey, *The*

Strategy of Minerals: A Study of the Mineral Factor in the World Position of America in War and in Peace, 108–9; Wall, *Andrew Carnegie*, 500–505; Misa, *Nation of Steel*, 22–28, 48–50; Aaron J. Ihde, *The Development of Modern Chemistry*, 466.

15. George Gilder quoted in Walter B. Wriston, "Technology and Sovereignty," 508.

16. Charles Singer et al., eds., *A History of Technology*, 64–65; R. Gordon, "The 'Kelly' Converter," 773–74; Tweedale, "Metallurgy," 210–14.

17. Geoffrey Tweedale, *Steel City: Entrepreneurship, Strategy, and Technology in Sheffield, 1743–1993*, 75–78, 95–98; Geoffrey Tweedale, "Sir Robert Abbot Hadfield F.R.S. (1858–1940) and the Discovery of Manganese Steel."

18. Tweedale, "Metallurgy," 214–16; John A. Vaccari et al., *Materials Handbook: An Encyclopedia for Managers, Technical Professionals, Purchasing and Production Managers, Technicians, and Supervisors*, 850–51.

19. Singer, *A History of Technology*, 64–65.

20. Geoffrey Tweedale, *Sheffield Steel and America: A Century of Commercial and Technological Interdependence, 1830–1930*, 69; Li and Wang, *Tungsten*, 255. For the purpose of Taylor's experiments and their connection with labor efficiency, consult Hugh G. J. Aitken, *Taylorism at Watertown Arsenal: Scientific Management in Action, 1908–1915*, 29–33. See also Edward Eyre Hunt, *Scientific Management Since Taylor: A Collection of Authoritative Papers*, 5–9.

21. Colin Fink, "Review of the Strategic Metals," 419–20; Tweedale, *Sheffield Steel*, 71–72; Tweedale, *Steel City*, 116–18.

22. Nathan Rosenberg, "The Direction of Technological Change: Inducement Mechanisms and Focusing Devices," 7–10.

23. Michael Sanderson, "The Professor as Industrial Consultant: Oliver Arnold and the British Steel Industry, 1900–1914," 587–88; Tweedale, "Metallurgy," 217–20; Sheffield University home page; WSJ, 17 Oct. 1916, 8; NYT, 18 Dec. 1929, 1; Frederick Betz, *Executive Strategy: Strategic Management and Information Technology*, 81–82.

24. Ralph D. Gray, *Alloys and Automobiles: The Life of Elwood Haynes*, 117–26, 145–57; Christopher Freeman and Luc Soete, *The Economics of Industrial Innovation*, 62; Tweedale, *Sheffield Steel*, 79–80.

25. Gray, *Alloys and Automobiles*, 126–43.

26. Leonard S. Reich, "Lighting the Path to Profit: GE's Control of the Electric Lamp Industry, 1892–1941," 306–18; Leonard S. Reich, *The Making of American Industrial Research: Science and Business at GE and Bell, 1876–1926*, 81.

27. W. Loach, "Developing the American Tungsten Industry"; Werner et al., "International Strategic Mineral Issues," 2; Frank L. Hess, "Tungsten in 1930," 181.

28. Li and Wang, *Tungsten*, 281.

29. Simon Ball, "The German Octopus: The British Metal Corporation and the Next War, 1914–1939," 452–62; Rondo E. Cameron, "Some French Contributions to the Industrial Development of Germany, 1840–1870"; Alfred D. Chandler Jr., "The Emergence of Managerial Capitalism"; Susan Becker, "The German Metal Traders Before 1914."

30. W. O. Henderson, "Germany's Trade With Her Colonies, 1884–1914"; William H. Dennis, *A Hundred Years of Metallurgy*, 24–27, 41–43, 269–70, 286–87; T. T. Read, "Historical Aspects of Mining and Metallurgical Engineering"; Cameron, "Some French Contributions," 296–304; Goran Ahlsrom, *Engineers and Industrial Growth: Higher Technical Education and the Engineering Profession During the Nineteenth and Early Twentieth Centuries: France, Germany, Sweden, and England*, 94–105; Arnold Pacey, *The Culture of Technology*, 19–20.

31. William O. Vanderburg, "Methods and Costs of Mining Ferberite Ore at the Cold Springs Mine, Nederland, Boulder Co., Colorado," 1–15; Vanderburg, "Mining and Milling Tungsten Ores," 2–5; Charles O. Hardy, "The Tungsten Market Situation in 1919"; J. W. Furness, "The Marketing of Tungsten Ores and Concentrates," ii, 1.

32. Edgar Crammond, "The Economic Relations of the British and German Empires"; M. J. Bonn, "The Nationalization of Capital," 254–57; Paul A. Papayoanou, "Interdependence, Institutions, and the Balance of Power: Britain, Germany, and World War I," 54–56; Steven E. Lobell, "Second Image Reversed Politics: Britain's Choice of Freer Trade or Imperial Preferences, 1903–1906, 1917–1923, 1930–1932," 677–78.

33. F. W. Taussig, *Tariff History of the United States*, 210–14, 256–59.

34. WSJ, 11 Nov. 1907, 6; 30 Nov. 1908, 1.

35. Taussig, *Tariff History*, 224; Abraham Berglund, "The Tariff Act of 1922," 22; Abraham Berglund, "The Ferroalloy Industries and Tariff Legislation," 251–53.

2 || TUNGSTEN IN WORLD WAR I

1. Alex Roland, "Science and War," 261–63; Marshall J. Bastable, "From Breechloaders to Monster Guns: Sir William Armstrong and the Invention of Modern Artillery, 1854–1880," 217–24; George Raudzens, "War-Winning Weapons: The Measurement of Technological Determinism in Military History," 417–24; Clive Trebilcock, "'Spin-off' in British Economic History: Armaments and Industry, 1760–1914," 483. Jared Diamond cautions us against assuming wars always advance technology in *Guns, Germs, and Steel*, 249–58. See also T. H. E. Travers, "The Offensive and the Problem of Innovation in British Military Thought, 1870–1915."

2. "Artillery: Construction and Use," in *The Times History and Encyclopaedia of the War*; WSJ, 12 Mar. 1917, 3; 14 Nov. 1917, 7; Bernard M. Baruch, *American Industry in the War: A Report of the War Industries Board (March 1921)*, 153–54; Berglund, "Ferroalloy Industries," 249.

3. Tweedale, *Steel City*, 71–78; Trebilcock, "'Spin-off,'" 483–85; Clive Trebilcock, "British Armaments and European Industrialization, 1890–1914," 256–58.

4. Richard J. Stoll, "Steaming in the Dark? Rules, Rivals, and the British Navy, 1860–1913," 268–73.

5. Thomas William Harvey, *Memoir of Hayward Augustus Harvey by His Sons*, 26–28; Karl Lautenschlager, "Technology and the Evolution of Naval Warfare," 12–14; Misa, *Nation of Steel*, 119–20.

6. Stuart D. Brandes, *Warhogs: A History of War Profits in America,* 113–14; Misa, *Nation of Steel.*

7. NYT, 12 May 1900, 6; Walter Millis, *The Martial Spirit: A Study of Our War With Spain,* 340; Walter Millis, *Arms and Men: A Study in American Military History,* 187–98; Michael Epkenhans, "Military-Industrial Relations in Imperial Germany, 1870–1914," 8–10; Papayoanou, "Interdependence," 55–56, 69–71; Misa, *Nation of Steel,* 118.

8. Archer Jones and Andrew J. Keogh, "The Dreadnought Revolution: Another Look," 124–29.

9. Strikwerda, "Troubled Origins of European Economic Integration"; Sidney Pollard, "Industrialization and the European Economy," 636–48; G. R. Searle, *A New England? Peace and War, 1886–1918,* 809; WSJ, 14 Nov. 1917, 7; John McDermott, "Trading With the Enemy: British Business and the Law During the First World War," 219.

10. Bonn, "The Nationalization of Capital," 261; NYT, 27 Dec. 1914, 4; WSJ, 18 Mar. 1916, 5; *Iron Trade Review* 61 (9 Aug. 1917): 293–94; Michael Pattison, "Scientists, Inventors, and the Military in Britain, 1915–19: The Munitions Inventions Department," 527–29; M. S. Birkett, "The Iron and Steel Trades During the War," 353–69; Ball, "German Octopus," 452–62.

11. WSJ, 18 Mar. 1916, 5; NYT, 3 May 1918, 17; Hew Strachan, *The First World War,* 1029. See also Dwight R. Messimer, *The Merchant U-boat: Adventures of the Deutschland, 1916–1918.*

12. NYT, 12 Oct. 1917, 3; 9 Nov. 1917, 9; 10 Nov. 1917, 4; 19 July 1918, 8; 23 July 1918, 5; CHS to GIE, 10 Oct. 1939. On the Red Scare, see Elizabeth Stevenson, *Babbitts and Bohemians: The American 1920s,* 58–62.

13. NYT, 16 June 1918, 48; WSJ, 17 July 1915, 5; David Stevenson, *Cataclysm: The First World War as Political Tragedy,* 195–97.

14. Donald M. Liddell, ed., *Handbook of Nonferrous Metallurgy: Recovery of the Metals,* 7–16; Sanderson, "Professor as Industrial Consultant," 591; Misa, *Nation of Steel,* 247–51; Gold et al., *Technological Progress,* 542–43.

15. Li and Wang, *Tungsten,* 163–64, 174–75, 192–201; Charles O. Hardy, "Marketing Tungsten Ores," 667.

16. WSJ, 14 Nov. 1917, 7; S. B. Ritchie, "Molybdenum in High-Speed Steel: The Elimination of Tungsten, a Strategic Material"; Segerstrom, "Strategic Minerals," 36–39.

17. NYT, 16 June 1918, 48; Tweedale, *Steel City,* 199–201; Tweedale, "Metallurgy," 221; Berglund, "Ferroalloy Industries," 257.

18. NYT, 11 Mar. 1913, 13; 21 June 1914, XX12; 6 Sept. 1914, X11; Roland Stromberg, "On Cherchez Le Financier: Comments on the Economic Interpretation of World War I," 435–37; Harry N. Scheiber, "World War I as Entrepreneurial Opportunity: Willard Straight and the American International Corporation," 493–95.

19. NYT, 16 Feb. 1915, 4; E. M. Bernstein, "War and the Pattern of Business Cycles," 528–29; Robert Sobel, *The Big Board: A History of the New York Stock Market,* 211–15. Wilson is quoted in Charles A. Beard and Mary R. Beard, *The Rise of American Civilization,* 628.

20. NYT, 3 Aug. 1916, 10; 26 Oct. 1916, 10; Tweedale, "Metallurgy," 220–21; Hope L. MacBride, "Export and Import Associations as Instruments of National Policy," 190–93.

21. NYT, 8 July 1915, 9; 25 July 1915, xx9; Gavin Wright, "The Origins of American Industrial Success, 1879–1940," 665.

22. Brandes, *Warhogs,* 135–36; WSJ, 14 Mar. 1916, 8; James Brown Scott, "The Black List of Great Britain and Her Allies," 834–39; Thomas A. Bailey, *A Diplomatic History of the American People,* 637–38.

23. WSJ, 17 July 1915, 1; 29 July 1916, 2; 12 Mar. 1917, 3; NYT, 11 May 1916; 7 Feb., 10 Apr. 1917; 6 Feb., 20 Nov. 1918.

24. WSJ, 6 Jan. 1916, 1; Hobert P. Sturm, "Webb-Pomerene Associations," 83–84; Mary A. Yeager, "Trade Protection as an International Commodity: The Case of Steel," 33–35.

25. Bailey, *Diplomatic History,* 638–41; John B. McMaster, *The United States in the World War,* 366–67; Arthur M. Schlesinger Jr., *The Crisis of the Old Order, 1919–1933,* 37–38; Herbert Hoover, *The Challenge to Liberty,* 108–11; Eric Goldman, *Rendezvous With Destiny: A History of Modern American Reform,* 237–38; Philip Read Bradley Jr., "A Mining Engineer in Alaska, Canada, the Western United States, Latin America, and Southeast Asia," 14.

26. Robert D. Cuff and Melvin I. Urofsky, "The Steel Industry and Price-Fixing During World War I," 292–95.

27. Baruch, *American Industry in the War,* 125–26; Samuel E. Morison and Henry S. Commager, *Growth of the American Republic,* 469–73; Robert D. Cuff, "Bernard Baruch: Symbol and Myth in Industrial Mobilization," 124–27; Cuff and Urofsky, "Steel Industry and Price-Fixing," 296–306; Robert Sobel, *The Age of Giant Corporations: A Microeconomic History of American Business, 1914–1992,* 10–12.

28. HSUS, 130, 143.

29. Brandes, *Warhogs,* 135–36; WSJ, 10 Apr. 1915, 6; 1 May 1916, 8; 15 May 1916, 6.

30. Dwight M. Lemmon and John V. N. Dorr, "Tungsten Deposits of the Atolia District, San Bernardino and Kern Counties, California," 207–9; Frank J. Wiebelt and Spangler Ricker, "Investigation of the Atolia Tungsten Mines, San Bernardino County, California," 2–3.

31. William Glenn Emminger, "Reminiscences," NC337.

32. NYT, 24 Sept. 1916, x12; Frank L. Hess, "Tungsten," USGSMR 1917 (1921), 931–33; Joseph M. Kurtak, *A Mine in the Sky: The History of California's Pine Creek Tungsten Mine and the People Who Were Part of It,* 18–33.

33. Frank L. Hess, "Tungsten," USGSMR 1915 (1917), 823–26; Frank L. Hess, "Tungsten," 1916 (1919), 789–94. For a table showing American and world tungsten production and prices, 1898–1930, see Hess, "Tungsten in 1930," 181.

34. Charles O. Hardy, *Wartime Control of Prices,* 119–40.

35. Furness, "Tungsten Ores and Concentrates," 19–20; WSJ, 17 July 1915, 1. The most important independents in the period were the Crucible Steel Company of America, with plants in New Jersey, Pennsylvania, and New York, and two Pennsylvania firms near the

rich Connellsville coal district in Latrobe, Vanadium Alloys Steel Co. and Latrobe Electric Steel Co. (Frank L. Hess, "Tungsten," USGSMR 1918 [1921], 986–88).

36. *Reese River Reveille,* 13 June, 17 June 1865, facsimile copy in Harold K. Stager and Joseph V. Tingley, "Tungsten Deposits in Nevada," 13–19. See also Francis Church Lincoln, *Mining Districts and Mineral Resources of Nevada,* 174–75; and Joseph V. Tingley, *Mining Districts of Nevada,* 85–86.

37. Donald I. Segerstrom, "A Brief History of the Mill City Tungsten Mines, Pershing County"; American Tungsten Association, "Tungsten, the Aristocrat of Rare Metals," 10–11; Paul F. Kerr, "Geology of the Tungsten Deposits Near Mill City, Nevada," 10–27.

38. C. H. Fry to WJL, 12 Apr. 1918; PTC Minute Book, 3 May 1918; WMRC, Reporter's Transcript, San Francisco, 24 July 1919, Claim No. 1018; D. Segerstrom, "Brief History"; M&SP 116 (Mar. 23, 1918): 421; Kerr, "Geology of the Tungsten Deposits," fig. 2.

39. NYT, 19 July 1916, 1; WJL to CHS, 20 May 1918; WMRC, 70; Hess, "Tungsten," USGSMR 1918 (1921), 974–75.

40. WSJ 25 June 1918, 6.

41. WJL to CHS, 4 May 1918; WMRC, 6. For Loring's career, see Ronald H. Limbaugh, "Making the Most of Experience: The Career of William J. Loring, Nevada Mining Engineer"; and Ronald H. Limbaugh, "'There Is a Game Against Us': W. J. Loring's Troubled Years as Bewick-Moreing Company's General Manager and Partner in Western Australia, 1905–1912."

42. "Tungsten's Meteoric Rise and Fall," *Iron Trade Review* 61 (9 Aug. 1917): 293–94; WSJ, 18 Mar. 1916, 5; 12 Mar. 1917, 3; 16 Aug. 1917, 8.

43. WJL to CHS, 4 May 1918; WMRC, 6–20; D. Segerstrom, "Brief History."

44. Mary Etta Segerstrom interview.

45. NYT, 13 Dec. 1917, 12; WSJ, 26 Feb. 1918, 9. On government encouragement to domestic producers, see WJL to CHS, 7 Aug. 1918; WJL to Local [Draft] Board of the County of Humboldt, Winnemucca, Nev., 26 Sept. 1918; H. Foster Bain to WJL, 29 Nov. 1918; and WSJ, 9 Nov. 1918, 6.

46. NYT, 27 Mar. 1918, 20; 12 Apr. 1918, 17; 1 May 1918, 24; 12 Sept. 1918, 13; 24 Nov. 1918, 20; Richard L. Lael, "The Pressure of Shortage: Platinum Policy and the Wilson Administration During World War I," 549–51.

47. WJL to CHS, 8 Mar. 1918; CHS to WJL, 3 June 1918.

48. PTC Minute Book, 3 Mar. 1918.

49. Ibid., 3 Mar., 18 May 1918; WJL to CHS, 19 Apr., 30 Apr., 4 May, 17 May 1918; WMRC, 23–43, 94–103.

50. WJL to CHS, 8 Mar., 17 May 1918; CHS to WJL, 24 May 1918.

51. CHS to WJL, 3 June 1918; WJL to C. W. Terry, 28 June 1918; WJL to Local [Draft] Board of the County of Humboldt, 26 Sept. 1918; PTC Minute Book, 18 June, 3 Aug., 20 Aug. 1918; WMRC, 50–94; WJL to CHS, 3 June 1924; CHS to Stockholders of PTC, 6 Nov. 1925; "Pacific Tungsten Company Is Sued," R-M, 10 Apr. 1925, 1.

52. WJL to Terry, 26 Apr. 1918; to CHS, 27 Apr. 1918; to J. H. Bell, 21 Aug. 1918; to Local Draft Board of the County of Humboldt, 26 Sept. 1918; CHS to WJL, 3 June, 8 June 1918.

53. CHS to Terry, 22 Apr. 1918; to WJL, 24 May, 3 June 1918; WJL to CHS, 30 Apr., 17 May, 20 May 1918; to Terry, 26 July 1918; to J. H. Bell, 4 Nov. 1918; Terry to CHS, 12 June 1918; PTC Minute Book, 20 Aug. 1918; Thiel Detective Service, Operative Reports, 7 Oct. 1918–Jan. 1919; NYT, 24 Sept. 1916, X12; WSJ, 25 June 1918, 6; Joseph Kurtak, "History of Pine Creek: A World Class Tungsten Deposit," 42–43.

54. PTC Minute Book, 20 Aug. 1918; WJL to L. A. Friedman, 14 Oct. 1918; to J. H. Bell, 4 Nov. 1918; "Pacific Tungsten Company Is Sued"; Hess, "Tungsten," USGSMR 1918 (1921), 974; Hardy, "Tungsten Market Situation."

55. "Tungsten's Meteoric Rise and Fall," 293–94.

56. Li and Wang, *Tungsten,* v–ix; Hardy, "Tungsten Market Situation"; George J. Young, "The Tungsten Mining Industry in 1919"; Charles H. Segerstrom, "The American Tungsten Industry," 53. Complaints against "coolie labor" in Asian tungsten mining first surfaced late in 1916. See EMJ 103 (21 Apr. 1917): 714–15.

57. NYT, 26 May 1918, X12.

58. NYT, 26 Oct. 1916, 10; 3 Oct. 1918, 1; William B. Colver, "Recent Phases of Competition in International Trade," 233–39; Scheiber, "World War I," 508–11.

59. MacBride, "Export and Import Associations," 190–97; Sturm, "Webb-Pomerene Associations," 83–84.

60. Felix Edgar Wormser, "The Importance of Foreign Trade in Copper and Other Metals," 75–76; Robert B. Pettengill, "The United States Copper Industry and the Tariff," 145.

61. Yeager, "Trade Protection," 33–36; Richard A. Lauderbaugh, *American Steel Makers and the Coming of the Second World War,* 121–41.

62. Wormser, "Importance of Foreign Trade," 66–69, 75–76; Yeager, "Trade Protection," 36; Pettengill, "United States Copper Industry," 143–56.

63. WSJ, 14 Nov. 1917.

3 ‖ FROM PACIFIC TUNGSTEN TO NEVADA-MASSACHUSETTS

1. G. Young, "Tungsten Mining Industry"; J. H. Bell to WJL, 13 Dec. 1918; WJL to Bell, 17 Dec., 27 Dec. 1918, 27 Aug. 1919; NYT, 14 June 1919, 15.

2. Sobel, *Age of Giant Corporations,* 25–75.

3. M&SP 120 (24 Apr. 1920): 621; G. Young, "Tungsten Mining Industry"; Hardy, "Tungsten Market Situation"; Hardy, "Marketing Tungsten Ores," 666–69.

4. U.S. Department of the Interior, *Annual Report of the Secretary of the Interior* (1920), 30–31; WJL to P. George Gow, 6 June 1919; AMC, "Bulletin to Members," 28 May, 4 June, 6 June 1919; U.S. Geological Survey, *Strategy of Minerals,* 358–60; WSJ, 12 July 1919, 7; U.S. House, Committee on Rules, "War Minerals Relief" [hearings on proposed amendment S.3641], 2–12.

5. H. Foster Bain to WJL, 29 Nov. 1918; WJL to EAC, 11 Feb. 1920; WJL to CHS, 2 Dec. 1918.

6. WMRC, Reporter's Transcript, San Francisco, 24 July 1919, Claim No. 1018; WJL to C. A. Hight, 18 May 1923; PEC, Ms. Notes, ca. 1 June 1924; "Payments Made to Judge John F. Davis by the Pacific Tungsten Company," typescript, ca. May 1925; CHS to Frederick B. Hyder, 21 Nov. 1935.

7. U.S. House, Committee on Rules, "War Minerals Relief" [hearings], 2–12; WSJ, 18 Nov. 1921, 11; 9 Dec. 1925, 15; PTC, WMRC Claim Chronology [two typescripts], 3 Feb. 1932. On Fall, see Francis Russell, *The Shadow of Blooming Grove: Warren G. Harding and His Times,* 491–500.

8. PTC, WMR Claim Chronology, 3 Feb. 1932. See the bibliography for citations to various War Minerals Relief legal cases in the 1920s and '30s.

9. PTC, WMRC Claim Chronology, 3 Feb. 1932; WSJ, 1 May 1936, 3; U.S. Department of the Interior, *Annual Report* (1935), 27.

10. U.S. Senate, Appropriations Committee, "Hearings on Supplemental Appropriation Bill"; E. K. Burlew to JCT, 25 June 1941; R. H. Yeatman to JCT, 21 Apr. 1943; JCT to CHS, 27 Apr. 1943. Ironically, some appeals from World War I were still pending when in 1944 Senator Richard Harless of Arizona introduced a new relief bill. Described by a mining lobbyist using a colloquialism heard more frequently in later decades, it was designed to "bail out" financially strapped producers who had worked to meet government metal needs in the early years of World War II (*American Metal Market* [New York] [18 Apr. 1944]: 1, 7).

11. Herbert Croly, *The Promise of American Life,* 191, 304–5; Richard Hofstadter, *The Age of Reform: From Bryan to F.D.R.,* 232–54; Cuff and Urofsky, "Steel Industry and Price-Fixing," 258–59.

12. "Thomas Hardy Quotations"; Reed Smoot, "Why a Protective Tariff?" 21.

13. NYT, 26 May 1918, X12; 11 June 1919, 5; 17 June 1919, 19; WSJ, 25 Sept. 1920, 4; WJL to Wolf Tongue Mining Co., 23 Nov. 1920; Benjamin B. Wallace, "Postwar Tariff Changes and Tendencies," 181; Yeager, "Trade Protection," 36. Yeager's study of steel tariffs is noteworthy for its empirical analysis and hindsight. Writing in the 1980s, a decade of triumphant "economic internationalism," as Senator Smoot earlier characterized free trade, she did not anticipate the rebirth of protectionist sentiment a decade later. On the global proliferation of antidumping laws and their impact on trade, see Thomas J. Prusa, "On the Spread and Impact of Antidumping."

14. U.S. Tariff Commission, "A Review of the Tungsten Industry."

15. M&SP 117 (28 Sept. 1918): 3; JFD to EAC, 31 May 1919.

16. J. J. Holmes Jr. to E. C. Voorheis, 21 June 1919. See also RCM to Frank L. Hess, 8 Oct. 1919.

17. NYT, 11 June 1919, 5; 13 June 1919, 13; 17 June 1919, 19; 21 June 1919, 6; 1 Aug. 1919, 28; 3 Aug. 1919, 8; WSJ, 19 June 1919, 6, 8.

18. JFD to EAC, 31 May 1919.

19. RCM to Frank L. Hess, 8 Oct. 1919; Hardy, "Tungsten Market Situation"; RCM to J. H. Holmes and JFD, 13 Sept. 1919.

20. AMC, "Bulletin to Members," 28 May 1919; "Tariff on Tungsten Is Vital to Continuation of Production," *Nevada Mining Press* 25 (Dec. 1919): 14–22, both in SC UOPWA; WSJ, 19 June 1919, 8; Wormser, "Importance of Foreign Trade," 67–69; Berglund, "Ferroalloy Industries," 259–68; Berglund, "Tariff Act of 1922," 22; NYT, 22 Aug. 1919, 6; 8 Dec. 1919, 6.

21. WJL to EAC, 23 Mar. 1920; Nelson Franklin to G. S. Wood, 27 Feb. 1920.

22. Frank W. Griffin to Tungsten Mines Co., 6 Mar. 1920.

23. WSJ, 25 Sept. 1920, 4; WJL to Wolf Tongue Mining Co., 23 Nov. 1920; William Anthony Lovett et al., *U.S. Trade Policy: History, Theory, and the W.T.O.*, 51–53.

24. WSJ, 16 Apr. 1921, 9; 14 Sept. 1921, 3; NYT, 8 Oct. 1919, 18; 29 Jan. 1922, 82; 4 Dec. 1922, 30; Nelson Franklin to Frank W. Griffin, 29 Sept. 1922; Hardy, "Tungsten Market Situation"; U.S. Senate, Committee on Finance, "Digest of Tariff Hearings . . . on the Bill H.R. 7456," 166–68, 171–72.

25. Susan C. Schwab, *Trade-offs: Negotiating the Omnibus Trade and Competitiveness Act,* 18–19; Lovett, *U.S. Trade Policy,* 51–53; Berglund, "Tariff Act of 1922," 23–27.

26. F. C. M[erritt] to WJL, 12 Dec. 1921; W. T. Connors to WJL, 22 July 1922; WJL to Connors, 31 July 1922; WJL to C. A. Hight, 18 May 1923.

27. WJL to E. A. Stent, 7 Sept., 1 Dec. 1921; Stent to WJL, 12 Sept. 1921; WJL to E. C. Voorheis, 22 Sept. 1921; WJL to F. W. Batchelder, 14 Aug. 1922; Ms. notes, PEC, Boston, ca. 1 June 1924; CHS to PEC, 18 Mar. 1925

28. D. Segerstrom, "Brief History"; WJL to EAC, 5 June 1920; Frederick Foster to First National Bank of Sonora, 12 Mar. 1923.

29. W. J. Loring, "Report on Properties of the Pacific Coast Gold Mining Corporation, Quartz, Calif." (typescript), 1 Oct. 1921, Archie Douglas Stevenot Papers, UOPWA; WJL to CHS, 30 Apr. 1918, 13 Mar., 5 July 1919; to EAC, 16 Dec. 1919; to Union Trust Co. of San Francisco, 5 July 1919; to W. B. Devereux Jr., 4 Feb. 1920; Phil. Huber to WJL, 20 Apr. 1920; C. H. Bell to CHS, 12 Sept., 27 Dec. 1919; CHS to WJL, 15 Dec. 1919. Segerstrom sold the property in the late 1930s. During the gold revival of the 1980s it was incorporated into the Sonora Mining Company's holdings.

30. WJL to CHS, 3 June 1918.

31. WJL to Henry Ford, 29 Sept. 1920.

32. WJL to F. W. Batchelder, 2 Nov. 1922; C. W. Poole, "Plan for Resuming Operations at Nevada-Humboldt and Pacific Tungsten Properties"; CHS to WJL, 6 Feb. 1924; C. W. Pool Lease, 18 Mar. 1924, in PTC Minute Book; PEC typescript notes, ca. 1 June 1924; CHS to PEC, 18 Mar. 1925; CHS to Geo. Wm. Sargent, 13 May 1925.

33. W. J. Loring, "Plan Regarding Pacific Tungsten"; WJL to CHS, 4 Aug. 1924; CHS to PEC, 26 Nov. 1924.

34. WJL to CHS, 31 May 1924, 18 Sept., 7 Oct. 1925, 30 June 1926, 15 Feb. 1927; WJL to First National Bank of Sonora, 22 Sept. 1924; PTC Minute Book, minutes of special meeting of board of directors, San Francisco, 29 Nov. 1924; Frank J. Solinsky to CHS, 5 Apr.,

25 Sept., 20 Nov. 1926; George S. Green to F. J. Solinsky, 18 May 1926; CHS to F. J. Solinsky, 2 Dec. 1926, 18 June 1927; CHS to WJL, 1 Mar. 1927; A. D. Stevenot to Donald Emery, 17 Oct. 1952. For Loring's later career, see Limbaugh, "Making the Most of Experience," 9–13.

35. Frederick Lewis Allen, *Only Yesterday: An Informal History of the Nineteen Twenties,* 159–85; Sobel, *Age of Giant Corporations,* 25–51; "Review of the First Quarter of the Year," *Review of Economic Statistics* 6 (Apr. 1924): 64–67; HSUS, 200.

36. Alfred D. Chandler Jr., "The Structure of American Industry in the Twentieth Century: A Historical Overview," 270–74; Sobel, *Age of Giant Corporations,* 72.

37. NYT, 18 Dec. 1929, 118; Sobel, *Age of Giant Corporations,* 70–71; Yeager, "Trade Protection," 34–36; Daniel R. Fusfield, "Joint Subsidiaries in the Iron and Steel Industry." For the vanadium oligopoly, see the opinions in a series of antitrust cases against Union Carbide decided in 1961 by the U.S. Court of Appeals, cited in the bibliography. On Alcoa's control of the aluminum industry, see George David Smith, *From Monopoly to Competition: The Transformation of Alcoa, 1888–1986;* and Raymond Vernon, *Two Hungry Giants: The United States and Japan in the Quest for Oil and Ores,* 40–41.

38. Li and Wang, *Tungsten,* 136–92; NYT, 2 Oct. 1920, 8; 18 Jan. 1921, 1; 9 Feb. 1921, 1; 5 Aug. 1921, 1; 8 Jan. 1922, 82; 4 July 1922, 1; 26 Apr. 1924, 3; 27 May 1924, 21; 21 Sept. 1924, 1; COH to Pacific Tungsten Leasing Co., 4 Sept. 1924.

39. Stefan Lorant, *The Presidency,* 556–61; COH to PTC, 16 June 1924.

40. PEC to CHS, 3 Sept., 12 Sept., 11 Nov. 1924, 2 Apr. 1925; C. W. Poole, "Report on PTC Operations," 14 Nov. 1924; CHS to PEC, 26 Nov., 1 Dec., 2 Dec., 3 Dec. 1924, 18 Mar. 1925; "Agreement Between 1st National Bank of Boston, National Shawmut Bank of Boston, First National Bank of Sonora, CHS, and George W. Johnson of Sonora" (typescript), 24 Dec. 1924; Minutes, Board of Directors, NMC, Boston, 18 Feb. 1925; R-M, 10 Apr. 1925, 1; Judgment (typescript), NMC v. PTC, 6th Judicial Dist., Humboldt Co., Nev., 25 Apr. 1925; Alfred Sutro to PEC, 8 June 1925; CHS to stockholders of PTC, 6 Nov. 1925; RMM, [List of Stockholders] NMC, 20 Nov. 1934.

41. CHS to Geo. Wm. Sargent, 13 May 1925; to PEC, 18 Mar. 1925.

42. CHS to PEC, 18 Mar. 1925.

43. CHS to officers and directors of NMC, 29 May 1925.

44. Misa, *Nation of Steel,* 254–82.

45. M&SP 116 (23 Mar. 1918): 421; G. Young, "Tungsten Mining Industry"; Hess, "Tungsten," USGSMR 1917 (1921), 931–94; Hess, "Tungsten," 1918 (1921), 986–88; M&SP 117 (28 Sept. 1918): 432; "Tungsten in Nevada," *Mining and Metallurgy* 10 (Feb. 1929): 97–98; CHS to WJL, 24 May 1918; PEC typescript notes, ca. 1 June 1924; COH to WJL, 25 Jan. 1924; to PTC, 16 June 1924; to Pacific Tungsten Leasing Co., 4 Sept. 1924; CHS to WJL, 1 Mar. 1927; to *Daily Metal Trade,* 21 Apr. 1938.

46. Li and Wang, *Tungsten,* 199–201; C. G. Roser to CHS, 28 June 1927.

47. COH to WJL, 29 Aug. 1924; Furness, "Tungsten Ores and Concentrates," 3–7, 19–20.

48. Charles H. Segerstrom, "Report as of 30 June 1926, to Officers and Directors of Nev-Mass Co."

49. JJH to Pacific Tungsten Leasing Co., 14 July 1924; Richard W. Haesler to author, 29 Sept. 2008.

50. NYT, 13 July 1958, 68.

51. NYT, 25 Nov. 1938, 23; 26 Aug. 1964, 39.

52. COH to WJL, 21 Mar. 1923; to Pacific Tungsten Leasing Co., 1 May, 2 June 1924; WJL to MH, 20 Mar., 31 Mar. 1923; Geo. Wm. Sargent to CHS, 1 May 1925; CHS to Sargent, 13 May 1925; CHS to officers and directors of NMC, 29 May 1925; William H. King, "Investigation of Nevada-Massachusetts Tungsten Deposits, Pershing County, Nev."

53. Y. S. Brenner, *Looking Into the Seeds of Time: The Price of Modern Development,* 72–73; Hess, "Tungsten in 1930," 181. Steffens is quoted in William E. Leuchtenburg, *The Perils of Prosperity, 1914–1932,* 202.

54. Segerstrom, "Report as of 30 June 1926"; CHS to WJL, 1 Mar. 1927; NMC Balance Sheets, 1925–29.

55. CHS, affidavit, sworn before notary public (undated typescript, Sonora, 1935); D. Segerstrom, "Brief History"; Ott F. Heizer, "Method and Cost of Mining Tungsten at the Nevada-Massachusetts Co. Mines, at Mill City, Nev."; Ott F. Heizer, "Concentration of Tungsten Ore by the Nevada-Massachusetts Co."

56. "Articles of Agreement and Association, Nevada-Massachusetts Exploration Co."

57. CHS to RGE, 23 Jan. 1935; TT, JJH and CHS, 23 Nov. 1936; CHS to PFK, 2 May 1938; Lincoln, *Mining Districts,* 203–4; Tingley, *Mining Districts of Nevada,* 64, 195.

58. "Quitclaim Deed, 20 September 1929, by Which O. F. Heizer, A. Ranson, and C. H. Segerstrom [Representing Nevada-Massachusetts Exploration Co.] . . . Release All Claims [to Cottonwood and Silver Dyke]"; Oscar H. Hershey, "Geological Report on the Silver Dyke Mine"; CHS to A. P. Rogers, 21 Mar. 1931; to Anne E. Beane, 24 May 1939; Paul F. Kerr, "The Tungsten Mineralization at Silver Dyke, Nevada," 9.

59. Henry J. Long, "Tungsten Carbide Cutting Tools"; "Why Tungsten Prices Have Increased," EMJ 128 (7 Sept. 1929): 398; NMC Balance Sheets, 1925–29.

4 ‖ TUNGSTEN AND THE GREAT DEPRESSION

1. "Tungsten in Nevada," *Mining and Metallurgy* 10 (Feb. 1929): 97–98.

2. Joan Hoff Wilson, *Herbert Hoover: Forgotten Progressive,* 31–43; Donald R. Stabile, "Herbert Hoover, the FAES, and the AF of L," 819; Edwin T. Layton, *The Revolt of the Engineers: Social Responsibility and the American Engineering Profession,* 190; Peri E. Arnold, "The 'Great Engineer' as Administrator: Herbert Hoover and Modern Bureaucracy," 338–41.

3. "Tungsten in Nevada," 97–98. A revised edition of Heizer's paper was published as "Method and Cost of Mining Tungsten."

4. NYT, 11 Jan. 1929, 1; WSJ, 14 Jan. 1929, 10; Abraham Berglund, "The Tariff Act of 1930," 467–69.

5. U.S. House, Committee on Ways and Means, "Tariff Act of 1929"; NYT, 11 Jan. 1929, 1; WSJ, 25 Jan. 1929, 5; 2 Apr. 1929, 5; Berglund, "Tariff Act of 1930," 472.

6. JFD to EAC, 31 May 1919; J. J. Holmes Jr. to Senator E. C. Voorheis, 21 June 1919; WJL to EAC, 23 Mar. 1920; to Wolf Tongue Mining Co., 23 Nov. 1920; Nelson Franklin to G. S. Wood, 27 Feb. 1920; to CHS, 31 Jan. 1930, 12 June 1931; to JJH, 13 July 1931.

7. "Arguments for a Tungsten Tariff Increase," EMJ 127 (2 Feb. 1929): 215; Hess, "Tungsten in 1930," 181; Loach, "American Tungsten Industry."

8. "Arguments for a Tungsten Tariff Increase," 215; NYT, 4 June 1929, 27; Berglund, "Tariff Act of 1930," 472.

9. U.S. House, Committee on Ways and Means, "Tariff Act of 1929," 3:32–45, 55, 96–104, 117–23; NYT, 29 May 1929, 1; 4 June 1929, 27; WSJ, 16 Aug. 1929, 2.

10. NYT, 12 Nov. 1929, 2; 25 Mar. 1930, 1; 13 June 1930, 1; 15 June 1930, 1; 16 June 1930, 1, 14; WSJ, 14 June 1930, 10; Berglund, "Tariff Act of 1930," 467–69, 478–79.

11. Schlesinger, *Crisis of the Old Order,* 234; Charles H. Segerstrom, "Herbert Hoover, the Man Leading the Way to Recovery"; Frank C. Jordan, *Statement of Vote at General Election Held on November 8, 1932.*

12. HSUS, 200; U.S. Bureau of Labor Statistics, *Wholesale Commodity Price Index,* as quoted in NYT, 1929–33; production of steel ingots, alloy steels, and automobiles, 1913–33, compiled from *Iron Age* 124 (29 Aug. 1929): 544–45; 131 (5 Jan. 1933): 93; (12 Jan. 1933): 110; and *Index* 15 (Sept. 1935): 180–83, 186–88.

13. Schlesinger, *Crisis of the Old Order,* 230–31.

14. HSUS, 912; Heizer, "Method and Cost of Mining Tungsten," 10; Charles H. Segerstrom, "[Annual Report] to the Offices and Directors of the Nevada-Massachusetts Co."

15. WSJ, 18 Mar. 1916, 5; Hess, "Tungsten," USGSMR 1918 (1921), 986–88; U.S. Senate, Committee on Finance, "Digest of Tariff Hearings," 166–68; U.S. Senate, Finance Committee, "Hearings … on the Tariff Act of 1929," 36–38, 45–46; NYT, 18 Dec. 1929, 1; JJH to CHS, 7 Jan. 1931; "Why Tungsten Prices Have Increased," EMJ 128 (7 Sept. 1929): 398; John N. Ingham, "Iron and Steel in the Pittsburgh Region: The Domain of Small Business," 108–9.

16. Li and Wang, *Tungsten,* 249–77; Ingham, "Iron and Steel," 107–16.

17. HSUS, 693; Iron and Steel Institute figures, as cited in NYT, 1 Jan. 1929, 1 Jan. 1931, 22 Sept 1932; "Alloy Steels," *Index* 15 (Sept. 1935): 180–83, 186–88; JJH to CHS, 7 Jan. 1931; CHS to C. G. Roser, 13 Mar. 1931.

18. CHS to [Charles G.] Roser, 20 Dec. 1930; to JJH, 2 Feb. 1931; to C. G. Roser, 21 Jan., 12 Feb. 1931; to OFH, 30 Apr. 1931; to RGE, 18 Apr. 1931; to Frank E. Launsberry, 3 July 1931; C. G. Roser to CHS, 11 Mar. 1931; JJH to CHS, 23 Apr. 1931.

19. NMC Balance Sheets, 1925–50; RMM to CHS, 21 Aug. 1933; Charles H. Segerstrom, "Semi-annual Report to Officers and Directors of the Nevada-Massachusetts Co."; CHS to Louis J. Hunter, 3 Apr. 1931; JJH to CHS, 23 Apr. 1931; C. G. Roser to CHS, 11 June 1931; NYT, 23 Apr. 1931, 40; David M. Kennedy, *Freedom From Fear: The American People in Depression and War, 1929–1945,* 65–71.

20. Charles H. Segerstrom, "Annual Report of Operations."

21. CHS to C. G. Roser, 26 Jan. 1932; JJH to CHS, 2 Feb. 1932; WSJ, 14 Dec. 1931, 5; Gerald D.

Nash, "Herbert Hoover and the Origins of the Reconstruction Finance Corporation," 462–68; Kennedy, *Freedom From Fear,* 75–82. Segerstrom's friend Archie Stevenot, a Mother Lode gold miner down on his luck, recognized the problem and offered Hoover's personal secretary some grassroots advice during the 1932 campaign: "If we could stimulate everything a little mining would come in for its share and with this feeling of better times, people would not be so much inclined to have a [leadership] change take place" (to George Hastings, 28 July 1932, Stevenot Papers, UOPWA).

22. NYT, 16 June 1930, 14; Schlesinger, *Crisis of the Old Order,* 426–28.

23. Kendalle E. Bailes, "The American Connection: Ideology and the Transfer of American Technology to the Soviet Union, 1917–1941," 429; NYT, 20 Mar. 1932, XX1; E. C. Ropes, "American-Soviet Trade Relations," 90–92.

24. JJH to CHS, 27 Feb. 1932.

25. Schlesinger, *Crisis of the Old Order,* 224–49; Kennedy, *Freedom From Fear,* 75–84; HSUS, 135, 912.

26. In his 1933 report to stockholders, Charles H. Segerstrom exaggerated the study's preliminary conclusions. He said Kerr's report will indicate that "the property at Mill City is the most important in the United States and ranks as the third in importance in the world, and that the possibilities for ore in depth are tremendous, and it is their belief that this property alone can supply the normal tungsten requirements of the United States" ("Annual Report of the Nevada-Massachusetts Company, Inc."). The published report actually said only that "it is estimated *by the operators* that in normal times, with reasonable prices, sufficient concentrates could be produced to care for the demands of the United States" (Kerr, "Geology of the Tungsten Deposits," 7 [emphasis added]).

27. CHS to JJH, 9 Mar. 1932.

28. "Memorandum of Agreement, 1 Jul 1932, Between Nevada-Massachusetts Co. and Molybdenum Corp. of America." A draft of the contract is contained in CHS to Louis J. Hunter, 12 June 1932.

29. CHS to Louis J. Hunter, 12 June 1932; to JJH, 25 June 1932; JJH to CHS, 9 Aug. 1932.

30. CHS to JJH, 25 June 1932; R. G. Emerson to CHS, 28 Feb. 1933; Wilson, *Herbert Hoover: Forgotten Progressive,* 39–40; Hoover, *The Challenge to Liberty,* 104–5, 108–11.

31. CHS to officers and directors of NMC, 23 Sept. 1932.

32. CHS to Louis J. Hunter, 12 June 1932; to JJH, 18 July 1932.

33. CHS to JJH, 22 Mar. 1932; to RMM, 24 Aug. 1934; JJH to CHS, 17 Aug. 1932; GIE to CHS, 22 Apr. 1937; PEC to CHSJR, 6 Aug. 1947; CHSJR to PEC, 21 Jan. 1950.

34. Segerstrom, "Annual Report of the Nevada-Massachusetts Company."

35. Kennedy, *Freedom From Fear,* 126–28; NYT, 5 Feb. 1933, 29; "The Buy American Act," approved 3 Mar. 1933, 47 *Stat.* 1520; Laurence A. Knapp, "The Buy American Act: A Review and Assessment," 430–32; M. S. Noorzoy, "'Buy American' as an Instrument of Policy," 100.

36. Arthur M. Schlesinger Jr., *The Coming of the New Deal,* 3–23; W. L. Crum, "Review of the First Quarter of 1933," 68.

37. Schlesinger, *Coming of the New Deal,* 195–252; CHS to J. P. Sievers, 4 Sept. 1934.

38. Henry A. Wallace, *America Must Choose: The Advantages and Disadvantages of Nationalism, of World Trade, and of a Planned Middle Course,* 26–33; Judith Goldstein, "The Impact of Ideas on Trade Policy: The Origins of U.S. Agricultural and Manufacturing Policies," 32–48.

39. Schlesinger, *Coming of the New Deal,* 87–102; Robert F. Himmelberg, *The Origins of the National Recovery Administration: Business, Government, and the Trade Association Issue, 1921–1933,* 181–212. On countervailing powers and the New Deal, see chapter 10 of John Kenneth Galbraith, *American Capitalism.*

40. Willard M. Kiplinger, "Industry Control Postscript No. 21, Emergency Service"; Thuman W. Arnold, *The Folklore of Capitalism,* 207–29.

41. E. R. Coombes to JFC, 9 June 1933.

42. "Tungsten and Tungsten Products Industries Code of Fair Competition Under the National Industrial Recovery Act."

43. [AMC], "Statement in Transmittal," [undated typescript with attached] "Code of Fair Competition" [typescript carbon copy, Aug. 1933]; AMC [Attendance List], Rare Metals and Non-Metals Division, 26 June 1933; Sylvia Snowiss, "Presidential Leadership of Congress: An Analysis of Roosevelt's First Hundred Days," 74; F. W. Taussig, "Necessary Changes in Our Commercial Policy," 397–98. See also Section 3(e) of the National Industrial Recovery Act, 90 *Stat.* 196–97.

44. CHS to Atkins Kroll & Co., 8 July 1933.

45. Ibid.; to T. H. McGraw, 7 Sept. 1933.

46. JFC to CHS, 14 July 1933; Schlesinger, *Coming of the New Deal,* 119–51.

47. Schlesinger, *Coming of the New Deal,* 114; CHS to Wm. Loach, 9 July 1933; to JGC, 9 Aug. 1933; to Wm. L. Esson, 28 Aug. 1933.

48. JFC to CHS, 10 July 1933; J. F. Callbreath, "The Industrial Recovery Act: Its Purposes and the Obligations of Industry Thereunder"; Abraham Berglund, "The Reciprocal Trade Agreements Act of 1934," 425; Lauderbaugh, *American Steel Makers,* 27–29.

49. JFC to CHS, 14 July 1933.

50. Ibid., 14 July, 27 July 1933; CHS to F. M. Beckett, 29 July 1933.

51. CHS to T. H. McGraw, 7 Sept. 1933; J. P. Sievers to CHS, 15 Sept. 1933; CHS to JJH, 7 Sept. 1933.

52. George H. Snyder, "Domestic Tungsten Ore Supply Becomes Increasingly Important"; Vanderburg, "Methods and Costs of Mining Ferberite Ore," 15.

53. Wm. Loach to JFC, 11 Aug. 1933; to CHS, 31 July, 18 Aug., 18 Sept. 1933; JGC to CHS, 3 Aug., 17 Aug., 2 Oct., 7 Oct. 1933.

54. Schlesinger, *Coming of the New Deal,* 116–20; CHS to Wm. Loach, 9 July, 7 Sept., 29 Sept. 1933, 14 Apr. 1934; to T. H. McGraw, 7 Sept. 1933; to JFC, 14 Oct. 1933; to ACD, 31 July 1933; to JGC, 9 Aug. 1933; to Dr. F. M. Beckett, 2 Oct. 1933.

55. Wm. Loach to CHS, 18 Sept., 25 Oct. 1933. At the Salt Lake meeting were representatives from the Nevada-Massachusetts Company (CHS, OFH), the Atolia Mining Co.

(A. V. Udell and David Atkins), the Borianna Mining Co. (J. P. Sievers), the Tungsten Production Co. (J. G. Clark), and the Wolf Tongue Mining Co. (Wm. Loach).

56. JJH to CHS, 25 Jan. 1934.

57. Michael P. Malone, *C. Ben Ross and the New Deal in Idaho,* 101–3; Butler Shaffer, *In Restraint of Trade: The Business Campaign Against Competition, 1918–1938,* 123–44; JFC to CHS, 30 Mar., 3 Apr. 1934; CHS to Wm. Loach, 14 Apr. 1934; telegram, H. W. Klipstein to Franklin D. Roosevelt, 31 Oct. 1933, copy in SC UOPWA.

58. CHS to Columbia Steel Company, 11 June 1934; *Schechter v. U.S.*

59. Hoover, *The Challenge to Liberty,* 111. On Hoover's role as elder statesman and postpresidential critic, see Brant Short, "The Rhetoric of the Post-Presidency."

5 || MINING AND MARKETING DURING THE NEW DEAL

1. CHS to ACD, 19 Sept. 1938.

2. B. W. Zachau to W. G. Emminger, 30 Mar. 1936.

3. Wm. Loach to CHS, 27 Mar. 1934.

4. Cordell Hull, "Should the U.S. Adopt a Reciprocity Tariff Policy? Pro," 140, 142, 146.

5. Taussig, "Necessary Changes," 91–100; Douglas A. Irwin, "Interests, Institutions, and Ideology in Securing Policy Change: The Republican Conversion to Trade Liberalization After Smoot-Hawley," 648–49. For a recent study that reinforces traditional views of the impact of the RTAA on liberalizing trade, see Karen E. Schnietz, "The Reaction of Private Interests to the 1934 Reciprocal Trade Agreements Act."

6. Paul H. Douglas, *America in the Market Place: Trade, Tariffs, and the Balance of Payments, 1933–45,* 88–90; Stephen Haggard, "The Institutional Foundations of Hegemony: Explaining the Reciprocal Trade Agreements Act of 1934," 91; Berglund, "Reciprocal Trade Agreements Act," 418.

7. Taussig, "Necessary Changes," 397–400; Goldstein, "Impact of Ideas on Trade Policy," 58–59.

8. Berglund, "Reciprocal Trade Agreements Act," 411–25.

9. NYT, 1 July 1934, 20; 3 Feb. 1935, 1.

10. NYT, 26 Jan. 1935, 36; 24 Feb. 1935, 21; 10 Mar. 1935, SM3; Irwin, "Interests," 648; Goldstein, "Impact of Ideas on Trade Policy," 64–67.

11. NYT, 15 Apr. 1934, N17; 4 Feb. 1935, 1; 11 Feb. 1935, 5.

12. J. P. Sievers to CHS, 9 Mar. 1934; JJH to CHS, 5 June 1934. Manganese data compiled from U.S. Geological Survey, "Historical Statistics for Mineral and Material Commodities in the United States, Manganese, 2006," by Thomas Goonan and Lisa Corathers, in *Data Series* 140.

13. OFH to P. A. McCarran, 20 Apr. 1933; McCarran to OFH, 20 May 1933.

14. CHS to RMM, 12 June 1934; JJH to CHS, 30 Mar. 1936.

15. CHS to GIE, 21 Nov. 1935; to MH, 19 Mar. 1938; U.S. Department of State, "The Trade Agreement With the United Kingdom Signed November 17, 1938," 153–68.

16. Michael Goldfield, "Worker Insurgency, Radical Organization, and New Deal

Labor Legislation," 1274; Kennedy, *Freedom From Fear*, 297–98; TT, JJH and CHS, undated [ca. 1 Sept. 1936].

17. NYT, 13 June 1938, 8; CHS to ERH, 12 May 1938; to GIE, 23 Nov. 1938; Kennedy, *Freedom From Fear*, 344–46.

18. Truman C. Bingham, "Economic Effects of the New Deal Tax Policy," 270–80; Eleanor Roosevelt, *Courage in a Dangerous World: The Political Writings of Eleanor Roosevelt*, 270; Mark H. Leff, *The Limits of Symbolic Reform: The New Deal and Taxation, 1933–1939*, 91–92, 164. See also CHS to ERH, 9 Mar. 1939.

19. Bingham, "Economic Effects," 279; CHS to GIE, 15 Jan. 1935.

20. TT, JJH and CHS, undated [ca. 20 Dec. 1937]; CHS to GIE, 9 Jan., 8 Feb. 1937, 25 May 1938; Segerstrom, "[Annual Report] to the Offices and Directors"; *Nevada-Massachusetts Co. v. Commissioner of Internal Revenue*; Emminger, "Reminiscences," NC337, 13.

21. CHS to ACD, 9 Apr. 1931; to T. H. McGraw, 7 Sept. 1933; to JGC, 9 Mar. 1934; to J. P. Sievers, 4 Sept. 1934; to ERH, 21 Jan. 1939; TT, JJH and CHS, 3 May 1937; JJH to CHS, 23 July, 15 Sept. 1937; "Tungsten Market Shows Strength as Supply Is Limited," *Reno Gazette*, 15 Sept. 1937, 6.

22. CHS to ACD, 19 Sept. 1938; to ERH, 21 Jan. 1939; F. W. Horton to CHS, 28 Sept. 1934; JJH to GIE, 9 Mar. 1937; TT, JJH to CHS, 1 July 1937; Fink, "Review of the Strategic Metals," 419–20; William L. Shirer, *The Rise and Fall of the Third Reich: A History of Nazi Germany*, 281–308; Arthur M. Schlesinger Jr., *The Politics of Upheaval*, 272–74, 333–34.

23. TT, JJH and CHS, 1 Apr., 30 Apr., 23 July 1937; CHS to GIE, 3 Apr. 1937; to B. W. Holeman, 5 Oct. 1938; to ERH, 21 Jan. 1939.

24. JJH to ACD, 12 Aug. 1935.

25. Weekly prices on the New York Curb, renamed the American Stock Exchange in 1952, were reported in the *New York Times*. See also Molycorp's financial statement in the *Times*, 22 Feb. 1939, 38; E. A. Lucas to JJH, 18 May 1933; JJH to CHS, 25 Jan. 1934; Nevada-Massachusetts Financial Statements, 1925–35; and WSJ, 27 Feb. 27 1935, 6.

26. JJH to CHS, 24 Mar. 1934; RMM to CHS, 9 May 1934; CHS to RMM, 6 Apr., 11 Apr., 7 May, 8 June 1934; to JJH, 11 Apr. 1934.

27. TT, JJH and CHS, 14 Oct., 31 Oct. 1936, 20 Apr. 1937; CHS to ACD, 9 Jan. 1937; to GIE, 12 Apr., 20 Aug. 1937; JJH to CHS, 15 Oct. 1934, 23 July 1937; ACD to CHS, 21 Aug. 1934; JJH to ACD, 21 Aug. 1936; J. V. Emmons, *The Molybdenum-Tungsten High Speed Steels Marketed Under the General Trade Name Mo-Max*; Li and Wang, *Tungsten*, 275–76.

28. JJH to ACD, 19 Apr. 1934; ACD to CHS, 28 Mar. 1934; JJH to CHS, 19 Dec. 1935, 25 Aug. 1936; TT, JJH and CHS, 1 Aug., 4 Aug., 1 Sept., 14 Oct. 1936, undated [ca. 15 Apr. 1937].

29. Robinson-Patman Antidiscrimination Act, U.S. Code, Title 15, Sec. 13–13b; TT, JJH and CHS, 3 Dec. 1937.

30. JJH to CHS, 13 Mar. 1935; CHS to GIE, 15 Sept., 25 Nov. 1936; NYT, 17 Nov. 1956, 21.

31. CHS to GIE, 8 May 1936.

32. *Mohave County Miner* [Kingman, Ariz.], 23 Apr. 1937; CHS to JJH, 28 Apr. 1931. See

also TT, JJH and CHS, 2 Feb. 1937; CHS to GIE, 5 Mar. 1937; Hubert W. Davis, "Tungsten," BOMMY 1943, 674.

33. CHS to GIE, 8 Sept. 1937.

34. CHS to RGE, 23 Jan. 1935; to JJH, 24 Sept. 1934; RGE to CHS, 15 Feb. 1935.

35. CHS to OFH, 11 Jan. 1936; to CGF, 21 May 1936; OFH to CHS, 14 Jan. 1936.

36. CHS to JJH, 16 Mar. 1936; to GIE, 16 Mar. 1936.

37. OFH to CHS, 10 May 1937; CHS to David Jiles, 27 Aug. 1937; "Nevada Tungsten Company Showing Great Ore Output" (unidentified clipping), Mines scrapbook 1941–42, SC UOPWA.

38. CHS to JJH, 31 Mar., 21 Apr. 1936; JJH to CHS, 6 Apr. 1936.

39. JJH to CHS, 24 Mar. 1934; TT, JJH and CHS, 3 Dec. 1937.

40. Production data from Nevada-Massachusetts corporate records, SC UOPWA. See also chart on page @@@.

41. Robert H. Ridgway, "Ferro-alloying Minerals: Domestic Productive Capacity Showed Important Increase for Tungsten, Molybdenum, and Vanadium Ore"; Kurtak, "History of Pine Creek."

42. CHS to RMM, 13 Apr., 24 Sept. 1934; to CGF, 21 May 1936; MI 49 (1940): 605–20.

43. OFH to employees of NMC, 28 Feb. 1935; CHS to GIE, 16 Apr. 1936, 17 Mar. 1942.

44. CHS to RMM, 24 Sept., 13 Nov. 1934; to GIE, 15 Jan., 8 Mar., 17 July, 4 Dec. 1935; Segerstrom, "[Annual Report] to the Offices and Directors."

45. CHS to JJH, 18 Apr. 1931; to RMM, 12 Aug., 3 Oct. 1933, 11 Apr., 24 Sept. 1934; to JJH, 6 Nov. 1933; to CGF, 11 June 1934; to GIE, 18 Nov. 1935, 14 Nov. 1939; OFH to CHS, 30 Apr. 1931; Heizer, "Method and Cost of Mining Tungsten," 4–9; Frank L. Hess, "Tungsten," BOMMY 1931–32 (1933), 275–76.

46. CHS to RMM, 24 Sept., 13 Nov. 1934; to GIE, 13 Nov. 1937, 11 Apr., 21 Aug. 1939.

47. CHS to RMM, 12 Aug. 1933, 6 Apr., 8 June, 24 Sept. 1934; to GIE, 17 Oct. 1935, 16 Mar. 1936, 13 Nov., 3 Dec. 1937, 21 Feb. 1938; to CGF, 21 May 1936; OFH to CHS, 19 Sept. 1933, 10 May 1937; to PFK, 16 Apr. 1935; [W. G. Emminger], Silver Dyke Report for week ending 16 Aug. 1931; JJH to ACD, 18 Jan. 1937; TT, JJH and CHS, 18 Nov. 1936; Kerr, "Tungsten Mineralization at Silver Dyke, Nevada," 9–14.

48. [W. G. Emminger], Silver Dyke Monthly reports, Oct. 1934–Jan. 1936; OFH to CHS, 8 Jan. 1936; B. W. Zachau to W. G. Emminger, 9 Mar., 30 Mar. 1936.

49. CHS to PFK, 5 Nov. 1937; to Robert H. Ridgway, 10 Mar. 1939; to JKG, 7 Jan. 1943; OFH to CHS, Mill City, 10 May, 3 Oct. 1937; TT, JJH and CHS, 14 Oct., 23 Nov. 1936, 28 Dec. 1937. Nevada-Massachusetts 1938 production figures for Oreana were 305 tons of concentrates for a total of 24,450 units of 60 percent WO_3. See also Tingley, *Mining Districts of Nevada,* 195; Frank L. Hess, "Tungsten," BOMMY 1934, 495; and Paul F. Kerr, "Tungsten Mineralization at Oreana, Nevada."

50. Data from periodic mining and milling production reports of the Nevada-Massachusetts Company, 1925–1950, in SC UOPWA.

51. OFH to CHS, 9 July 1933; JJH to RMM, 24 Aug. 1933; CHS to RMM, 3 Oct., 7 Oct. 1933; Heizer, "Concentration of Tungsten Ore," 835–40.

52. CHS to CGF, 11 June 1934; to RMM, 8 June 1934; George Crerar to CHS, 17 May 1935.

53. Jeremy Mouat, "The Development of the Flotation Process: Technological Change and the Genesis of Modern Mining, 1898–1911"; Ralph W. Birrell, "The Role of Minerals Separation Ltd. in the Development of the Flotation Process." For the legal decision undergirding Minerals Separation's American patent control, see *Minerals Separation v. Hyde,* 242 U.S. 261 (1916). One important case Minerals Separation lost involved differential copper flotation. See *Minerals Separation v. Magma Copper Co.*

54. Edward H. Nutter to WJL, 1 Nov. 1916; James L. Brooks to WJL, 16 Nov., 7 Dec. 1916; WJL to CHS, 13 Mar. 1919; CHS to GIE, 10 Sept. 1935.

55. E. S. Leaver and M. B. Royer, "Flotation for Recovery of Scheelite from Slimed Material," 1–12.

56. R. S. Dean et al., "Flotation of Scheelite"; Li and Wang, *Tungsten,* 104–7.

57. CHS to GIE, 25 Nov. 1936; to CGF, 17 Mar. 1942.

58. George Crerar to CHS, 29 Sept. 1939. Crerar's assumption was correct, but slightly off the mark. Minerals Separation did call again in 1942, this time demanding a settlement from Segerstrom for using a patented reagent to float gold from the Malvina Mine near his home in Tuolumne County. But by then the government had forced a shutdown of all "nonessential" mining operations for the duration. With a lawyer's appreciation for the art of courteous dismissal, Segerstrom offered to settle for "a few dollars" if there was anything left after disposal of the property (CHS to Minerals Separation North American Corp., 17 Mar. 1942).

59. George Crerar to CHS, 17 May 1935; CHS to JJH, 5 Aug. 1935; Lincoln, *Mining Districts,* 97–98; Paul F. Kerr, "Tungsten-Bearing Manganese Deposit at Golconda, Nevada," 1360–61; Tingley, *Mining Districts of Nevada,* 99; Li and Wang, *Tungsten,* 143.

60. J. Carson Adkerson to W. H. Phillips, 9 May 1935; NYT, 26 June 1935, 2.

61. Hiland Batcheller, "Economic Significance of Special Alloy Steels," 316; P. M. Heldt, "Tungsten Has Played a Prominent Role in the Development of the Automotive Industry," 816; Liddell, *Handbook of Nonferrous Metallurgy,* 629.

62. Heldt, "Tungsten Has Played a Prominent Role," 816; Gregory J. Comstock, "Tungsten Carbide: The First Product of a New Metallurgy"; Karl Schroeter, "Inception and Development of Hard Metal Carbides"; Karl Schroeter, "Analysis of Hard Metal Carbide Theory"; *U.S. v. General Electric Co. et al.*; Gabriel Kolko, "American Business and Germany, 1930–1941," 726–27. For examples of other "breakthrough" technologies, see Dianne Newell, *Technology on the Frontier: Mining in Old Ontario,* 3–5, 18–27; Donald C. Jackson, *Building the Ultimate Dam: John S. Eastwood and the Control of Water in the West,* 130–32.

63. Robert W. Geehan, "Tungsten," 5–6.

64. CHS to JJH, 5 Aug. 1935.

65. TT, JJH and CHS, 2 Mar. 1937.

66. CHS to CGF, 21 May 1936; to GIE, 15 Sept. 1936; to Henry C. Carlisle, 24 Sept. 1942; TT, JJH and CHS, 14 Oct. 1936; OFH to CHS, 10 May, 3 Oct. 1937.

67. CHS to JJH, 14 Sept. 1936; TT, JJH and CHS, 8 Mar. 1937.

68. CHS to George Crerar, 28 July 1939; TT, JJH and CHS, 8 Mar., 17 Mar. 1937; Leaver and Royer, "Flotation for Recovery of Scheelite," 21–24; Frank L. Hess, "Rare Metals and Minerals"; Fink, "Review of the Strategic Metals," 419–20; Geehan, "Tungsten," 2–10.

69. Alan M. Bateman to CHS, 21 Oct. 1942; Li and Wang, *Tungsten,* 136–43; John B. Huttl, "Unique Golconda Deposit Yields Its Tungsten."

70. CGF to CHS, 17 Nov. 1936; TT, JJH and CHS, 23 Nov. 1936; CHS to CGF, 17 Mar. 1942; MI 49 (1940): 609.

71. JJH to CHS, 16 Sept. 1935; CHS to GIE, 5 Sept., 21 Sept., 17 Oct., 20 Nov. 1935, 12 Apr., 4 Nov. 1937, 25 May 1938; GIE to CHS, 1 June 1938.

72. Leaver and Royer, "Flotation for Recovery of Scheelite," 21–24; Hess, "Rare Metals and Minerals."

73. NYT, 2 Nov. 1937, 45; 19 Nov. 1937, 42; 2 Dec. 1937, 39; 3 Dec. 1937, 42.

74. WSJ, 3 Jan. 1938, 6; NYT, 30 Apr. 1938, 2; Andrew H. Bartels, "The Office of Price Administration and the Legacy of the New Deal, 1939–1946," 7–8.

75. TT, CHS and JJH, 17 Sept. 1940; Gene M. Gressley, "Thurman Arnold, Antitrust, and the New Deal."

76. *U.S. v. General Electric Co. et al.*

77. JJH to ACD, 16 Nov. 1937; TT, JJH and CHS, undated [ca. 20 Dec. 1937]; CHS to Clifford L. Ach, 17 Jan. 1938; to GIE, 24 Mar. 1938; to ERH, 23 Aug. 1938; to B. W. Holeman, 5 Oct. 1938; GIE to CHS, 6 Apr. 1938; Lauderbaugh, *American Steel Makers,* 51–64.

78. GIE to CHS, 23 Mar., 6 Apr. 1938; CHS to Anne E. Beane, 2 May 1938; to GIE, 25 May, 14 June 1938; to ACD, 19 Sept. 1938; to B. W. Holeman, 5 Oct. 1938.

79. CHS to ACD, 10 Dec. 1937; to PFK, 25 May 1938; Kerr, "Tungsten-Bearing Manganese Deposit at Golconda, Nevada," 1361. For the Segerstrom mansion, see Sharon Marovich, "The C. H. Segerstrom Estate: A Haven for Family and Friends on Knowles Hill."

80. CHS to JJH, 25 Mar., 28 July 1939; to GIE, 26 Oct. 1939, 11 June 1940; P. Joralemon, "California's Foothill Gold Belt: Some Famous Lode Mines Are Showing Signs of Renewed Interest."

81. CHS to GIE, 25 May, 23 Nov. 1938.

6 || STRATEGIC METALS AT THE START OF WORLD WAR II

1. TT, JJH and CHS, 30 Mar. 1939.

2. For the isolationist response to the administration's refusal to enforce the neutrality laws during the Sino-Japanese war, see Robert A. Divine, *The Illusion of Neutrality,* 200–204.

3. Roland Stromberg, "American Business and the Approach of War, 1935–1941," 58–64, 72–78; Stromberg, "On Cherchez Le Financier"; E. Bernstein, "War and Business

Cycles," 524–25; "Business Stands Against War," unidentified clipping attached to James H. McGraw Jr. to CHS, 4 Oct. 1939.

4. David Hinshaw, *The Home Front,* 17–18; *U.S. v. General Electric Co. et al.;* Kolko, "American Business and Germany," 718–28. See also Arnold A. Offner, "Appeasement Revisited: The United States, Great Britain, and Germany, 1933–1940."

5. NYT, 24 Nov. 1935, E11; Robert H. Whealey, *Hitler and Spain: The Nazi Role in the Spanish Civil War,* 17–21; John Toland, *The Rising Sun: The Decline and Fall of the Japanese Empire, 1936–1945,* 58–63; Walter A. Radius, "United States Trade and the Sino-Japanese War"; Stephen G. Craft, "Peacemakers in China: American Missionaries and the Sino-Japanese War, 1937–1941," 575–85.

6. Douglas Miller, *You Can't Do Business With Hitler,* 167.

7. WSJ, 1 Sept. 1937, 2; OFH to CHS, 3 Dec. 1937; CHS to GIE, 31 Dec. 1937.

8. Paul A. C. Koistinen, "The 'Industrial-Military Complex' in Historical Perspective: The Interwar Years," 828.

9. Chester W. Wright, "American Economic Preparations for War, 1914–1917 and 1939–1941," 161–63; Koistinen, "'Industrial-Military Complex,'" 823–26.

10. Robert H. Ferrell, *Peace in Their Time: The Origins of the Kellogg-Briand Pact,* 21–30, 266–69; John W. Killigrew, *The Impact of the Great Depression on the Army;* Koistinen, "'Industrial-Military Complex,'" 828–35.

11. Harry B. Yoshpe, "Economic Mobilization Planning Between the Two World Wars," 199–203.

12. NYT, 6 July 1933, 1; 10 July 1933, 3; "Military Expenditures by U.S. Government, 1919–1940," SAUS, 1934, 162–63; 1940, 168–69; "Congress Considers Bills Affecting National Defense," 74; Yoshpe, "Economic Mobilization," 203–4.

13. ERH to CHS, 25 Sept. 1934; CHS to ERH, 24 Oct. 1934.

14. Koistinen, "'Industrial-Military Complex,'" 826.

15. NYT, 15 May 1932, E6; *Biographical Directory of Congress;* C. Elizabeth Raymond, *George Wingfield: Owner and Operator of Nevada,* 174–75; Don W. Driggs and Leonard E. Goodall, *Nevada Politics and Government: Conservatism in an Open Society,* 51–52.

16. James G. Scrugham to OFH, 20 May 1933; NYT, 31 Dec. 1933, E7; WSJ, 10 Mar. 1934, 1; Schlesinger, *Coming of the New Deal,* 195–252; Betty Glad, *Key Pittman: The Tragedy of a Senate Insider,* 210–12.

17. WSJ, 10 Mar. 1934, 1. Scrugham's frustrations and motives appear in his statement published in U.S. House, Committee on Military Affairs, "Hearings," 43–46.

18. CHS to ERH, 4 Sept. 1934.

19. NYT, 14 Oct. 1934, 29; B. Friedman to NMC, 24 Oct. 1934.

20. CR (25 Apr. 1935): 6404; statement of Capt. F. A. Daubin in U.S. House, Committee on Military Affairs, "Hearings," 190–93; Schlesinger, *Coming of the New Deal,* 289–96.

21. CR (Apr.–May 1935): 6224, 6232, 6404, 7832–35; J. Carson Adkerson to W. H. Phillips, 9 May 1935; NYT, 26 June 1935, 2.

22. From Baruch's "Report to the President," 24 Dec. 1919, and "Taking the Profit Out of War," reprinted in *American Industry in the War*, by Baruch, 8, 373–477.

23. Bruce Catton, *The War Lords of Washington*, 100–102; Koistinen, "'Industrial-Military Complex,'" 837.

24. AIMME news release, 19 Feb. 1935, in SC UOPWA; Schlesinger, *Coming of the New Deal*, 350–53.

25. NYT, 29 Jan. 1922, 82; WSJ, 22 May 1937, 4; Charles K. Leith, "Mineral Resources and Peace"; Charles K. Leith, "The Struggle for Mineral Resources."

26. Arthur W. Schatz, "The Anglo-American Trade Agreement and Cordell Hull's Search for Peace, 1936–1938," 85; AIMME news release, 19 Feb. 1935.

27. Charles K. Leith, "Conservation of Certain Mineral Resources," 243–45; AIMME news release, 19 Feb. 1935.

28. Ernest R. May, "The Development of Political-Military Consultation in the United States," 171–72.

29. *Biographical Directory of Congress*; WSJ, 19 May 1937, 4; William W. Lockwood, "Washington Works on Barter Scheme."

30. JJH to CHS, 30 Mar. 1936; statement of John R. Murdock in U.S. House, Committee on Military Affairs, "Supplies for the Armed Forces," 2–10.

31. U.S. House, Committee on Military Affairs, "Hearings," 43–46, 193.

32. U.S. House, Committee on Military Affairs, "Supplies for the Armed Forces," 2–10, 23–38; U.S. House, Subcommittee of the Committee on Appropriations, "Hearings . . . on the Military Establishment Appropriation Bill for 1938," 781–82.

33. For an example of problems in economic geology, see G. P. Glasby, "Deep Seabed Mining: Past Failures and Future Prospects."

34. JJH to CHS, 1 Apr., 23 July 1937; American Tungsten Association, "Tungsten: The Aristocrat of Rare Metals"; CHS to GIE, 21 Feb. 1938.

35. Snyder, "Domestic Tungsten Ore Supply"; WSJ, 22 May 1937, 4; "Tungsten Market Shows Strength as Supply Is Limited," *Reno Gazette*, 15 Sept. 1937, 6; CR (28 Mar. 1939): 3412.

36. Kennedy, *Freedom From Fear*, 361.

37. CR (25 Jan. 1938): 1093; JJH to CHS, 31 Jan. 1938.

38. Harold L. Ickes to Senator Morris Sheppard, 9 Apr. 1938, in U.S. Senate, Subcommittee of the Committee on Military Affairs, "Hearing . . . on S. 3460 . . . Part 2," 61. See also Part 1, 3–41.

39. "The Buy American Act of [3 Mar] 1933," 47 *Stat.* 1520; Public No. 117 (the Thomas Bill), 76th Cong., 1st sess., Senate 572 (1939), both reprinted in U.S. House, Subcommittee on Mines and Mining of the Committee on Public Lands, "Hearings [on Strategic and Critical Minerals and Metals]," 981–82, 1235–36, hereafter cited as "Hearings 1948"; NYT, 25 Feb. 1939, 1; U.S. House, Committee on Military Affairs, "Hearings," 193.

40. NYT, 21 Jan. 1939, 7; 25 Feb. 1939, 1.

41. WSJ, 14 Feb. 1939, 2; ERH to CHS, 11 Mar. 1939; E. W. Pehrson, "Stockpiling: Development of Submarginal Resources: Conservation and Substitution," 6–8.

42. WSJ, 27 Apr. 1938, 17; ERH to CHS, 28 Jan., 5 Apr. 1939.

43. CR (25 Apr. 1939): 4760, 4778–79; WSJ, 9 June 1939, 6; 12 Sept. 1939, 4; "American Manganese Producers Association."

44. NYT, 11 Nov. 1938, 39; CHS to GIE, 10 Oct. 1939.

45. Kennedy, *Freedom From Fear,* 466–79. Churchill's famous exhortation was first heard on a BBC radio broadcast on 9 Feb. 1941. Part of the speech is reprinted on the Churchill Memorial Web site of Westminster College.

46. "Hearings 1948," 1135–36, 1300–1301.

47. WSJ, 19 Sept. 1941, 2; 25 Sept. 1941, 1; 3 Nov. 1941, 2; 31 Mar. 1943, 4; "Hearings 1948," 984–85; *American Metal Market* (New York) (18 Apr. 1944): 1, 7.

48. Luther Gulick, "War Organization of the Federal Government," 1167–72; U.S. War Production Board, *Industrial Mobilization for War: History of the War Production Board and Predecessor Agencies,* 6–11; Koistinen, "'Industrial-Military Complex,'" 836–37; Robert Higgs, "Private Profit, Public Risk: Institutional Antecedents of the Modern Military Procurement System in the Rearmament Program of 1940–1941," 171–72.

49. NYT, 30 Dec. 1940, 6; Hinshaw, *The Home Front,* 58–59; Matthew J. Dickinson, *Bitter Harvest: FDR, Presidential Power, and the Growth of the Presidential Branch,* 204–20; John W. Jeffries, *Wartime America: The World War II Home Front,* 17–18. The Wallace quote is cited in Arthur M. Schlesinger Jr., *The Imperial Presidency,* 409.

50. Herman Miles Somers, *Presidential Agency: OWMR, the Office of War Mobilization, and Reconversion,* 6–24.

51. WSJ, 30 June 1941, 3; Studs Terkel, *"The Good War": An Oral History of World War Two,* 321–22; Lauderbaugh, *American Steel Makers,* 42–51; Bartels, "Office of Price Administration," 10.

52. WSJ, 2 Jan. 1941, 1; 17 Feb. 1941, 1; 8 Mar. 1941, 1; 25 Apr. 1941, 11; 14 July 1941, 1; 18 July 1941, 1; 28 Nov. 1941, 2; Calvin L. Christman, "Donald Nelson and the Army: Personality as a Factor in Civil-Military Relations During World War II"; Lauderbaugh, *American Steel Makers,* 79–80.

53. WSJ, 18 July 1941, 1; E. Cary Brown, "Accelerated Depreciation: A Neglected Chapter in War Taxation"; C. Rudolf Peterson, "The Statutory Evolution of the Excess Profits Tax," 24–25; Sobel, *Age of Giant Corporations,* 21–23; Higgs, "Private Profit, Public Risk," 170–71; Kennedy, *Freedom From Fear,* 622–24.

54. Harland Prechel, "Steel and the State: Industry Politics and Business Policy Formation, 1940–1989," 653–54.

55. Higgs, "Private Profit, Public Risk," 170.

56. CHS to ERH, 12 May 1938; JJH to CHS, 10 June 1940; NYT, 18 Nov. 1938, 1.

57. WSJ, 28 Mar. 1942, 5; Kennedy, *Freedom From Fear,* 637–41.

58. CHS to RMM, 11 Apr. 1934; ACD to CHS, 9 Apr. 1937; TT, JJH and CHS, 20 Apr. 1937; JJH to CHS, 23 July 1937; MH to CHS, 26 Dec. 1939; Ritchie, "Molybdenum in High-Speed

Steel"; Fink, "Review of the Strategic Metals," 419–20; WSJ, 1 Mar. 1941, 2; 25 Mar. 1941, 1; 13 June 1941, 3; 2 Sept. 1941, 3; 6 Dec. 1941, 9; Paul B. Coffman, "The Rise of a New Metal: The Growth and Success of the Climax Molybdenum Company."

59. JJH to GIE, 19 Mar. 1941; CHS to GIE, 1 Nov. 1941; A. I. Henderson to Hiram W. Johnson, 4 Dec. 1941; Leonard Caruana and Hugh Rockoff, "A Wolfram in Sheep's Clothing: Economic Warfare in Spain, 1940–1944," 101.

60. MH to CHS, 26 Dec. 1939.

61. TT, JJH and CHS, 1 July 1937, 13 Feb. 1940; "The Tungsten Situation," Metal Trade, 11 Feb. 1941 (clipping), in SC UOPWA; Charles H. Segerstrom, "War's Effect on Tungsten"; Edward S. Mason, "American Security and Access to Raw Materials," 152.

62. Shirer, Rise and Fall, 839–40; Arthur N. Young, China's Nation-Building Effort, 1927–1937: The Financial and Economic Record, 367–68; George F. Botjer, A Short History of Nationalist China, 1919–1949, 154; Douglas L. Wheeler, "The Price of Neutrality: Portugal, the Wolfram Question, and World War II," 108; Christian Leitz, "Nazi Germany's Struggle for Spanish Wolfram During the Second World War," 72–88. On the German prewar stockpile question, see Fink, "Review of the Strategic Metals," 419–20; Li and Wang, Tungsten, x–xi; "Tungsten Saga," Business Week (6 Jan. 1945): 28–32; and Offner, "Appeasement Revisited," 375–76.

63. EMJ 140 (Nov. 1939): 29–30; WSJ, 26 Sept. 1940, 1; 22 Oct. 1940, 3; 13 Feb. 1941, 2; 17 Dec. 1942, 7; Kung-Ping Wang, "Mineral Resources of China: With Special Reference to the Non-ferrous Metals," 625–26; Carl A. DeFrancesco and Kim B. Shedd, "Tungsten Statistics."

64. Li and Wang, Tungsten, 83–87.

65. Donald G. Stevens, "World War II Economic Warfare: The United States, Britain, and Portuguese Wolfram," 543–44; Caruana and Rockoff, "Wolfram in Sheep's Clothing," 102–17.

66. NYT, 4 Sept. 1941, 10; Edgar S. Furniss Jr., "American Wartime Objectives in Latin America," 377–79.

67. NYT, 5 May 1941, 8; 22 May 1941, 5; Li and Wang, Tungsten, 69–73; Mining Record 55 (18 May 1944): 1–2; "Hearings 1948," 1136–37.

68. Robert C. Arnold to CHS, 22 Apr. 1943; JJH to CHS, 30 Mar. 1944; Dun & Bradstreet Report on Haesler Metal & Ore Corp., 18 Aug. 1959.

69. ERH to CHS, 11 Mar. [1939]; NYT, 29 Sept. 1939, 6; Hess, "Rare Metals and Minerals," 9; WSJ, 20 Sept. 1940, 2; Segerstrom, "War's Effect on Tungsten."

70. In 1938, for example, CHS complained bitterly about the effect of Molybdenum Corporation's trading tactics on the tungsten tariff. Just at the "very moment" that the tariff on foreign ore brought the import price to $24 a unit, he told a friend, "Moly is bidding all over the UAS [sic] trying to buy ore at about an average of $17.50 and thus take no account of the duty and if the miners sell at this price it shows clearly they need no duty and it will be cut without any doubt" (TT, CHS to JJH, 19 Feb. 1938).

71. CHS to J. B. Coryell, 21 Jan. 1937; TT, CHS to JJH, 15 Feb. 1938; CHS to Clifford L. Ach, 17 Mar. 1938.

72. Van Rensselaer Lansingh, [trade letter], Los Angeles, 27 Jan. 1938.

73. U.S. Department of State, "Trade Agreement"; JDC to CHS, 17 Nov. 1938; CHS to JDC, 23 Nov. 1938, SC UOPWA.

74. JCT to CHS, 1 July 1941; CHS to Pat McCarran, 22 Oct. 1941; "Hearings 1948," 1074–76, 1157–61; U.S. Tariff Commission, *Tungsten Ores and Concentrates: Report to the President on Investigation No. 120 Under Section 336, Title III of the Tariff Act of 1930*, 4.

75. NYT, 4 Sept. 1941, 10; WSJ, 7 Sept. 1943, 6.

76. WSJ, 10 Oct. 1939, 4. On protectionist theory, see Taussig, *Tariff History*, 1–15.

77. CHS to B. W. Holeman, 5 Oct. 1938; JJH to ERH, 13 June, 19 June 1939; to Paymaster General of the Navy, 20 Sept. 1939; ERH to CHS, 30 Aug. 1939; CHS to GIE, 21 Aug. 1939; TT, CHS and JJH, 19 Sept. 1939.

78. TT, CHS and JJH, 16 Oct., 17 Oct. 1939; Hardy, *Wartime Control of Prices*, 54–55; WSJ, 10 Feb. 1943, 4; Schlesinger, *The Politics of Upheaval*, 509–10.

79. TT, CHS and JJH, 17 Oct. 1939; CHS to GIE, 7 Feb. 1941.

80. CHS to GIE, 14 Nov. 1939; Kennedy, *Freedom From Fear*, 426–34.

81. CHS to GIE, 11 Apr. 1939. On the Watterson brothers' role in the Owens Valley water wars, see Abraham Hoffman, *Vision or Villainy: Origins of the Owens Valley–Los Angeles Water Controversy*.

82. CHS to GIE, 11 Apr. 1939.

83. Ibid., 16 Sept. 1939.

84. Nevada-Massachusetts corporate records, 1941–45, SC UOPWA; Kurtak, *Mine in the Sky*, appendix, 198.

85. Stager and Tingley, "Tungsten Deposits in Nevada," 46; CHS to JJH, 25 Mar. 1939; TT, CHS and OFH, ca. Dec. 1942.

86. CHS to GIE, 11 June 1940, 3 Sept. 1941.

87. W. G. Stewart to CHS, 23 Jan. 1941.

88. CHS to ACD, 30 Nov. 1940, 2 Jan. 1941; to JJH, 6 Jan., 8 Jan., 5 Feb., 7 Feb., 23 May 1941; to GIE, 8 Jan., 5 Feb., 19 Mar., 29 Mar., 19 Apr., 10 June, 3 Sept. 1941; George F. Yoran to Metal & Ore Corp., 29 Jan. 1941; JJH to CHS, 21 Jan., 30 Jan., ca. 1 Feb., 4 Mar., 4 Apr., 21 May, 23 May, 23 June, 7 Aug. 1941; to Navy Dept., Bureau of Supplies and Accounts, 4 Apr., 19 May, 23 May 1941; M. A. Norcross to Metal & Ore Corp., 29 May 1941; OFH to CHS, 22 Apr. 1941; JCT to CHS, 26 Apr. 1941; E. R. Stettinius Jr., "General Metals Order No. 1," 1 May 1941; Office of Production Mobilization, "Preference Rating Certificate," re: Navy Contract No. 68968, 16 June 1941, all in SC UOPWA.

89. CHS to GIE, 11 Sept. 1939.

7 || AMERICAN WARTIME METAL POLICY AND PRACTICE

1. Percy W. Bidwell, "A Trade Policy for National Defense"; WSJ, 2 Sept. 1941, 4.

2. Kennedy, *Freedom From Fear*, 476–79.

3. WSJ, 14 Jan. 1942, 2; 22 Jan. 1942, 3; 27 Apr. 1942, 1; Gulick, "War Organization,"

1167–72; Higgs, "Private Profit, Public Risk," 172; Jim F. Heath, "American War Mobilization and the Use of Small Manufacturers, 1939–1943," 302–4.

4. U.S. War Production Board, *Industrial Mobilization for War,* 208–11; Christman, "Donald Nelson and the Army"; J. K. Galbraith, quoted in *"The Good War,"* by Terkel, 323.

5. Donald M. Nelson to President Roosevelt, 17 Apr. 1943, reprinted in "Hearings 1948," 1464–67; Barton J. Bernstein, "The Debate on Industrial Reconversion: The Protection of Oligopoly and Military Control of the Economy," 160–61; Robert D. Cuff, "From the Controlled Materials Plan to the Defense Materials System, 1942–1953," 1; Kennedy, *Freedom From Fear,* 620–28; Heath, "American War Mobilization," 295; Higgs, "Private Profit, Public Risk," 186–90.

6. WSJ, 18 Feb. 1943, 3; NYT, 29 May 1943, 1; 13 June 1943, E6; Hinshaw, *The Home Front,* 46–47; Kennedy, *Freedom From Fear,* 629–30.

7. WSJ, 25 Mar. 1941, 1; 13 Nov. 1942, 1; Hinshaw, *The Home Front,* 62–64.

8. WSJ, 18 Feb. 1943, 3; Eliot Janeway, *The Struggle for Survival: A Chronicle of Economic Mobilization in World War II,* 312–17; Sir Richard Clarke, *Anglo-American Economic Collaboration in War and Peace, 1942–1949,* 16–17; Robert D. Cuff, "Ferdinand Eberstadt, the National Security Resources Board, and the Search for Integrated Mobilization Planning, 1947–1948," 38–39.

9. Higgs, "Private Profit, Public Risk," 184.

10. J. S. McGrath, "International Aspects of War Mineral Procurement," 26; WSJ, 14 Nov. 1945, 1.

11. Seymour E. Harris, "Subsidies and Inflation," 564.

12. WSJ, 10 Aug. 1940, 1; 25 Sept. 1941, 1; McGrath, "International Aspects," 27–28; Hubert W. Davis, "Tungsten," BOMMY 1942, 675–76; Norwood B. Melcher, "Manganese," BOMMY 1944, 581–84; Edwin K. Jenckes and Alice P. van Siclen, "Vanadium," 642–44; *Business Week* (6 Jan. 1945): 28–32; "Hearings 1948," 988–93, 1153–57; Tyler Priest, *Global Gambits: Big Steel and the U.S. Quest for Manganese,* 136–40; Stephen R. Niblo, *War, Diplomacy, and Development: The United States and Mexico, 1938–1954,* 89–91; Furniss, "American Wartime Objectives," 373–79; Herbert A. Franke and M. E. Trought, "Bauxite and Aluminum 1941," 675; McGrath, "International Aspects," 26–28.

13. Carl A. Hatch, "The Truman Committee Reports on Small Business." On problems of small business during mobilization, see Heath, "American War Mobilization."

14. Ickes reviewed and reiterated his 1941 testimony during a hearing before the U.S. Senate, Special Committee to Study and Survey Problems of Small Business Enterprises, 13 Jan. 1943, in "Hearings 1948," 1450–63.

15. E. W. Pehrson, "Review of the Mineral Industries in 1941," xxiii; T. H. Miller and H. M. Meyer, "Copper," 131–33; WSJ, 30 Mar. 1942, 2; 13 June 1942, 6.

16. Norwood B. Melcher, "Manganese and Manganiferous Ores," 583–86; Pehrson, "Review, 1941," xv–xxiii; McGrath, "International Aspects," 26.

17. WSJ, 31 Mar. 1941, 3; U.S. Senate, Subcommittee of the Committee on Military

Affairs, "Hearings . . . Relative to Strategic and Critical Materials and Minerals," 35–40; "Hearings 1948," 895, 1471–94.

18. The statistical information and the accompanying charts for the ferrous and non-ferrous metals industry were compiled from the Bureau of Mines series of *Mineral Yearbooks* (BOMMY) for 1940–46.

19. H. Davis, "Tungsten," BOMMY 1942, 673; 1944, 652–54; Norwood B. Melcher and John Hozik, "Manganese," 597.

20. WSJ, 1 Oct. 1941, 4; 10 Oct. 1941, 2; 18 Oct. 1941, 8; 10 Dec. 1941, 2.

21. Leon Henderson, "The Consumer and Competition," 263–68; Henderson testimony before the House Committee on Banking and Currency, 7 Aug. 1941, excerpted in *Congressional Digest* 20 (Oct. 1941): 236; WSJ, 1 Oct. 1941, 4.

22. WSJ, 26 June 1941, 9; 18 July 1941, 1.

23. WSJ, 20 Sept. 1941, 2.

24. WSJ, 10 Sept. 1941, 3; NYT, 21 Sept. 1941, 35; 25 Sept. 1941, 1.

25. WSJ, 10 Sept. 1941, 3; 28 Nov. 1941, 2; NYT, 31 Oct. 1941, 40; Bartels, "Office of Price Administration," 8–9.

26. WSJ, 1 Dec. 1941, 11; 15 May 1942, 6; 16 July 1942, 5; NYT, 29 Apr. 1942, 15; 18 July 1942, 19; James J. O'Leary, "A General Wage Ceiling," 24–25; Bartels, "Office of Price Administration," 11–12; John Perry Miller, "Military Procurement Policies: World War II and Today," 458–65; Jeffries, *Wartime America,* 27–29.

27. NYT, 6 Aug. 1941, 10; 4 Oct. 1942, 46; WSJ, 10 Oct. 1941, 2; 13 Jan. 1942, 4; 7 May 1943, 1; E. W. Pehrson, "Review of the Mineral Industries in 1942," 21–22; T. Miller and Meyer, "Copper," 153; Alfred L. Ransome and Esther B. Miller, "Zinc," 192; Alfred L. Ransome, "Lead," 167; A. L. Ransome and B. A. Estill, "Zinc," 188; "Hearings 1948," 1300–1301, 1307–11.

28. Pehrson, "Review, 1942," 21–22; Frederick Betz Jr., "Chromite," 632; Norwood B. Melcher, "Manganese," BOMMY 1943, 607, 620–21; C. E. Nighman and Alice P. van Siclen, "Vanadium," 661–62; Jenckes and van Siclen, "Vanadium," 642–44.

29. Pehrson, "Review, 1941," xv–xix; NYT, 7 July 1942, 27; WSJ, 19 Nov. 1942, 1; HDS to NMC, 16 Nov. 1942; CHS to GFS, 27 Nov. 1942; John D. Sumner to CHS, 22 Feb. 1943, attached to OPA press release, 22 Feb. 1943.

30. MH to W. H. Phillips, 31 Dec. 1942; to CHS, 25 Jan. 1943.

31. WSJ, 31 Mar. 1943, 4; 13 Dec. 1943, 1; H. DeWitt Smith to CHS, 14 May 1943; CHS to GFS, 14 Mar. 1944; JDC to CHS, 18 Jan. 1945; Hubert W. Davis, "Tungsten," BOMMY 1944, 652–54.

32. WSJ, 14 May 1942, 3; 14 Nov. 1942, 2; Bartels, "Office of Price Administration," 14–16.

33. WSJ, 8 Sept. 1941, 1; 1 Dec. 1941, 11; 28 Mar. 1942, 5; 7 Apr. 1942, 3; 13 May 1942, 4; 15 May 1942, 6; 16 July 1942, 5; 14 Nov. 1942, 2; 10 Dec. 1941, 2; 11 Dec. 1942, 6; NYT, 18 July 1942, 19; Harris, "Subsidies and Inflation," 561–65; Bartels, "Office of Price Administration," 10–16; Hinshaw, *The Home Front,* 79–80; J. Miller, "Military Procurement Policies," 458–65; Kennedy, *Freedom From Fear,* 627–28; Jeffries, *Wartime America,* 27–29.

34. Janeway, *Struggle for Survival,* 327–31; Terkel, *"The Good War,"* 323; Bartels, "Office of Price Administration," 14–16.

35. NYT, 15 Sept. 1940, 30; Richard A. Lester, "War Controls of Materials, Equipment, and Manpower: An Experiment in Economic Planning," 207–14; John G. Clifford, "Grenville Clark and the Origins of Selective Service," 17–19.

36. Lewis B. Hershey, "Procurement of Manpower in American Wars," 22; Albert A. Blum and J. Douglas Smyth, "Who Should Serve: Pre–World War II Planning for Selective Service," 380–401; Jordan A. Schwarz, "Review: Hershey's Draft," *Reviews in American History* 14 (Mar. 1986): 134–35.

37. L. Hershey, "Procurement of Manpower," 21; George Q. Flynn, "Conscription and Equity in Western Democracies, 1940–75," 6–9.

38. James E. Pate, "Mobilizing Manpower," 154–55; Paul A. C. Koistinen, "Mobilizing the World War II Economy: Labor and the Industrial-Military Alliance," 450–2; George Q. Flynn, *The Mess in Washington: Manpower Mobilization in World War II,* 14.

39. NYT, 19 Apr. 1942, 1; 21 Apr. 1942, 22; Pate, "Mobilizing Manpower," 155–59; Flynn, *Mess in Washington,* 9–35; Kent Carter, "Total War: The Federal Government and the Home Front," 226.

40. Francis E. Rourke, "The Department of Labor and the Trade Unions," 661–72; Lester, "War Controls of Materials," 207–14; Janeway, *Struggle for Survival,* 325–27; Koistinen, "Mobilizing the World War II Economy," 452–63.

41. Pate, "Mobilizing Manpower," 156–57; Koistinen, "Mobilizing the World War II Economy," 454–55; Flynn, *Mess in Washington,* 65–75; Jeffries, *Wartime America,* 25–26.

42. WSJ, 26 June 1941, 9; 12 July 1941, 2; 18 July 1941, 1; NYT, 22 Mar. 1942, 44; 8 Sept. 1942, 25; 17 Sept. 1942, 1; SAUS, 1940–50; Bartels, "Office of Price Administration," 8–9.

43. WSJ, 25 Sept. 1942, 6.

44. NYT, 4 Oct. 1942, 46; William H. Davis, "Aims and Policies of the National War Labor Board," 145–46. Historic CPI charts can be found online.

45. Richard E. Lingenfelter, *The Hardrock Miners: A History of the Mining Labor Movement in the American West, 1863–1893,* 23–33; Alan Derickson, *Workers' Health, Workers' Democracy: The Western Miners' Struggle, 1891–1925,* 155–80, 196–201; Mark Wyman, *Hard Rock Epic: Western Miners and the Industrial Revolution, 1860–1910,* 182–200; OFH to Matt Murphy, 25 Oct. 1942.

46. NYT, 17 July 1942, 1; W. Davis, "Aims and Policies," 145–46; SAUS, 1944.

47. NYT, 12 Aug. 1942, 15; 17 Sept. 1942, 1; 17 Oct. 1942, 1.

48. WSJ, 16 July 1942, 1; 8 Sept. 1942, 2; NYT, 8 Sept. 1942, 25; 13 Sept. 1942, 42; 17 Oct. 1942, 1.

49. NYT, 11 Oct. 1942, E6; 24 Oct. 1942, 7; SAUS, 1939–45.

50. NYT, 17 Sept. 1942, 1; 13 Sept. 1942, 29; 11 Oct. 1942, F1; WSJ, 21 Nov. 1942, 3; Otey M. Scruggs, "The United States, Mexico, and the Wetbacks, 1942–1947," 152.

51. NYT, 21 Oct. 1942, 1; 19 Nov. 1942, 1; WSJ, 21 Oct. 1942, 1; 8 Dec. 1942, 4; 1 May 1943, 3; Albert A. Blum, "The Fight for a Young Army," 81–84.

52. WSJ, 10 Feb. 1943, 4; NYT, 3 Mar. 1943, 1; Paul McNutt, "Is America's War Manpower Policy Striking the Proper Balance Between Civilian and Armed Forces? Pro," 87–88; Hinshaw, *The Home Front,* 79–80; U.S. War Production Board, *Industrial Mobilization for War,* 703–4.

53. John D. Bradley et al., "Yellow Pine Mine"; T. Miller and Meyer, "Copper," 131–33; WSJ, 9 Sept. 1943, 2. Annual mineral production data were compiled from U.S. Bureau of Mines *Minerals Yearbooks,* 1941–45, and from U.S. Geological Survey, *Mineral Commodity Statistics.* Monthly copper production data, 1942–50, were recently compiled by the U.S. Geological Survey and released upon request.

54. EMJ 140 (Nov. 1939): 29–30; "Hearings 1948," 989; Pehrson, "Stockpiling," 8; Alfred E. Eckes, *The United States and the Global Struggle for Minerals,* 94–95, 99–103; Priest, *Global Gambits,* 136–40.

55. D. Miller, *You Can't Do Business With Hitler,* 109–32; Eckes, *Global Struggle,* 99–102, 107–8.

56. Charles H. Behre Jr., "Mineral Resources and the Atlantic Charter"; Kennedy, *Freedom From Fear,* 496.

57. U.S. Senate, Subcommittee of the Committee on Military Affairs, "Hearings . . . Relative to Strategic and Critical Materials and Minerals," 53; Eckes, *Global Struggle,* 99–102.

58. WSJ, 17 Feb. 1943, 8; 7 Sept. 1943, 6; "Hearings 1948," 1464–67; Eckes, *Global Struggle,* 113–14.

59. WSJ, 6 Jan. 1944, 1; "Hearings 1948," 1443–48; Kennedy, *Freedom From Fear,* 587–88. For conflicting views on the postwar impact of metals overexpansion, see Eckes, *Global Struggle,* 112–14, 159–56; Alan M. Bateman, "Our Future Dependence on Foreign Minerals," 26.

60. James Scrugham, Extracts from Speech, CR (7 Oct. 1941), copy in SC UOPWA.

61. WSJ, 4 Jan. 1943, 11; 17 Feb. 1943, 8.

62. "Hearings 1948," 1067–99.

63. WSJ, 19 June 1943, 1; 26 Aug. 1943, 1; "Hearings 1948," 989.

64. WSJ, 4 Oct. 1943, 5; 13 Dec. 1943, 1; 29 Jan. 1944, 2; "Hearings 1948," 1103–4, 1464–71; Public Law 520, 79th Cong., 60 *U.S. Statutes* 596; U.S. Office of Technology Assessment, "An Assessment of Alternative Economic Stockpiling Policies," 222–23.

65. WSJ, 4 Jan. 1943, 11; 28 Oct. 1943, 3; NYT, 27 Aug. 1943, 19; statement, Donald M. Nelson to the president, 17 Apr. 1943, in "Hearings 1948," 1464–67.

8 || NEVADA-MASSACHUSETTS IN WORLD WAR II

1. CHS to JCT, 5 Sept. 1941.

2. JDC to CHS, 29 Aug. [1941]; CHS to GIE, 3 Sept. 1941, 13 Nov. 1941; to Senator Pat McCarran, 22 Oct. 1941; to H. K. Masters, 4 Dec. 1941; McCarran to CHS, 4 Sept. 1941.

3. CHS to GIE, 13 Nov. 1941.

4. JJH to Clifford L. Ach, 26 June 1941; TT, CHS and CHSJR, 21 July 1941; CHS to GIE,

27 Sept. 1941; to W. R. Fablinger, 29 Apr. 1942; to GFS, 9 July 1942; NMC Contract P 326 with Metals Reserve, 18 Oct. 1941, amended 27 Mar. 1942; Pehrson, "Review, 1941," xxiii.

5. CHS to GIE, 19 Apr., 10 June, 3 Sept. 1941, 17 Jan. 1942; to J. K. Galbraith, 16 Dec. 1941; to Geo. W. Malone, 28 Apr. 1942.

6. CHS to GIE, 29 Apr., 21 May 1940; to T. H. McGraw, 5 Sept. 1940; to IBJ, 17 Mar. 1942; Segerstrom, "War's Effect on Tungsten."

7. CHS to GIE, 29 Mar., 14 Oct. 1941; MI 49 (1940): 605–20; MW 3 (Dec. 1941): 36; Hubert W. Davis, "Tungsten," BOMMY 1941, 648–49.

8. CHS to GIE, 19 Apr., 1 Nov. 1941; to CGF, 17 Mar. 1942; TT, CHS and CHSJR, 21 July 1941; IBJ to CHS, 9 Mar. 1942.

9. CHS to MH, 22 June, 3 July 1942; to GFS, 9 July 1942.

10. CHS to GIE, 10 June, 14 Oct. 1941; to GFS, 9 May 1944.

11. IBJ to HDS, 30 Apr. 1943.

12. CHS to GIE, 19 Apr., 10 June, 3 Sept., 1 Nov. 1941, 17 Jan. 1942; to IBJ, 17 Mar. 1942; to HDS, 2 Sept. 1943.

13. CHS to J. B. Heilman, 7 Apr. 1942; to GFS, 4 Aug. 1944.

14. CHS to JJH, 7 Aug. 1941; to JCT, 5 Sept. 1941; to G. Temple Bridgeman, 13 May 1943; "Agreement Between Rare Metals Co. and John Pedro"; *Milford News,* 23 Apr. 1942, 1, 8; NMC Production Reports, 1925–50; Palomar Scheelite Mine report and Old Hickory Mine report, in "War Production Board Mine and Smelter Production Report," 15 June 1944 (ms. on printed form), all in SC UOPWA.

15. TT, CHS and OFH, ca. Dec. 1942; "Agreement and Lease Between CHS and Rare Metals Corp.," 20 Feb. 1943; CHS to G. Temple Bridgman, 22 June 1942, 13 May 1943; to GFS, 7 Sept. 1944; to Rare Metals Corp., 10 Nov. 1944.

16. WSJ, 28 Mar. 1942, 5.

17. CHS to MKS, 1 Sept. 1942; to GFS, 23 Sept. 1942; HDS to NMC, 16 Nov. 1942; TT, CHS and JJH, 23 Nov. 1942; TT, CHS to MKS, 29 Dec. 1942.

18. CHS to Henry C. Carlisle, 24 Nov. 1942; Carlisle to CHS, 2 Dec. 1942.

19. TT, CHS to JJH, 2 Mar. 1943.

20. CHS to MH, 29 Jan. 1943; to E. Franklin Hatch, 30 Mar. 1943; to HDS, 13 May 1943; to PEC, 6 Aug. 1943; to JKG, 23 Sept. 1943; IBJ to HDS, 30 Apr. 1943.

21. TT, CHS to HDS, 5 May 1943; CHS to IBJ, 10 May 1943; to HDS, 13 May 1943; to JKG, 2 Sept. 1943; affidavit, CHS, 2 Apr. 1945.

22. WSJ, 26 Aug. 1943, 1; NYT, 27 Aug. 1943, 19; Henry C. Carlisle to CHS, 6 Oct. 1943.

23. WSJ, 26 Aug. 1943, 1; 9 Sept. 1943, 2; TT, CHS to CHSJR, 20 Oct. 1943.

24. T. Miller and Meyer, "Copper," 131–33; BOMMY 1944 (1946), 141; Ransome and Estill, "Zinc," 187–88; A. L. Ransome and J. H. Schaum, "Lead," 170; H. Davis, "Tungsten," BOMMY 1944, 652–54; Jenckes and van Siclen, "Vanadium," 642–46; John H. Weitz and Mary E. Trought, "Bauxite and Aluminum," 669–71; Edwin K. Jenckes and Katharine D. Wildensteiner, "Chromite," 614; Helena M. Meyer and Alethea W. Mitchell, "Mercury," 702–4; CHS to GFS, 24 Oct. 1944; to ACD, 10 May 1945; WSJ, 28 Oct. 1943, 3.

25. CHS to MKS, 12 Nov. 1943.

26. CHS to HDS, 2 Sept. 1943; to JKG, 23 Sept., 27 Nov. 1943; to WJL, 29 Nov. 1943; D. Segerstrom, "Brief History"; CHS to GFS, 19 Jan. 1944.

27. CHS to GIE, 11 Apr. 1939; to J. R. Van Fleet, 14 Dec. 1943; to WJL, 4 Jan. 1944.

28. CHS to Worthen Bradley, 20 Apr. 1937; P. R. Bradley to CHS, 26 Apr. 1937; TT, JJH and CHS, 26 Apr. 1937; TT, CHS and JJH, 2 Nov. 1937; MI 49 (1940): 607.

29. J. Bradley et al., "Yellow Pine Mine"; H. Davis, "Tungsten," BOMMY 1942, 673; 1943, 676; 1944, 652; 1945, 660.

30. ACD to CHS, 29 Jan. 1944; CHS to GFS, 7 Apr. 1944.

31. COH to Pacific Tungsten Leasing Co., 1 May 1924; CHS to Paul M. Tyler, 28 Dec. 1930; TT, JJH and CHS, 1 Apr. 1937; ACD to CHS, 15 June 1937; TT, CHS and JJH, 8 Sept. 1937; CHS to ACD, 19 Sept. 1938; Ingham, "Iron and Steel," 112.

32. TT, CHS and CHSJR, ca. 1 May 1944.

33. Ibid., ca. 1 May 1944; CHS to GFS, 7 Apr. 1944.

34. CHS to ACD, 17 Aug. 1944; Hubert W. Davis, "Tungsten," BOMMY 1945, 660.

35. Kurtak, *Mine in the Sky,* 91–100; H. Davis, "Tungsten," BOMMY 1945, 660, 664; 1946, 1196.

36. John J. Judge to Walter A. Janssen, 8 Feb. 1941; JCT to Cordell Hull, 2 Apr. 1941; TT, CHS and JJH, 2 Apr. 1941; ca. 29 Apr. 1941; Bailey, *Diplomatic History,* 791.

37. TT, JJH and CHS, 17 Apr., 23 Apr., 30 Apr. 1940.

38. CHS to GIE, 7 Feb. 1941; to JJH, 27 Feb. 1942; TT, CHS and JJH, 5 Feb. 1942.

39. J. H. East and Russell R. Trengove, "Investigation of Nightingale Tungsten Deposit, Pershing County, Nev."; Jack A. Crowley, "Garnet and Clinozoisite From the Nightingale Mining District."

40. Norman C. Stines, "Memorandum on Nightengale [sic] and Starr [sic] Tungsten Ore Zones in Western Pershing County, Nevada"; Ward C. Smith and Philip W. Guild, "Tungsten Deposits of the Nightingale District Pershing County, Nevada," 41–56.

41. JJH to CHS, 2 Oct. 1931; CHS to ACD, 31 July, 9 Aug. 1933.

42. CHS to JGC, 7 Sept. 1933, 9 Mar., 29 Mar. 1934; OFH to CHS, 19 Sept. 1933; TT, CHS to JJH, 23 Oct., 6 Nov. 1933; JGC to ACD, 7 Feb. 1934; to CHS, 2 Feb., 3 Mar., 12 Mar. 1934.

43. C. G. Clark to CHS, 12 Mar. 1934.

44. R. D. George, "Nightingale and Eastern Star Tungsten Properties, Pershing Co. NV"; JGC to CHS, 21 Nov. 1939. Two months after George's report, another geologist estimated Nightingale's total ore reserves at more than one million tons. See Stines, memorandum, 5 July 1939.

45. CHS to JGC, 28 Nov. 1939; JGC to CHS, 6 Dec. 1939; MH to E. A. Lucas, 21 Feb. 1940.

46. E. A. Lucas to JGC, 17 May 1940.

47. V. R. Lansingh to E. A. Lucas, 7 June 1940.

48. V. R. Lansingh to CHS, 1 Mar. 1941; CHS to V. R. Lansingh, 3 Mar. 1941; to MH, 22 June 1942.

49. "Confidential Report, Boulder Credit Bureau for Retail Credit Bureau, Sonora,

Calif."; "[Confidential Credit Report on] Gold, Silver and Tungsten, Inc., Boulder, Co.";
MH to Henry Schultheis, 22 Oct. 1942.

50. Wm. M. Kearney to CHS, 7 Jan. 1943; CHS to Kearney, 29 Jan. 1943.

51. MH to CHS, 16 July 1943; CHS to Wm. M. Kearney, 16 July 1943; to JKG, 2 Sept. 1943; James A. Adams to Metals Reserve Co., 2 Sept. 1943. The carbon copies of the letters written by "Adams" and Segerstrom are in the company files and show identical formatting and mechanical characteristics.

52. Wm. M. Kearney to James A. Adams, 10 Sept. 1943; to CHS, 12 Oct. 1943, 7 Aug. 1944; MH to Henry R. Schultheis, 18 Oct. 1943.

53. CHS to Wm. M. Kearney, 10 Nov. 1944; to OFH, 12 May 1945; L. E. Davis and Warren C. Fischer, "Nevada" (1956), 712.

54. CHS to GIE, 3 Sept. 1941.

55. TT, JJH and CHS, 7 Oct. 1941; TT, CHS and CHSJR, undated [ca. June 1943]. See chart in chapter 5.

56. Ruth G. Weintraub and Rosalind Tough, "Federal Housing and World War II," 155–56, 160–61.

57. CHS to GIE, 6 Mar. 1942; to GFS, 9 July, 1 Sept. 1942.

58. CHS to GFS, 20 Aug., 1 Sept., 23 Sept. 1942; to MKS, 29 Dec. 1942; to MH, 29 Jan. 1943; NYT, 9 Sept. 1942, 25.

59. TT, CHS to Edwin K. Jenckes, 4 Dec. 1941; CHS to GIE, 14 Oct. 1941; to PFK, 13 Mar. 1942; to GFS, 24 Oct. 1944, 10 Mar., 20 July 1945; to OFH, 12 May 1945; W. W. Mutter to CHS, 8 Nov. 1941. Labor and technology conflicts are common themes in mining history. See Wyman, *Hard Rock Epic*, 84–117; Larry Lankton, *Cradle to Grave: Life, Work, and Death at the Lake Superior Copper Mines*, 99–109; Logan Hovis and Jeremy Mouat, "Miners, Engineers, and the Transformation of Work in the Western Mining Industry, 1880–1930."

60. CHS to GFS, 1 Sept., 5 Nov. 1942; to MKS, 29 Dec. 1942; to E. Franklin Hatch, 30 Mar. 1943.

61. IBJ to HDS, 30 Apr. 1943.

62. T. Miller and Meyer, "Copper," 131–33; E. W. Pehrson, "Review of the Mineral Industries in 1944," 8; *Nevada State Journal*, 4 Aug. 1943 (clip), in SC UOPWA; CHS to PEC, 6 Aug. 1943; to E. Franklin Hatch, 30 Aug. 1943; Chas. F. Mulford to CHS, 24 July 1943; Hinshaw, *The Home Front*, 79–80; Jeffries, *Wartime America*, 23–26.

63. CHS to E. Franklin Hatch, 30 Aug. 1943; to CHSJR, 20 Oct. 1943; WSJ, 28 Oct. 1943, 3.

64. Pehrson, "Review, 1944," 8–9; Blum and Smyth, "Who Should Serve," 366–76.

65. WSJ, 3 Feb. 1943, 3.

66. TT, JJH and CHS, 16 Jan., 26 Jan. 1943; Richard W. Haesler to author, 20 Oct. 2008.

67. WSJ, 3 Feb. 1943, 3; L. A. Levensaler to CHS, 22 Mar. 1944; CHS, Selective Service Affidavit (Ms. on printed form), 2 Apr. 1945.

68. CHS to Selective Service Board, Los Angeles, 6 May 1941, 16 Mar. 1944; CHSJR to Selective Service Local Board 238, 8 Nov. 1941, 19 June, 15 Dec. 1942, 21 July 1943, 15 Jan. 1944.

69. Undated TT, CHSJR to CHS, ca. June 1943.

70. Undated TT, CHS to CHSJR, ca. June 1943; CHS to Avery D. Stitser, 11 July 1944.

71. JDC to CHS, 18 Jan. 1945.

72. CHS to A. George Keating, 20 Jan. 1945; to L. A. Levensaler, 15 Feb. 1945; MH to CHS, 9 Mar. 1945; Clinton S. Golden to CHS, 19 Apr. 1945; H. Davis, "Tungsten," BOMMY 1945, 661; Edwin K. Jenckes, "Molybdenum," 638.

73. R-M, 28 June 1945, 1, in SC UOPWA; Morison and Commager, *American Republic,* 780–82.

74. CHS to GFS, 12 Sept. 1945; to GIE, 4 Jan. 1946.

75. CHS to GIE, Sonora, 4 Jan. 1946.

76. Glenn J. Degner to CHS, 29 Oct. 1943; CHS to Degner, 9 Nov. 1943.

77. CHS to GFS, 22 Aug. 1945.

EPILOGUE

1. CHS obituary in *Tuolumne Independent,* 6 Aug. 1946, 1–2.

2. D. Segerstrom, "Brief History"; "Agreement Between Nevada Massachusetts Company . . . and District 50 United Mine Workers of America . . . Local No. 13747, Effective Jan. 1, 1957."

3. Robert E. Wallace, "Nevada," 670; L. E. Davis and Warren C. Fischer, "Nevada" (1955), 712–13; L. Davis and Fischer, "Nevada" (1956), 758–59; R. B. Maurer and Robert E. Wallace, "Nevada," 606; Robert G. Reeves and Victor E. Kral, "Geology and Iron Ore Deposits of the Buena Vistal Hills, Churchill and Pershing Counties, Nevada"; "Agreement of Sale, Between Chas. H. Segerstrom Jr. and John M. Heizer, Partnership of the Wolfram Co., Seller, and Space Metals Inc., Buyer."

4. Merrill D. Cronwall to CHSJR, 22 Feb. 1958; Howmet Corporation history; Emminger, "Reminiscences," NC337.

5. Arthur C. Meisinger, "Nevada," 490; V. Anthony Cammarota Jr., "Nevada" (1970), 456; Anthony V. Cammarota Jr., "Nevada" (1971), 466.

6. The Stock Piling Act of 1946, Public Law 520; Stockpiling Hearings Before the Committee on Military Affairs, Feb. 5 and 26, 1946; S. H. Williston, testimony, 20 May 1948, all three reprinted in "Hearings 1948," 889, 1177–1209, 1232–35; E. W. Pehrson, "Our Mineral Resources and Security"; NYT, 24 July 1946, 1; Claude Barbier, *The Economics of Tungsten,* 78; Eckes, *Global Struggle,* 136–45; Vernon, *Two Hungry Giants,* 77–78; Charlotte Twight, "The Political Economy of the National Defense Stockpile," 778–83.

7. WSJ, 29 July 1950, 2; 16 Aug. 1950, 2; 2 Nov. 1950, 3; 5 Jan. 1951, 1; 23 July 1953, 13; 16 July 1956, 12; NYT, 22 Aug. 1956, 16; Melvyn P. Leffler, "The American Conception of National Security and the Beginnings of the Cold War, 1945–48," 369–80.

8. WSJ, 31 Dec. 1965, 1; NYT, 28 Nov. 1976, 184.

9. Eckes, *Global Struggle,* 163–73, 209–15.

10. Ibid., 221, 230–36.

11. NYT, 24 Sept. 1964, 57; 15 Mar. 1973, 89; 2 Oct. 1976, F31; 11 Apr. 1982, E4; 28 Aug.

1986, D3; WSJ, 31 Dec. 1965, 1; 16 Mar. 1981, 7; 15 Apr. 1981, 1; 24 Apr. 1981, 4; 26 June 1981, 37; 7 Aug. 1981, 26; 2 Oct. 1981, 29; 12 Oct. 1981, 16; National Research Council, Committee on Assessing the Need for a Defense Stockpile, *Managing Materials for a 21st Century Military*, 5–14; Eckes, *Global Struggle*, 248–50.

  12. Raymond F. Mikesell, *Nonfuel Minerals: Foreign Dependence and National Security*, 1; National Research Council, Committee on Assessing the Need for a Defense Stockpile, *Managing Materials for a 21st Century Military*, box 1–1, 6–4. For current stockpile statutory provisions, see 50 U.S. Code 98. For current stockpile inventory, see Defense National Stockpile Center.

  13. For Buy American legal provisions and exceptions, see 41 U.S. Code 10a; and "Federal Acquisition Regulation." For an example of MOU principles and practice in international relations, see John H. McNeill, "International Agreements: Recent U.S.-U.K. Practice Concerning the Memorandum of Understanding." Interest groups continue to argue over Buy American laws, as seen most recently in the congressional debates in February 2009 over the Obama administration's $900 billion economic stimulus package.

BIBLIOGRAPHY

A NOTE ON SOURCES

As a retired historian living on a remote section of the northern California coast, I have taken full advantage of the Internet resources available, especially full-text reproductions from Google Print, Jstor, Lexis, Ebsco, the Library of Congress Thomas Web site, and other online databases. Some are open to anyone with an Internet hookup; others require academic affiliation. Even snippets of longer text have brought to light obscure publications and provided useful leads to further research. This book is a good example of how the "fair use" guidelines of the 1978 copyright law can be used both for the advancement of learning and for the benefit of scholarly authors whose obscure or limited-edition monographs find larger audiences. To avoid redundancy, online citations are listed in the bibliography but not in endnotes.

For background chapters and national politics, as well as macroeconomic analysis, I have relied almost exclusively on secondary sources and published government documents. In order to concentrate on telling the Nevada-Massachusetts story, I preferred to follow the lead of those who know these peripheral subjects best rather than "reinvent the wheel."

MANUSCRIPT COLLECTIONS AND UNPUBLISHED SOURCES

The chief source of information on Charles H. Segerstrom and the Nevada-Massachusetts Company is the Segerstrom Collection at the Holt-Atherton Library, University of the Pacific. Accumulated over a period of years from family deposits, it totals some five hundred linear feet of business records, family and business correspondence, scrapbooks, maps, diagrams, charts, photographs, and publications, covering multiple family and business enterprises from 1910 to the 1960s. Segerstrom was heavily invested and directly involved in the management of at least eight different mining companies during his long career, but spent most of his time and resources on Nevada-Massachusetts and its subsidiaries. They represent the bulk of the collection.

The Segerstrom Collection is not fully organized and cataloged, although a preliminary organization has been completed and a finding aid prepared. Researching the collection will challenge the patience and endurance of any scholar, but the friendly help of the

Holt-Atherton staff is considerable consolation. All references to Nevada-Massachusetts, and all Segerstrom and related correspondence, unless otherwise indicated, are located in the Segerstrom Collection. The standard citation is SC UOPWA.

Manuscripts in the Segerstrom Collection

Among the Segerstrom papers are the following manuscripts and published sources:

"Agreement Between Nevada Massachusetts Company . . . and District 50 United Mine Workers of America . . . Local No. 13747, Effective January 1, 1957."

"Agreement Between Rare Metals Co. and John Pedro." February 1944.

"Agreement of Sale [draft], Between Chas. H. Segerstrom Jr. and John M. Heizer, Partnership of the Wolfram Co., Seller, and Space Metals Inc., Buyer." Reno, April 1969.

"American Institute of Mining and Metallurgical Engineers." Engineering News Bureau. News release, February 19, 1935.

"American Manganese Producers Association." News release (typescript), August 7, 1939.

American Tungsten Association. "Tungsten, the Aristocrat of Rare Metals." July 1937.

"Articles of Agreement and Association, Nevada-Massachusetts Exploration Co." Undated typescript, ca. 1928.

Callbreath, J. F. "The Industrial Recovery Act: Its Purposes and the Obligations of Industry Thereunder." Ms. (draft), pencil date November 11, 1933.

"[Confidential Credit Report on] Gold, Silver and Tungsten, Inc., Boulder Co." July 31, 1942.

"Confidential Report, Boulder Credit Bureau for Retail Credit Bureau, Sonora Calif." May 13, 1942.

Conover, J. D. "Address Before the 26th American Zinc Institute, 1944." Excerpted in American Metal Market (April 18, 1944).

Emmons, J. V. The Molybdenum-Tungsten High Speed Steels Marketed Under the General Trade Name Mo-Max. Cleveland: Cleveland Twist Drill Co. [1937].

George, R. D. "Nightingale and Eastern Star Tungsten Properties, Pershing Co. NV." Typescript, Geology Department, University of Colorado, May 15, 1939.

Hershey, Oscar H. "Geological Report on the Silver Dyke Mine." Typescript, June 3, 1930.

Kiplinger, Willard M. "Industry Control Postscript No. 21, Emergency Service." Kiplinger Washington Letter, October 21, 1933.

Loring, W. J. "Plan Regarding Pacific Tungsten." Ms. (draft), November 14, 1923.

"Memorandum of Agreement, 1 Jul 1932, Between Nevada-Massachusetts Co., and Molybdenum Corp. of America." Typescript.

Pacific Tungsten Company Minute Book, 1918–24.

Poole, C. W. "Plan for Resuming Operations at Nevada-Humboldt and Pacific Tungsten Properties." Typescript, October 8, 1923.

"Quitclaim Deed, 20 September 1929, by Which O. F. Heizer, A. Ranson, and C. H.

Segerstrom [Representing Nevada-Massachusetts Exploration Co.] . . . Release All Claims [to Cottonwood and Silver Dyke]."

Segerstrom, Charles H. "Annual Report of Operations." Typescript, February 1, 1932.

———. "Annual Report of the Nevada-Massachusetts Company, Inc." Typescript, May 26, 1933.

———. "[Annual Report] to the Offices and Directors of the Nevada-Massachusetts Co." Typescript, July 28, 1937.

———. "Herbert Hoover, the Man Leading the Way to Recovery." Typescript, March 11, 1932.

———. "Report as of 30 June 1926, to Officers and Directors of Nev-Mass Co." Typescript, July 26, 1926.

———. "Semi-annual Report to Officers and Directors of the Nevada-Massachusetts Co." Typescript, March 30, 1931.

Segerstrom, Donald I. "A Brief History of the Mill City Tungsten Mines, Pershing County." Typescript, 1971.

"Tungsten and Tungsten Products Industries Code of Fair Competition Under the National Industrial Recovery Act." Typescript draft, ca. July 1, 1933.

U.S. Tariff Commission. "A Review of the Tungsten Industry." Unpaged draft, May 1918.

War Minerals Relief Commission. PTC claim chronology. Typescript, February 3, 1932.

———. Reporter's transcript, San Francisco, July 24, 1919, Claim No. 1018.

OTHER UNPUBLISHED DOCUMENTS AND COLLECTIONS

Dun & Bradstreet. "Report on Haesler Metal and Ore Corp., 18 August 1959." Personal copy in possession of Richard W. Haesler.

Emminger, William Glenn. "Reminiscences, 1965." Typescript. University of Nevada–Reno Special Collections.

Haesler, Richard W. Correspondence with author, 2008.

Segerstrom, Charles Homer, Sr. and Jr. Papers. 1916–57? Nevada Historical Society.

Stevenot, Archie Douglas. Papers. 1918–66. Holt-Atherton Library, University of the Pacific.

Stines, Norman C. "Memorandum on Nightengale [sic] and Starr [sic] Tungsten Ore Zones in Western Pershing County, Nevada." Typescript, Seattle, July 5, 1939. University of Nevada–Reno Special Collections.

GOVERNMENT DOCUMENTS

Betz, Frederick, Jr. "Chromite." In U.S. Bureau of Mines, Minerals yearbook 1942. Year 1942 (1943), 631–44. Available at http://digicoll.library.wisc.edu/EcoNatRes/.

Biographical Directory of the U.S. Congress. Available at http://bioguide.congress.gov/.

Cammarota, V. Anthony, Jr. "Nevada." In Minerals Yearbook Area Reports: Domestic, 1970 (1970), 455–69ff. Available at http://digicoll.library.wisc.edu/EcoNatRes/.

———. "Nevada." In *Minerals Yearbook Area Reports: Domestic, 1971* (1971), 465–78. Available at http://digicoll.library.wisc.edu/EcoNatRes/.

Congressional Record. April–May 1935; January 25, 1938; April 25, 1939.

Davis, Hubert W. "Tungsten." In U.S. Bureau of Mines, Minerals yearbook 1941. Year 1941 (1943), 643–53. Available at http://digicoll.library.wisc.edu/EcoNatRes/.

———. "Tungsten." In U.S. Bureau of Mines, Minerals yearbook 1942. Year 1942 (1943), 673–84. Available at http://digicoll.library.wisc.edu/EcoNatRes/.

———. "Tungsten." In U.S. Bureau of Mines, Minerals yearbook 1943. Year 1943 (1945), 670–82. Available at http://digicoll.library.wisc.edu/EcoNatRes/.

———. "Tungsten." In U.S. Bureau of Mines, Minerals yearbook 1944. Year 1944 (1946), 652–66. Available at http://digicoll.library.wisc.edu/EcoNatRes/.

———. "Tungsten." In U.S. Bureau of Mines, Minerals yearbook 1945. Year 1945 (1947), 660–73. Available at http://digicoll.library.wisc.edu/EcoNatRes/.

———. "Tungsten." In U.S. Bureau of Mines, Minerals yearbook 1946. Year 1946 (1948), 1192–1204. Available at http://digicoll.library.wisc.edu/EcoNatRes/.

Davis, L. E., and Warren C. Fischer. "Nevada." In U.S. Bureau of Mines, *Minerals Yearbook Area Reports, 1955* (1958), 673–716. Available at http://digicoll.library.wisc.edu/EcoNatRes/.

———. "Nevada." In U.S. Bureau of Mines, *Minerals Yearbook Area Reports, 1956* (1958), 721–69ff. Available at http://digicoll.library.wisc.edu/EcoNatRes/.

Dean, R. S., et al. "Flotation of Scheelite." *Report of Investigations* (U.S. Bureau of Mines Progress Reports, Metallurgical Division) 3331 (January 1937): 41–42.

DeFrancesco, Carl A., and Kim B. Shedd. "Tungsten Statistics." U.S. Geological Survey Web site, http://minerals.usgs.gov/minerals/pubs/commodity/.

East, J. H., and Russell R. Trengove. "Investigation of Nightingale Tungsten Deposit, Pershing County, Nev." *Report of Investigations* (U.S. Bureau of Mines) 4678 (April 1950).

Franke, Herbert A., and M. E. Trought. "Bauxite and Aluminum 1941." In U.S. Bureau of Mines, Minerals yearbook 1941. Year 1941 (1943), 655–84. Available at http://digicoll.library.wisc.edu/EcoNatRes/.

Furness, J. W. "The Marketing of Tungsten Ores and Concentrates." *Trade Information Bulletin* (U.S. Bureau of Foreign and Domestic Commerce), no. 643 (1929).

Geehan, Robert W. "Tungsten." In *Mineral Facts and Problems.* Bulletin 556. Washington, D.C.: U.S. Bureau of Mines, 1955.

Heizer, Ott F. "Method and Cost of Mining Tungsten at the Nevada-Massachusetts Co. Mines, at Mill City, Nev." *Information Circular* (U.S. Bureau of Mines) 6284 (June 1930).

Hess, Frank L. "Tungsten." In U.S. Bureau of Mines, *Mineral Resources of the United States, 1915.* Pt. 1, *Metals* (1917), 823–30.

———. "Tungsten." In U.S. Bureau of Mines, *Mineral Resources of the United States, 1916.* Pt. 1, *Metals* (1919), 789–96.

———. "Tungsten." In U.S. Bureau of Mines, *Mineral Resources of the United States, 1917.* Pt. 1, *Metals* (1921), 931–95.

———. "Tungsten." In U.S. Bureau of Mines, *Mineral Resources of the United States, 1918.* Pt. 1, *Metals* (1921), 973–1026.

———. "Tungsten in 1930." In U.S. Bureau of Mines, *Mineral Resources of the United States, 1930.* Pt. 1, *Metals* (1932), 179–207.

———. "Tungsten." In U.S. Bureau of Mines, Minerals yearbook 1932–33. Year 1931–32 (1933), 271–79. Available at http://digicoll.library.wisc.edu/EcoNatRes/.

———. "Tungsten." In U.S. Bureau of Mines, Minerals yearbook 1934. Year 1934 (1935), 491–98. Available at http://digicoll.library.wisc.edu/EcoNatRes/.

Jenckes, Edwin K. "Molybdenum." In U.S. Bureau of Mines, Minerals yearbook 1945. Year 1945 (1947), 637–48. Available at http://digicoll.library.wisc.edu/EcoNatRes/.

Jenckes, Edwin K., and Alice P. van Siclen. "Vanadium." In U.S. Bureau of Mines, Minerals yearbook 1944. Year 1944 (1946), 642–51. Available at http://digicoll.library.wisc .edu/EcoNatRes/.

Jenckes, Edwin K., and Katharine D. Wildensteiner. "Chromite." In U.S. Bureau of Mines, Minerals yearbook 1944. Year 1944 (1946), 602–18. Available at http://digicoll.library .wisc.edu/EcoNatRes/.

Jordan, Frank C. *Statement of Vote at General Election Held on November 8, 1932.* Sacramento: State Printing Office, 1932.

Kelly, T. D., and G. R. Matos, comps. "Historical Statistics for Mineral and Material Commodities in the United States." *USGS Data Series* 140. Available at http://pubs.usgs .gov/.

Kerr, Paul F. "Geology of the Tungsten Deposits Near Mill City, Nevada." *University of Nevada Bulletin* 28 (March 15, 1934).

———. "Tungsten-Bearing Manganese Deposit at Golconda, Nevada." *Bulletin of the Geological Society of America* 51 (September 1, 1940).

———. "Tungsten Mineralization at Oreana, Nevada." *Economic Geology* 33 (June–July 1938): 390–425.

———. "The Tungsten Mineralization at Silver Dyke, Nevada." *University of Nevada Bulletin* 30 (June 15, 1936).

King, William H. "Investigation of Nevada-Massachusetts Tungsten Deposits, Pershing County, Nev." *Report of Investigations* (U.S. Bureau of Mines) 4634 (February 1950).

Larson, L. P., et al. "Availability of Tungsten at Various Prices From Resources in the United States." *Information Circular* (U.S. Bureau of Mines) 8500 (1971).

Leaver, E. S., and M. B. Royer. "Flotation for Recovery of Scheelite From Slimed Material." *U.S. Bureau of Mines Technical Paper* 585 (1938): 1–24.

Lemmon, Dwight M., and John V. N. Dorr. "Tungsten Deposits of the Atolia District, San Bernardino and Kern Counties, California." *U.S. Geological Survey Bulletin* 922-H (1940).

Maurer, R. B., and Robert E. Wallace. "Nevada." In U.S. Bureau of Mines, *Minerals Yearbook Area Reports, 1952* (1955), 572–609. Available at http://digicoll.library.wisc.edu/ EcoNatRes/.

McGrath, J. S. "International Aspects of War Mineral Procurement." In U.S. Bureau of Mines, Minerals yearbook 1942. Year 1942 (1943), 25–34. Available at http://digicoll.library.wisc.edu/EcoNatRes/.

Meisinger, Arthur C. "Nevada." In U.S. Bureau of Mines, *Minerals Yearbook Area Reports: Domestic, 1969* (1969), 481–96. Available at http://digicoll.library.wisc.edu/EcoNatRes/.

Melcher, Norwood B. "Manganese." In U.S. Bureau of Mines, Minerals yearbook 1943. Year 1943 (1945), 605–23. Available at http://digicoll.library.wisc.edu/EcoNatRes/.

———. "Manganese." In U.S. Bureau of Mines, Minerals yearbook 1944. Year 1944 (1946), 581–601. Available at http://digicoll.library.wisc.edu/EcoNatRes/.

———. "Manganese and Manganiferous Ores." In U.S. Bureau of Mines, Minerals yearbook 1941. Year 1941 (1943), 583–601. Available at http://digicoll.library.wisc.edu/EcoNatRes/.

Melcher, Norwood B., and John Hozik. "Manganese." In U.S. Bureau of Mines, Minerals yearbook 1945. Year 1945 (1947), 590–606. Available at http://digicoll.library.wisc.edu/EcoNatRes/.

Meyer, Helena M., and Alethea W. Mitchell. "Mercury." In U.S. Bureau of Mines, Minerals yearbook 1945. Year 1945 (1947), 702–23. Available at http://digicoll.library.wisc.edu/EcoNatRes/.

Miller, T. H., and H. M. Meyer. "Copper." In U.S. Bureau of Mines, Minerals yearbook 1942. Year 1942 (1943), 127–61. Available at http://digicoll.library.wisc.edu/EcoNatRes/.

Nevada. State Mine Inspector. "Biennial Reports to the State Legislature." In *Appendix to Journals of State and Assembly.* Carson City: State Printing Office, 1926–41.

———. "Mining Reports, 1923–1948." Typescript. Nevada State Archives, Carson City.

Nighman, C. E., and Alice P. van Siclen. "Vanadium." In U.S. Bureau of Mines, Minerals yearbook 1943. Year 1943 (1945), 660–69. Available at http://digicoll.library.wisc.edu/EcoNatRes/.

———. "Vanadium." In U.S. Bureau of Mines, Minerals yearbook 1944. Year 1944 (1946). Available at http://digicoll.library.wisc.edu/EcoNatRes/.

Pehrson, E. W. "Review of the Mineral Industries in 1941." In U.S. Bureau of Mines, Minerals yearbook 1941. Year 1941 (1943), ix–xxv. Available at http://digicoll.library.wisc.edu/EcoNatRes/.

———. "Review of the Mineral Industries in 1942." In U.S. Bureau of Mines, Minerals yearbook 1942. Year 1942 (1943), 1–23. Available at http://digicoll.library.wisc.edu/EcoNatRes/.

———. "Review of the Mineral Industries in 1944." In U.S. Bureau of Mines, Minerals yearbook 1944. Year 1944 (1946), 1–26. Available at http://digicoll.library.wisc.edu/EcoNatRes/.

Ransome, Alfred L. "Lead." In U.S. Bureau of Mines, Minerals yearbook 1944. Year 1944 (1946), 156–75. Available at http://digicoll.library.wisc.edu/EcoNatRes/.

Ransome, Alfred L., and B. A. Estill. "Zinc." In U.S. Bureau of Mines, Minerals yearbook 1943. Year 1943 (1945), 186–209. Available at http://digicoll.library.wisc.edu/ EcoNatRes/.

Ransome, Alfred L., and Esther B. Miller. "Zinc." In U.S. Bureau of Mines, Minerals yearbook 1944. Year 1944 (1946), 176–201. Available at http://digicoll.library.wisc.edu/ EcoNatRes/.

Ransome, Alfred L., and J. H. Schaum. "Lead." In U.S. Bureau of Mines, Minerals yearbook 1943. Year 1943 (1945), 168–85. Available at http://digicoll.library.wisc.edu/ EcoNatRes/.

Reeves, Robert G., and Victor E. Kral. "Geology and Iron Ore Deposits of the Buena Vistal Hills, Churchill and Pershing Counties, Nevada." *Nevada Bureau of Mines Bulletin* (Carson City) 53 (1956): 18–20. Available at http://www.nbmg.unr.edu.

Shedd, Kim B. "Tungsten Statistics and Information." In *Mineral Commodity Summaries*, 182–83. Washington, D.C.: U.S. Geological Survey, January 2008. Available at http://minerals.usgs.gov/minerals/pubs/.

Smith, Ward C., and Philip W. Guild. "Tungsten Deposits of the Nightingale District Pershing County, Nevada." *U.S. Geological Survey Bulletin* (Washington, D.C.) (1942): 936-B.

Stager, Harold K., and Joseph V. Tingley. "Tungsten Deposits in Nevada." *Nevada Bureau of Mines and Geology Bulletin* (Reno) 105 (1988).

Tingley, Joseph V. *Mining Districts of Nevada.* Report 47. Reno: Nevada Bureau of Mines and Geology, 1998. Available at http://www.nbmg.unr.edu/.

U.S. Bureau of the Census. *Historical Statistics of the United States, Colonial Times to 1970.* Bicentennial Edition. Washington, D.C.: Government Printing Office, 1975. Available at http://www2.census.gov/prod2/statcomp/documents/CT1970p1-01.pdf.

———. *Statistical Abstracts of the United States.* 1934, 1940–50. Available at http://www.census.gov/.

U.S. Code. *Supplies for the Armed Forces in Time of an Emergency: Hearings Before the Committee on Military Affairs . . . on H.R. 1608, a Bill to Provide for the Common Defense by Acquiring Certain Commodities Essential to the Manufacture of Supplies for the Armed Forces in Time of Emergency . . .* [April–June 1937]. 75th Cong., 1st sess. Washington, D.C.: Government Printing Office, 1938.

———. Title 15, Section 13–13b; Title 47, Section 1520; Title 50, Section 98a; Title 60, Section 596. Available at http://straylight.law.cornell.edu/uscode/.

U.S. Department of State. "The Trade Agreement With the United Kingdom Signed November 17, 1938." *Press Releases* 19, no. 477, supp. A, Publication 1252 (1938).

U.S. Department of the Interior. *Annual Report[s] of the Secretary of the Interior . . .* Washington, D.C.: Government Printing Office, 1920, 1935.

U.S. General Services Administration, Department of Defense, and National Aeronautics and Space Administration. "Federal Acquisition Regulation (FAR)" (March 2005). Available at http://www.acquisition.gov/far/.

U.S. Geological Survey. *Commodity Statistics and Information.* Available at http://minerals.usgs.gov/minerals/pubs/.

———. *Mineral Commodity Statistics. Data Series* 140. Available at http://pubs.usgs.gov.

———. *The Strategy of Minerals: A Study of the Mineral Factor in the World Position of American in War and in Peace.* Edited by George Otis Smith. New York: Appleton, 1919. Available at http://books.google.com/books.

U.S. House. Committee on Military Affairs. "Hearings . . . on H.R. 2969, 3320, 2556, 2643, 1987, 987, and 4373, to Provide for the Common Defense by Acquiring Stocks of Strategic and Critical Raw Materials in Time of National Emergency . . ." [February–March 1939]. 76th Cong., 1st sess. (1939).

———. "Supplies for the Armed Forces in Time of Emergency: Hearings . . . on H.R. 1608, a Bill to Provide for the Common Defense by Acquiring Certain Commodities Essential to the Manufacture of Supplies for the Armed Forces in Time of Emergency . . ." [April–June 1937]. 75th Cong., 1st sess. (1938).

U.S. House. Committee on Rules. "War Minerals Relief" [hearings, February 21, 1927]. 69th Cong., 2nd sess. (1927).

U.S. House. Committee on Ways and Means. "Tariff Act of 1929" [hearings, January–February 1929]. 18 vols. 71st Cong., 1st sess. (1929).

U.S. House. Subcommittee of the Committee on Appropriations. "Hearings . . . on the Military Establishment Appropriation Bill for 1938." 75th Cong., 1st sess. [March–April 1937] (1937).

U.S. House. Subcommittee on Mines and Mining of the Committee on Public Lands. "Hearings [on Strategic and Critical Minerals and Metals]." Pt. 4, "Preliminary Review of the Problems of the Tungsten and Mercury Mining Industries, March 31 and May 20, 1948." Pt. 5, "Stockpiling, May 3 and 6, 1948." 80th Cong., 2nd sess. (1948).

U.S. Office of Technology Assessment. "An Assessment of Alternative Economic Stockpiling Policies." August 1976.

U.S. Senate. Appropriations Committee. "Hearings on Supplemental Appropriation Bill" [March 25–April 10, 1941]. 77th Cong., 1st sess. (1941).

U.S. Senate. Committee on Finance. "Digest of Tariff Hearings . . . on the Bill H.R. 7456." 67th Cong., 1st sess. (1922).

U.S. Senate. Finance Committee. "Hearings . . . on the Tariff Act of 1929" [June 14–July 18, 1929]. Vol. 3. 71st Cong., 1st sess. (1929).

U.S. Senate. Subcommittee of the Committee on Military Affairs. "Hearing . . . on S. 3460 . . . Part 1 [8 March 1938]; Part 2 [May 3, 1938]." 75th Cong., 3rd sess. (1938).

———. "Hearings . . . Relative to Strategic and Critical Materials and Minerals" [May 15–July 1, 1941]. 77th Cong., 1st sess. (1941).

U.S. Tariff Commission. *Tungsten Ores and Concentrates: Report to the President on Investigation No. 120 Under Section 336, Title III of the Tariff Act of 1930.* February 1958.

U.S. War Production Board. *Industrial Mobilization for War: History of the War Production Board and Predecessor Agencies.* Vol. 1. Washington, D.C.: Government Printing Office, 1947.

Vanderburg, William O. "Methods and Costs of Mining Ferberite Ore at the Cold Springs Mine, Nederland, Boulder Co., Colorado." *Information Circular* (U.S. Bureau of Mines) 6673 (December 1932).

———. "Mining and Milling Tungsten Ores." *Information Circular* (U.S. Bureau of Mines) 6852 (September 1935).

Wallace, Robert E. "Nevada." In U.S. Bureau of Mines, *Minerals Yearbook Area Reports, 1953* (1956), 629–74. Available at http://digicoll.library.wisc.edu/EcoNatRes/.

Weitz, John H., and Mary E. Trought. "Bauxite and Aluminum." In U.S. Bureau of Mines, Minerals yearbook 1944. Year 1944 (1946), 666–96. Available at http://digicoll.library.wisc.edu/EcoNatRes/.

Werner, Anthony P. D., et al. "International Strategic Mineral Issues, Summary Report— Tungsten." *U.S. Geological Survey Circular* 930-O (1998). Available at http://pubs.usgs.gov/.

Wiebelt, Frank J., and Spangler Ricker. "Investigation of the Atolia Tungsten Mines, San Bernardino County, California." *Report of Investigations* (U.S. Bureau of Mines) 4627 (1950).

COURT REPORTS

Minerals Separation v. Hyde. U.S. District Court, District of Montana, July 28, 1913, 207 F. 956.

Minerals Separation v. Hyde. 242 U.S. 261 (1916).

Minerals Separation v. Magma Copper Co. 280 U.S. 400 (1930) (LEXIS 759).

Nevada-Massachusetts Co. v. Commissioner of Internal Revenue. U.S. Court of Appeals, 9th Circuit, 128 F.2d 347 (May 19, 1942); 1942 U.S. App. (LEXIS 3579).

Schechter v. U.S. 295 U.S. 495.

Union Carbide and Carbon Corporation and Vanadium Corporation of America v. Nisley et al.; Union Carbide and VCA v. Wade et al.; Union Carbide and VCA v. Balsley et al. 300 F.2d 561.

U.S. ex rel Chestatee v. Wilbur. 60 App. D.C. 62, 47 F.2d 424; *Wilbur v. U.S. ex rel Vindicator* and *Wilbur v. U.S. ex rel Chestatee,* 283 U.S. 817.

U.S. ex rel Chestatee v. Wilbur. 61 App. D.C. 324; 62 F.2d 863.

U.S. v. General Electric Co. et al. U.S. District Court for the Southern District of New York, 80 F. Supp. 989 (October 8, 1948).

Wilbur v. U.S. ex rel Chestatee. 61 App. D.C. 212; 59 F.2d 887.

Wilbur v. U.S. ex rel Chestatee. 287 U.S. 588.

Wilbur v. U.S. ex rel Chestatee. 288 U.S. 97.

Wilbur v. U.S. ex rel Chestatee. 288 U.S. 590.

Wilbur v. U.S. ex rel Vindicator and *Wilbur v. U.S. ex rel Chestatee.* 284 U.S. 231.

Work v. U.S. ex rel. Rives and *Work v. U.S. ex rel. Chestatee Pyrites and Chemical Corp.* 267 U.S. 187.

SERIALS AND PERIODICALS

American Metal Market [New York]

Business Week

Chicago Tribune

Congressional Digest

Engineering and Mining Journal

Index

International Directory of Company Histories

Iron Age

Iron Trade Review

Kelly's Customs Tariffs of the World

Las Vegas Review-Journal

Milford [Utah] News

Mineral Industry

Mining and Metallurgy

Mining and Scientific Press

Mining Record [Denver]

Mining World

Mohave County Miner [Kingman, Ariz.]

New York Times, 1900–1956

Reno Gazette

Review-Miner [Lovelock, Nev.]

Review of Economic Statistics

The Times History and Encyclopaedia of the War [London]

Tuolumne Independent

Wall Street Journal, 1907–50

Wikipedia

ORAL HISTORIES

Bradley, Philip Read, Jr. "A Mining Engineer in Alaska, Canada, the Western United States, Latin America, and Southeast Asia." Oral history conducted by Eleanor Swent, in 1986 and 1988. Regional Oral History Office, the Bancroft Library, University of California, Berkeley, 1988. Available at http://bancroft.berkeley.edu/ROHO/projects/mining/.

Segerstrom, Mary Etta. Interview with author. Sonora, Calif., July 15, 2003.

OTHER WEB SITES

Churchill Memorial Web Site of Westminster College. Available at http://www.churchillmemorial.org/.

Consumer Pride Index summary data. Available at http://inflationdata.com.

Defense National Stockpile Center. Available at https://www.dnsc.dla.mil/.

"Federal Acquisition Regulation." March 2005. Available at http://www.acquisition.gov/far/.

Howmet Corporation history. Available at http;//www.fundinguniverse.com/.

"Incandescent Light Bulb." *Wikipedia*. Available at http://en.wikipedia.org/wiki/.

Sheffield University home page. http://www.shef.ac.uk/about/history.html.

"Thomas Hardy Quotations." http://www.brainyquote.com/quotes/.

OTHER SOURCES

Agricola, Georgius. *De Re Metallica*. Translated from the first Latin edition of 1556 by Herbert Clark Hoover and Lou Henry Hoover. 1910. Reprint, New York: Dover, 1950.

Ahlsrom, Goran. *Engineers and Industrial Growth: Higher Technical Education and the Engineering Profession During the Nineteenth and Early Twentieth Centuries: France, Germany, Sweden, and England*. London: Croom Helm, 1982.

Aitken, Hugh G. J. *Taylorism at Watertown Arsenal: Scientific Management in Action, 1908–1915*. Cambridge: Harvard University Press, 1960.

Allen, Frederick Lewis. *Only Yesterday: An Informal History of the Nineteen Twenties*. New York: Harper, 1931.

Arnold, Peri E. "The 'Great Engineer' as Administrator: Herbert Hoover and Modern Bureaucracy." *Review of Politics* 43, no. 3 (1980): 329–48. Available at http://links .jstor.org.

Arnold, Thuman W. *The Folklore of Capitalism*. New Haven: Yale University Press, 1937.

Bailes, Kendalle E. "The American Connection: Ideology and the Transfer of American Technology to the Soviet Union, 1917–1941." *Comparative Studies in Society and History* 23 (July 1981): 421–48. Available at http://links.jstor.org.

Bailey, Thomas A. *A Diplomatic History of the American People*. New York: Appleton-Century-Crofts, 1946.

Balke, W. C. "The Story of Tungsten." *Journal of the Western Society of Engineers* 27 (August 1922): 223–32.

Ball, Simon. "The German Octopus: The British Metal Corporation and the Next War, 1914–1939." *Enterprise and Society* 5, no. 3 (2004): 451–89. Available at http://muse .jhu.edu.

Barbier, Claude. *The Economics of Tungsten*. London: Metal Bulletin Books, 1971.

Barnett, Harold J. "The Changing Relation of Natural Resources to National Security." *Economic Geography* 34 (July 1958): 189–201. Available at http://links.jstor.org.

Bartels, Andrew H. "The Office of Price Administration and the Legacy of the New Deal, 1939–1946." *Public Historian* 5 (Summer 1983): 5–29. Available at http://links.jstor.org.

Baruch, Bernard M. *American Industry in the War: A Report of the War Industries Board (March 1921)*. New York: Prentice-Hall, 1941.

Bastable, Marshall J. "From Breechloaders to Monster Guns: Sir William Armstrong and the Invention of Modern Artillery, 1854–1880." *Technology and Culture* 33 (April 1993): 213–47. Available at http://links.jstor.org.

Batcheller, Hiland. "Economic Significance of Special Alloy Steels." *Mining and Metallurgy* 12 (July 1931): 312–18.

Bateman, Alan M. "Our Future Dependence on Foreign Minerals." *Annals of the American Academy of Political and Social Science* 281 (May 1952): 25–32. Available at http://links.jstor.org.

Beard, Charles A., and Mary R. Beard. *The Rise of American Civilization.* Rev. ed. New York: Macmillan, 1935.

Becker, Susan. "The German Metal Traders Before 1914." In *The Multinational Traders,* edited by Geoffrey Jones, 66–85. London: Routledge, 1998.

Behre, Charles H., Jr. "Mineral Resources and the Atlantic Charter." *Geographical Review* 33 (July 1943): 482–86. Available at http://links.jstor.org.

Berglund, Abraham. "The Ferroalloy Industries and Tariff Legislation." *Political Science Quarterly* 36 (June 1921): 245–73. Available at http://links.jstor.org.

———. "The Reciprocal Trade Agreements Act of 1934." *American Economic Review* 25 (September 1935): 411–25. Available at http://links.jstor.org.

———. "The Tariff Act of 1922." *American Economic Review* 13 (March 1923): 14–33. Available at http://links.jstor.org.

———. "The Tariff Act of 1930." *American Economic Review* 20 (September 1930): 467–79. Available at http://links.jstor.org.

Bernstein, Barton J. "The Debate on Industrial Reconversion: The Protection of Oligopoly and Military Control of the Economy." *American Journal of Economics and Sociology* 26 (April 1967): 159–72. Available at http://links.jstor.org.

Bernstein, E. M. "War and the Pattern of Business Cycles." *American Economic Review* 30 (September 1940): 524–35. Available at http://links.jstor.org.

Betz, Frederick. *Executive Strategy: Strategic Management and Information Technology.* New York: Wiley, 2001. Available at http://books.google.com/books.

Bidwell, Percy W. "A Trade Policy for National Defense." *Foreign Affairs* 19 (January 1941): 282–96. Available at http://web.ebscohost.com/.

Bingham, Truman C. "Economic Effects of the New Deal Tax Policy." *Southern Economic Journal* 3 (January 1937): 270–80. Available at http://links.jstor.org.

Birkett, M. S. "The Iron and Steel Trades During the War." *Journal of the Royal Statistical Society* 83 (May 1920): 351–400. Available at http://links.jstor.org.

Birrell, Ralph W. "The Role of Minerals Separation Ltd. in the Development of the Flotation Process." *Australian Mining History Monographs,* no. 7 (Victoria: Zlota Press) (2000): 7–40.

Blum, Albert A. "The Fight for a Young Army." *Military Affairs* 18 (Summer 1954): 81–85. Available at http://links.jstor.org.

Blum, Albert A., and J. Douglas Smyth. "Who Should Serve: Pre–World War II Planning for Selective Service." *Journal of Economic History* 30 (June 1970): 379–404. Available at http://links.jstor.org.

Boklund, Uno. "Scheele, Carl Wilhelm." In *Dictionary of Scientific Biography,* edited by Charles Coulston Gillespie, 11:147–48. New York: Charles Scribner's Sons, 1980.

Bonn, M. J. "The Nationalization of Capital." *Annals of the American Academy of Political and Social Science* 68 (November 1916): 252–63. Available at http://links.jstor.org.

Botjer, George F. *A Short History of Nationalist China, 1919–1949.* New York: G. P. Putnam's Sons, 1979.

Bradley, John D., et al. "Yellow Pine Mine." *Engineering and Mining Journal* 144 (April 1943): 60–66.

Brandes, Stuart D. *Warhogs: A History of War Profits in America.* Lexington: University Press of Kentucky, 1997. Available at http://books.google.com/books.

Brenner, Y. S. *Looking Into the Seeds of Time: The Price of Modern Development.* New Brunswick, N.J.: Transaction, 1998. Available at http://books.google.com/books.

Brown, E. Cary. "Accelerated Depreciation: A Neglected Chapter in War Taxation." *Quarterly Journal of Economics* 57 (August 1943): 630–45. Available at http://links.jstor.org.

Brownlee, W. Elliot. Review of *The Limits of Symbolic Reform: The New Deal and Taxation, 1933–1939,* by Mark H. Leff. *Reviews in American History* 14 (March 1986): 121–26. Available at http://books.google.com/books.

Cameron, Rondo E. "Some French Contributions to the Industrial Development of Germany, 1840–1870." *Journal of Economic History* 16 (September 1956): 281–321. Available at http://links.jstor.org.

Carter, Kent. "Total War: The Federal Government and the Home Front." *Prologue* 23 (Fall 1991): 224–29.

Caruana, Leonard, and Hugh Rockoff. "A Wolfram in Sheep's Clothing: Economic Warfare in Spain, 1940–1944." *Journal of Economic History* 63 (March 2003): 100–126.

Catton, Bruce. *The War Lords of Washington.* New York: Harcourt, Brace, 1948.

Ceruzzi, Paul C. "Moore's Law and Technological Determinism: Reflections on the History of Technology." *Technology and Culture* 46 (July 2005): 584–93. Available at http://web.ebscohost.com/.

Chamberlain, John. *The Enterprising Americans: A Business History of the United States.* New York: Harper and Row, 1963.

Chandler, Alfred D., Jr. "The Emergence of Managerial Capitalism." *Business History Review* 58 (Winter 1984): 498–501. Available at http://links.jstor.org.

———. "The Structure of American Industry in the Twentieth Century: A Historical Overview." *Business History Review* 43 (Autumn 1969): 255–98. Available at http://links.jstor.org.

Christman, Calvin L. "Donald Nelson and the Army: Personality as a Factor in Civil-Military Relations During World War II." *Military Affairs* 37 (October 1973): 81–83. Available at http://links.jstor.org.

Clarke, Sir Richard. *Anglo-American Economic Collaboration in War and Peace, 1942–1949.* Oxford: Clarendon Press, 1982.

Clifford, John G. "Grenville Clark and the Origins of Selective Service." *Review of Politics* 35 (January 1973): 17–40. Available at http://links.jstor.org.

Coffman, Paul B. "The Rise of a New Metal: The Growth and Success of the Climax Molybdenum Company." *Journal of Business of the University of Chicago* 19 (January 1937): 30–45. Available at http://links.jstor.org.

Colver, William B. "Recent Phases of Competition in International Trade." *Annals of the American Academy of Political and Social Science* 83 (May 1919): 233–48. Available at http://links.jstor.org.

Comstock, Gregory J. "Tungsten Carbide: The First Product of a New Metallurgy." *Iron Age* 126 (November 13, 1930): 1381–83.

"Congress Considers Bills Affecting National Defense." *Congressional Digest* 17 (March 1938): 75–78.

Craft, Stephen G. "Peacemakers in China: American Missionaries and the Sino-Japanese War, 1937–1941." *Journal of Church and State* 41 (Summer 1999): 575–91. Available at http://web.ebscohost.com/.

Crammond, Edgar. "The Economic Relations of the British and German Empires." *Journal of the Royal Statistical Society* 77 (July 1914): 777–94. Available at http://links.jstor.org.

Croly, Herbert. *The Promise of American Life.* 1909. Reprint, New York: Capricorn Books, 1964.

Crowley, Jack A. "Garnet and Clinozoisite From the Nightingale Mining District." *Rocks and Minerals* (March 2000). Available at HiBeam Encyclopedia, http://www.encyclopedia.com/.

Crum, W. L. "Review of the First Quarter of 1933." *Review of Economic Statistics* 15 (May 15, 1933): 68–74. Available at http://links.jstor.org.

Cuff, Robert D. "Bernard Baruch: Symbol and Myth in Industrial Mobilization." *Business History Review* 43 (Summer 1969): 115–33. Available at http://links.jstor.org.

———. "Ferdinand Eberstadt, the National Security Resources Board, and the Search for Integrated Mobilization Planning, 1947–1948." *Public Historian* 7 (Autumn 1985): 37–52. Available at http://links.jstor.org.

———. "From the Controlled Materials Plan to the Defense Materials System, 1942–1953." *Military Affairs* 51 (January 1987): 1–6. Available at http://links.jstor.org.

Cuff, Robert D., and Melvin I. Urofsky. "The Steel Industry and Price-Fixing During World War I." *Business History Review* 44 (Autumn 1970): 291–306. Available at http://links.jstor.org.

Davis, William H. "Aims and Policies of the National War Labor Board." *Annals of the American Academy of Political and Social Science* 224 (November 1942): 141–46. Available at http://links.jstor.org.

Dennis, William H. *A Hundred Years of Metallurgy.* Chicago: Aldine Publishing, 1964.

Derickson, Alan. *Workers' Health, Workers' Democracy: The Western Miners' Struggle, 1891–1925.* Ithaca: Cornell University Press, 1988.

Diamond, Jared. *Guns, Germans, and Steel: The Fates of Human Societies.* New York: W. W. Norton, 1999.

Dickinson, Matthew J. *Bitter Harvest:* FDR, *Presidential Power, and the Growth of the Presidential Branch.* Cambridge: Cambridge University Press, 1997.

Divine, Robert A. *The Illusion of Neutrality.* Chicago: University of Chicago Press, 1962.

Douglas, Paul H. *America in the Market Place: Trade, Tariffs, and the Balance of Payments, 1933–45.* New York: Holt, Rinehart, and Winston, 1966.

Driggs, Don W., and Leonard E. Goodall. *Nevada Politics and Government: Conservatism in an Open Society.* Lincoln: University of Nebraska Press, 1996.

Dunn, John M. "American Dependence on Materials Imports the World-Wide Resource Base." *Journal of Conflict Resolution* 4 (March 1960): 106–22. Available at http://links .jstor.org.

Dunn, Peter J. *The Story of Franklin and Sterling Hill.* Washington, D.C.: Smithsonian Institution, 1997.

Durant, Will, and Ariel Durant. *The Age of Voltaire.* Pt. 9 of *The Story of Civilization.* New York: Simon and Schuster, 1965.

Eckes, Alfred E. *The United States and the Global Struggle for Minerals.* Austin: University of Texas Press, 1979.

Eggert, Gerald G. *The Iron Industry in Pennsylvania.* Pennsylvania History Studies no. 25. Harrisburg: Pennsylvania Historical Association, 1994.

Epkenhans, Michael. "Military-Industrial Relations in Imperial Germany, 1870–1914." *War in History* 10, no. 1 (2003): 1–26. Available at http://links.jstor.org.

Ferrell, Robert H. *Peace in Their Time: The Origins of the Kellogg-Briand Pact.* New Haven: Yale University Press, 1968.

Fetherling, Douglas. *The Gold Crusades: A Social History of Gold Rushes, 1849–1929.* Toronto: University of Toronto Press, 1997. Available at http://books.google.com/ books.

Fink, Colin G. "Review of the Strategic Metals." *Metals and Alloys* 12 (October 1940): 419–20.

Flynn, George Q. "Conscription and Equity in Western Democracies, 1940–75." *Journal of Contemporary History* 33 (January 1998): 5–20. Available at http://links.jstor.org.

———. *The Mess in Washington: Manpower Mobilization in World War II.* Westport, Conn.: Greenwood Press, 1979.

Freeman, Christopher, and Francisco Louçà. *As Time Goes By: From the Industrial Revolutions to the Information Revolution.* Oxford: Oxford University Press, 2001.

Freeman, Christopher, and Luc Soete. *The Economics of Industrial Innovation.* Cambridge: MIT Press, 1997. Available at http://books.google.com/books.

Frenkel, Stephen. "Geography, Empire, and Environmental Determinism." *Geographical Review* 82 (April 1992): 143–53. Available at http://links.jstor.org.

Furniss, Edgar S., Jr. "American Wartime Objectives in Latin America." *World Politics* 2 (April 1950): 373–89. Available at http://links.jstor.org.

Fusfield, Daniel R. "Joint Subsidiaries in the Iron and Steel Industry." *American Economic Review* 48 (May 1958): 578–87. Available at http://links.jstor.org.

Gago, Ramon. "The New Chemistry in Spain." *Osiris,* 2nd ser., 4 (1988): 169–92. Available at http://links.jstor.org.

Galbraith, John Kenneth. *American Capitalism.* 1952. Reprint, New Brunswick, N.J.: Transaction, 1993.

Gessing, Paul. "'Buy American' Provisions in the Defense Authorization Bill: If You Thought $250 Toilet Seats Were Outrageous . . ." In National Taxpayers Union *Issue Briefs.* Available at http://www.ntu.org/.

Glad, Betty. *Key Pittman: The Tragedy of a Senate Insider.* New York: Columbia University Press, 1986.

Glasby, G. P. "Deep Seabed Mining: Past Failures and Future Prospects." *Marine Georesources and Geotechnology* 20 (2002): 161–76. Available at http://web.ebscohost.com/.

Gold, Bela, Wm. S. Peirce, Gerhard Rosegger, and Mark Perlman. *Technological Progress and Industrial Leadership: The Growth of the U.S. Steel Industry, 1900–1970.* Lexington, Mass.: D. C. Heath, 1984.

Goldfield, Michael. "Worker Insurgency, Radical Organization, and New Deal Labor Legislation." *American Political Science Review* 83 (December 1989): 1257–82. Available at http://links.jstor.org.

Goldman, Eric. *Rendezvous With Destiny: A History of Modern American Reform.* 25th anniversary ed. New York: Vantage Books, 1977.

Goldstein, Judith. "The Impact of Ideas on Trade Policy: The Origins of U.S. Agricultural and Manufacturing Policies." *International Organization* 43 (Winter 1989): 31–71. Available at http://links.jstor.org.

Gordon, John Steele. "The Armor-Plate Scandal." *American Heritage* 45 (July–August 1994).

Gordon, Robert B. "The 'Kelly' Converter." *Technology and Culture* 33 (October 1992): 769–79. Available at http://links.jstor.org.

Gray, Ralph D. *Alloys and Automobiles: The Life of Elwood Haynes.* Indianapolis: Indiana Historical Society, 1979.

Gressley, Gene M. "Thurman Arnold, Antitrust, and the New Deal." *Business History Review* 38 (Summer 1964): 214–31. Available at http://links.jstor.org.

Gulick, Luther. "War Organization of the Federal Government." *American Political Science Review* 38 (December 1944): 1166–79. Available at http://links.jstor.org.

Haggard, Stephen. "The Institutional Foundations of Hegemony: Explaining the Reciprocal Trade Agreements Act of 1934." *International Organization* 42 (Winter 1988): 91–119. Available at http://links.jstor.org.

Hardy, Charles O. "Marketing Tungsten Ores." *Engineering and Mining Journal* 113 (April 22, 1922): 666–69.

———. "The Tungsten Market Situation in 1919." *Engineering and Mining Journal* 109 (January 17, 1920): 210–11.

——. *Wartime Control of Prices.* Washington, D.C.: Brookings Institution, 1940.

Harris, Seymour E. "Subsidies and Inflation." *American Economic Review* 33 (September 1943): 557–72. Available at http://links.jstor.org.

Harvey, Thomas William. *Memoir of Hayward Augustus Harvey by His Sons.* 1900. Reprint, Whitefish, Mont.: Kessinger Publishing, 2004. Available at http://books .google.com/books.

Hatch, Carl A. "The Truman Committee Reports on Small Business." *Congressional Digest* 21 (February 1942): 43–47.

Heath, Jim F. "American War Mobilization and the Use of Small Manufacturers, 1939–1943." *Business History Review* 46 (Autumn 1972): 295–319. Available at http://links .jstor.org.

Heizer, Ott F. "Concentration of Tungsten Ore by the Nevada-Massachusetts Co." AIME *Transactions* 112 (1934): 833–40.

Heldt, P. M. "Tungsten Has Played a Prominent Role in the Development of the Automotive Industry." *Automotive Industries* 78 (June 18, 1938): 810–17.

Henderson, Leon. "The Consumer and Competition." *Annals of the American Academy of Political and Social Science* 183 (January 1936): 263–71. Available at http://links .jstor.org.

Henderson, W. O. "Germany's Trade With Her Colonies, 1884–1914." *Economic History Review* 9 (November 1938): 1–16. Available at http://links.jstor.org.

Hershey, Lewis B. "Procurement of Manpower in American Wars." *Annals of the American Academy of Political and Social Science* 241 (September 1945): 15–25. Available at http://links.jstor.org.

Hess, Frank L. "Rare Metals and Minerals." *Mining and Metallurgy* 19 (January 1938): 5–9.

Higgs, Robert. "Private Profit, Public Risk: Institutional Antecedents of the Modern Military Procurement System in the Rearmament Program of 1940–1941." In *The Sinews of War: Essays on the Economic History of World War II,* edited by Geofrey T. Mills and Hugh Rockoff. Ames: Iowa State University Press, 1993.

Himmelberg, Robert F. *The Origins of the National Recovery Administration: Business, Government, and the Trade Association Issue, 1921–1933.* New York: Fordham University Press, 1976.

Hinshaw, David. *The Home Front.* New York: G. P. Putnam's Sons, 1943.

Hiorns, Arthur H. *Principles of Metallurgy.* London: Macmillan, 1895.

Hoffman, Abraham. *Vision or Villainy: Origins of the Owens Valley–Los Angeles Water Controversy.* College Station: Texas A&M University Press, 1981.

Hofstadter, Richard. *The Age of Reform: From Bryan to F.D.R.* New York: Alfred A. Knopf, 1965.

Hoover, Herbert. *The Challenge to Liberty.* New York: Charles Scribner's Sons, 1934.

Hovis, Logan, and Jeremy Mouat. "Miners, Engineers, and the Transformation of Work in the Western Mining Industry, 1880–1930." *Technology and Culture* 37, no. 3 (1996): 429–56.

Hughes, Thomas Parke. "Thomas Alva Edison and the Rise of Electricity." In *Technology in America: A History of Individuals and Ideas,* edited by Carroll W. Pursell Jr., 117–28. Cambridge: MIT Press, 1981.

Hull, Cordell. "Should the U.S. Adopt a Reciprocity Tariff Policy? Pro." *Congressional Digest* 12 (May 1933). Available at http://web.ebscohost.com/.

Hunt, Edward Eyre. *Scientific Management Since Taylor: A Collection of Authoritative Papers.* New York: McGraw-Hill, 1924.

Huttl, John B. "Unique Golconda Deposit Yields Its Tungsten." *Engineering and Mining Journal* 146 (August 1945): 79–81.

Ihde, Aaron J. *The Development of Modern Chemistry.* 1970. Reprint, New York: Harper and Row, 1984.

Ingham, John N. "Iron and Steel in the Pittsburgh Region: The Domain of Small Business." *Business and Economic History* 20 (1991): 107–16.

Irwin, Douglas A. "Interests, Institutions, and Ideology in Securing Policy Change: The Republican Conversion to Trade Liberalization After Smoot-Hawley." *Journal of Law and Economics* 42 (October 1999): 643–73. Available at http://links.jstor.org.

Jackson, Donald C. *Building the Ultimate Dam: John S. Eastwood and the Control of Water in the West.* Lawrence: University Press of Kansas, 1995.

Janeway, Eliot. *The Struggle for Survival: A Chronicle of Economic Mobilization in World War II.* New Haven: Yale University Press, 1951.

Jeffries, John W. *Wartime America: The World War II Home Front.* Chicago: Ivan R. Dee, 1996.

Jones, Archer, and Andrew J. Keogh. "The Dreadnought Revolution: Another Look." *Military Affairs* 49 (July 1985): 124–31. Available at http://links.jstor.org.

Joralemon, P. "California's Foothill Gold Belt: Some Famous Lode Mines Are Showing Signs of Renewed Interest." *Mining Engineering* 39 (July 1987): 493.

Kelly, Alfred H., et al. *The American Constitution: Its Origins and Development.* 6th ed. New York: W. W. Norton, 1983.

Kennedy, David M. *Freedom From Fear: The American People in Depression and War, 1929–1945.* New York: Oxford University Press, 1999.

Kerr, Paul F. "Tungsten-Bearing Manganese Deposit at Golconda, Nevada." *Bulletin of the Geological Society of America* 51 (September 1, 1940): 1359–90.

Killigrew, John W. *The Impact of the Great Depression on the Army.* 1969. Reprint, Ann Arbor: University of Michigan Press, 2006.

Klein, Cornelius, and Cornelius S. Hurlbut Jr. *Manual of Mineralogy* (after James D. Dana). 21st ed. New York: John Wiley and Sons, 1977.

Knapp, Laurence A. "The Buy American Act: A Review and Assessment." *Columbia Law Review* 61 (March 1962): 430–62. Available at http://links.jstor.org.

Koistinen, Paul A. C. "The 'Industrial-Military Complex' in Historical Perspective: The Interwar Years." *Journal of American History* 56 (March 1970): 819–39. Available at http://links.jstor.org.

———. "Mobilizing the World War II Economy: Labor and the Industrial-Military Alliance." *Pacific Historical Review* 42 (November 1973): 443–78. Available at http://links. jstor.org.

Kolko, Gabriel. "American Business and Germany, 1930–1941." *Western Political Quarterly* 15 (December 1962): 713–28. Available at http://links.jstor.org.

Kurtak, Joseph M. "History of Pine Creek: A World Class Tungsten Deposit." *Mining Engineering* 50 (December 1998): 42–47.

———. *A Mine in the Sky: The History of California's Pine Creek Tungsten Mine and the People Who Were Part of It.* Anchorage: privately printed, 1998.

Lael, Richard L. "The Pressure of Shortage: Platinum Policy and the Wilson Administration During World War I." *Business History Review* 56 (Winter 1982): 545–58. Available at http://links.jstor.org.

Lankton, Larry. *Cradle to Grave: Life, Work, and Death at the Lake Superior Copper Mines.* New York: Oxford University Press, 1991.

Lauderbaugh, Richard A. *American Steel Makers and the Coming of the Second World War.* Ann Arbor: UMI Research Press, 1980.

Lautenschlager, Karl. "Technology and the Evolution of Naval Warfare." *International Security* 8 (Autumn 1983): 3–51. Available at http://links.jstor.org.

Layton, Edwin T. *The Revolt of the Engineers: Social Responsibility and the American Engineering Profession.* 1971. Reprint, Baltimore: Johns Hopkins University Press, 1986.

Le Billon, Philippe. "The Geopolitical Economy of 'Resource Wars.'" *Geopolitics* 9 (Winter 2004): 1–28. Available at http://web.ebscohost.com/.

Leff, Mark H. *The Limits of Symbolic Reform: The New Deal and Taxation, 1933–1939.* Cambridge: Cambridge University Press, 1984. Available at http://books.google.com/books.

Leffler, Melvyn P. "The American Conception of National Security and the Beginnings of the Cold War, 1945–48." *American Historical Review* 89 (April 1984): 346–81. Available at http://links.jstor.org.

Leith, Charles K. "Conservation of Certain Mineral Resources." Pt. 3 of *The Foundations of National Prosperity: Studies in the Conservation of Permanent National Resources,* edited by Richard T. Ely et al. New York: Macmillan, 1920.

———. "Mineral Resources and Peace." *Foreign Affairs* 16 (April 1938): 515–24. Available at http://web.ebscohost.com/.

———. "Mineral Resources in Their International Relations." *Proceedings of the American Philosophical Society* 91 (February 25, 1947): 83–87. Available at http://links.jstor.org.

———. "The Struggle for Mineral Resources." *Annals of the American Academy of Political and Social Science* 204 (July 1939): 42–48. Available at http://links.jstor.org.

Leitz, Christian. "Nazi Germany's Struggle for Spanish Wolfram During the Second World War." *European History Quarterly* 25 (January 1995): 71–92.

Lester, Richard A. "War Controls of Materials, Equipment, and Manpower: An Experiment in Economic Planning." *Southern Economic Journal* 9 (January 1943): 197–216. Available at http://links.jstor.org.

Leuchtenburg, William E. *The Perils of Prosperity, 1914–1932*. Chicago: University of Chicago Press, 1958.

Li, K. C., and Chung Yu Wang. *Tungsten: Its History, Geology, Ore-Dressing, Metallurgy, Chemistry, Analysis, Applications, and Economics*. New York: Reinhold, 1943.

Liddell, Donald M., ed. *Handbook of Nonferrous Metallurgy: Recovery of the Metals*. 2nd ed. New York: McGraw-Hill, 1945.

Limbaugh, Ronald H. "Making Old Tools Work Better: Pragmatic Adaptation and Innovation in Gold-Rush Technology." In *A Golden State: Mining and Economic Development in Gold Rush California*, edited by James J. Rawls and Richard J. Orsi, 24–51. Berkeley and Los Angeles: University of California Press, 1999.

———. "Making the Most of Experience: The Career of William J. Loring, Nevada Mining Engineer." *Mining History Journal* 1 (1994): 9–26.

———. "'There Is a Game Against Us': W. J. Loring's Troubled Years as Bewick-Moreing Company's General Manager and Partner in Western Australia, 1905–1912." *Journal of Australasian Mining History* 2 (September 2004): 90–114.

Lincoln, Francis Church. *Mining Districts and Mineral Resources of Nevada*. 1923. Reprint, n.p.: Stanley Paher, 1982.

Lingenfelter, Richard E. *The Hardrock Miners: A History of the Mining Labor Movement in the American West, 1863–1893*. Berkeley and Los Angeles: University of California Press, 1974.

Loach, W. "Developing the American Tungsten Industry." *Mining Congress Journal* 20 (September 1934): 27–28.

Lobell, Steven E. "Second Image Reversed Politics: Britain's Choice of Freer Trade or Imperial Preferences, 1903–1906, 1917–1923, 1930–1932." *International Studies Quarterly* 43 (December 1999): 671–93. Available at http://links.jstor.org.

Lockwood, William W. "Washington Works on Barter Scheme." *Far Eastern Survey* 8 (May 24, 1939): 131–32. Available at http://links.jstor.org.

Long, Henry J. "Tungsten Carbide Cutting Tools." *Iron Age* 123 (May 23, 1929): 1414–16.

Lorant, Stefan. *The Presidency*. New York: Macmillan, 1953.

Lovett, William Anthony, et al. *U.S. Trade Policy: History, Theory, and the W.T.O.* 2nd ed. New York: M. E. Sharpe, 2004. Available at http://books.google.com/books.

MacBride, Hope L. "Export and Import Associations as Instruments of National Policy." *Political Science Quarterly* 57 (June 1942): 189–213. Available at http://links.jstor.org.

Malone, Michael P. C. *Ben Ross and the New Deal in Idaho*. Seattle: University of Washington Press, 1970.

Marcus, Alan I., and Howard P. Segal. *Technology in America: A Brief History*. San Diego: Harcourt Brace Jovanovich, 1989.

Marovich, Sharon. "The C. H. Segerstrom Estate: A Haven for Family and Friends on Knowles Hill." *Chispa* 41 (October–December 2001): 1463–69.

Mason, Edward S. "American Security and Access to Raw Materials." *World Politics* 1 (January 1949): 147–60. Available at http://links.jstor.org.

May, Ernest R. "The Development of Political-Military Consultation in the United States." *Political Science Quarterly* 70 (June 1955): 161–80. Available at http://links.jstor.org.

McDermott, John. "Trading With the Enemy: British Business and the Law During the First World War." *Canadian Journal of History* 32 (August 1997). Available at http://web.ebscohost.com/.

McMaster, John B. *The United States in the World War.* New York: D. Appleton, 1918. Available at http://books.google.com/books.

McNeill, John H. "International Agreements: Recent U.S.-UK Practice Concerning the Memorandum of Understanding." *American Journal of International Law* 88 (October 1994): 821–26. Available at http://links.jstor.org.

McNutt, Paul V. "Is America's War Manpower Policy Striking the Proper Balance Between Civilian and Armed Forces? Pro." *Congressional Digest* 22 (March 1943): 87–88. Available at http://web.ebscohost.com/.

Merkley, Jeff. "Merkley Plan to Create Family-Wage Jobs: End Outsourcing, Buy American." From the Jeff Merkley, Democrat for U.S. Senate, Web site, http://www.jeffmerkley.com/.

Messimer, Dwight R. *The Merchant U-boat: Adventures of "The Deutschland," 1916–1918.* Annapolis: Naval Institute Press, 1988.

Mikesell, Raymond F. *Nonfuel Minerals: Foreign Dependence and National Security.* Ann Arbor: University of Michigan Press, 1987.

Miller, Douglas. *You Can't Do Business With Hitler.* 1941. Reprint, Boston: Little, Brown, 1942.

Miller, E. Willard. "The Industrial Development of the Allegheny Valley of Western Pennsylvania." *Economic Geography* 19 (October 1943). Available at http://links.jstor.org.

Miller, John Perry. "Military Procurement Policies: World War II and Today." *American Economic Review* 42 (May 1952): 453–78. Available at http://web.ebscohost.com/.

Millis, Walter. *Arms and Men: A Study in American Military History.* New York: G. P. Putnam's Sons, 1956.

———. *The Martial Spirit: A Study of Our War With Spain.* Boston: Houghton Mifflin, 1931.

Misa, Thomas J. *A Nation of Steel: The Making of Modern America, 1865–1925.* Baltimore: Johns Hopkins University Press, 1995.

Molander, Earl A. "Historical Antecedents of Military-Industrial Criticism." *Military Affairs* 40 (April 1976): 59–63. Available at http://links.jstor.org.

Morison, Samuel E., and Henry S. Commager. *The Growth of the American Republic.* Vol. 2. New York: Oxford University Press, 1950.

Mouat, Jeremy. "The Development of the Flotation Process: Technological Change and the Genesis of Modern Mining, 1898–1911." *Australian Economic History Review* 36 (March 1996): 3–26.

Nash, Gerald D. "Herbert Hoover and the Origins of the Reconstruction Finance Corporation." *Mississippi Valley Historical Review* 46 (December 1949). Available at http://links.jstor.org.

National Research Council, Committee on Assessing the Need for a Defense Stockpile. *Managing Materials for a 21st Century Military.* Washington, D.C.: National Academies Press, 2007. Available at http://www.nap.edu/.

Newell, Dianne. *Technology on the Frontier: Mining in Old Ontario.* Vancouver: University of British Columbia Press, 1986.

Niblo, Stephen R. *War, Diplomacy, and Development: The United States and Mexico, 1938–1954.* Wilmington, Del.: Scholarly Resources, 1995.

Noorzoy, M. S. "'Buy American' as an Instrument of Policy." *Canadian Journal of Economics* 1 (February 1968): 96–105. Available at http://links.jstor.org.

Offner, Arnold A. "Appeasement Revisited: The United States, Great Britain, and Germany, 1933–1940." *Journal of American History* 64 (September 1977): 373–93. Available at http://links.jstor.org.

O'Leary, James J. "A General Wage Ceiling." *Southern Economic Journal* 9 (July 1942): 24–32. Available at http://links.jstor.org.

Pacey, Arnold. *The Culture of Technology.* Cambridge: MIT Press, 1983. Available at http://books.google.com/books.

Papayoanou, Paul A. "Interdependence, Institutions, and the Balance of Power: Britain, Germany, and World War I." *International Security* 20 (Spring 1996): 42–76. Available at http://links.jstor.org.

Parr, J. Gordon. "The Sinking of the *Ma Robert*: An Excursion Into Mid-19th-Century Steelmaking." *Technology and Culture* 13 (April 1972): 209–25. Available at http://links.jstor.org.

Partington, J. R. *A History of Chemistry.* Vol. 3. London: Macmillan, 1962.

Pate, James E. "Mobilizing Manpower." *Social Forces* 22 (December 1943): 154–62. Available at http://links.jstor.org.

Pattison, Michael. "Scientists, Inventors, and the Military in Britain, 1915–19: The Munitions Inventions Department." *Social Studies of Science* 13 (November 1983): 521–68. Available at http://links.jstor.org.

Pehrson, E. W. "Our Mineral Resources and Security." *Foreign Affairs* 23 (July 1945): 644–57. Available at http://web.ebscohost.com/.

———. "Stockpiling: Development of Submarginal Resources: Conservation and Substitution." Typescript, Industrial College of the Armed Forces, February 11, 1946. Available at http://www.ndu.edu/library/.

Peterson, C. Rudolf. "The Statutory Evolution of the Excess Profits Tax." *Law and Contemporary Problems* 10 (Winter 1943): 3–27. Available at http://links.jstor.org.

Pettengill, Robert B. "The United States Copper Industry and the Tariff." *Quarterly Journal of Economics* 46 (December 1931): 141–57. Available at http://links.jstor.org.

Pollard, Sidney. "Industrialization and the European Economy." *Economic History Review* 26, no. 4 (1973). Available at http://links.jstor.org.

Prechel, Harland. "Steel and the State: Industry Politics and Business Policy Formation,

1940–1989." *American Sociological Review* 55 (October 1990): 648–68. Available at http://links.jstor.org.

Priest, Tyler. *Global Gambits: Big Steel and the U.S. Quest for Manganese.* Westport, Conn.: Praeger, 2003.

Prusa, Thomas J. "On the Spread and Impact of Antidumping." *Canadian Journal of Economics* 34 (August 2001): 591–611. Available at http://links.jstor.org.

Radius, Walter A. "United States Trade and the Sino-Japanese War." *Far Eastern Survey* 7 (January 5, 1938): 2–7. Available at http://links.jstor.org.

Raudzens, George. "War-Winning Weapons: The Measurement of Technological Determinism in Military History." *Journal of Military History* 54 (October 1990): 403–34. Available at http://links.jstor.org.

Raymond, C. Elizabeth. *George Wingfield: Owner and Operator of Nevada.* Reno: University of Nevada Press, 1992.

Read, T. T. "Historical Aspects of Mining and Metallurgical Engineering." *Journal of Engineering Education* 24 (November 1933): 229–58.

Reich, Leonard S. "Lighting the Path to Profit: GE's Control of the Electric Lamp Industry, 1892–1941." *Business History Review* 66 (Summer 1992): 305–34. Available at http://links.jstor.org.

———. *The Making of American Industrial Research: Science and Business at GE and Bell, 1876–1926.* Cambridge: Cambridge University Press, 1985. Available at http://books.google.com/books.

Ridgway, Robert H. "Ferro-Alloying Minerals: Domestic Productive Capacity Showed Important Increase for Tungsten, Molybdenum, and Vanadium Ore." *Mining Congress Journal* 24 (February 1938): 41–44.

Ritchie, S. B. "Molybdenum in High-Speed Steel: The Elimination of Tungsten, a Strategic Material." *Army Ordinance* 11 (July–August 1930): 12–19.

Roland, Alex. "Science and War." *Osiris* 1 (1985): 247–72. Available at http://links.jstor.org.

Roosevelt, Eleanor. *Courage in a Dangerous World: The Political Writings of Eleanor Roosevelt.* New York: Columbia University Press, 1999. Available at http://books.google.com/books.

Ropes, E. C. "American-Soviet Trade Relations." *Russian Review* 3 (Autumn 1943): 89–94. Available at http://links.jstor.org.

Rosenberg, Nathan. "The Direction of Technological Change: Inducement Mechanisms and Focusing Devices." *Economic Development and Cultural Change* 18 (October 1969): 1–24. Available at http://links.jstor.org.

Rourke, Francis E. "The Department of Labor and the Trade Unions." *Western Political Quarterly* 7 (December 1954): 656–72. Available at http://links.jstor.org.

Russell, Francis. *The Shadow of Blooming Grove: Warren G. Harding and His Times.* New York: McGraw-Hill, 1968.

Sanderson, Michael. "The Professor as Industrial Consultant: Oliver Arnold and the

British Steel Industry, 1900–1914." *Economic History Review* 31 (November 1978): 585–600. Available at http://links.jstor.org.

Schatz, Arthur W. "The Anglo-American Trade Agreement and Cordell Hull's Search for Peace, 1936–1938." *Journal of American History* 57 (June 1970): 85–103. Available at http://links.jstor.org.

Scheiber, Harry N. "World War I as Entrepreneurial Opportunity: Willard Straight and the American International Corporation." *Political Science Quarterly* 84 (September 1969): 486–511. Available at http://links.jstor.org.

Schell, John D. "Tungsten Alloy and Cancer in Rats: Link to Childhood Leukemia?" *Environmental Health Perspectives* (December 1, 2005). Available at http://www .encyclopedia.com .

Schlesinger, Arthur M., Jr. *The Coming of the New Deal.* Vol. 2 of *The Age of Roosevelt.* Boston: Houghton Mifflin, 1959.

———. *The Crisis of the Old Order, 1919–1933.* Vol. 1 of *The Age of Roosevelt.* Boston: Houghton Mifflin, 1957.

———. *The Cycles of American History.* Boston: Houghton Mifflin, 1986.

———. *The Imperial Presidency.* Boston: Houghton Mifflin, 1973.

———. *The Politics of Upheaval.* Vol. 3 of *The Age of Roosevelt.* Boston: Houghton Mifflin, 1960.

Schnietz, Karen E. "The Reaction of Private Interests to the 1934 Reciprocal Trade Agreements Act." *International Organization* 57 (Winter 2003): 213–33. Available at http://links.jstor.org.

Schroeter, Karl. "Analysis of Hard Metal Carbide Theory." *Iron Age* 133 (February 22, 1934): 21–23.

———. "Inception and Development of Hard Metal Carbides." *Iron Age* 133 (February 1, 1934): 27–29.

Schwab, Susan C. *Trade-offs: Negotiating the Omnibus Trade and Competitiveness Act.* Boston: Harvard Business School Press, 1994. Available at http://books.google.com/books.

Schwarz, Jordan A. "Review: Hershey's Draft." *Reviews in American History* 14 (March 1986): 133–37. Available at http://links.jstor.org.

Scott, James Brown. "The Black List of Great Britain and Her Allies." *American Journal of International Law* 10 (October 1916): 832–43. Available at http://links.jstor.org.

Scruggs, Otey M. "The United States, Mexico, and the Wetbacks, 1942–1947." *Pacific Historical Review* 30 (May 1961): 149–64. Available at http://links.jstor.org.

Searle, G. R. *A New England? Peace and War, 1886–1918.* Oxford: Oxford University Press, 2004. Available at http://books.google.com/books.

Segerstrom, Charles H. "The American Tungsten Industry." *Mining Congress Journal* 23 (January 1937): 50–53.

———. "Operations in Milford District." *Mining and Contracting Review* (Salt Lake City) 44 (May 31, 1942): 22–27.

———. "Strategic Minerals." *Mining Congress Journal* 25 (February 1940): 36–39.

———. "War's Effect on Tungsten." Address before the American Mining Congress, September 19, 1940. Reprinted in *Mining Congress Journal* 27 (February 1941): 38–40.

Shafer, Michael. "Mineral Myths." *Foreign Policy* 47 (Summer 1982): 154–71. Available at http://links.jstor.org.

Shaffer, Butler. *In Restraint of Trade: The Business Campaign Against Competition, 1918–1938.* Lewisburg, Pa.: Bucknell University Press, 1997.

Sheffield University, home page at http://www.shef.ac.uk.

Shirer, William L. *The Rise and Fall of the Third Reich: A History of Nazi Germany.* New York: Simon and Schuster, 1960.

Short, Brant. "The Rhetoric of the Post-Presidency." *Presidential Studies Quarterly* 21 (Spring 1991): 333–50.

Singer, Charles, et al., eds. *A History of Technology.* Vol. 5, *The Late Nineteenth Century, c. 1850 to c. 1900.* Oxford: Clarendon Press, 1958.

Smith, Cyril Stanley. "The Discovery of Carbon in Steel." *Technology and Culture* 5 (Spring 1964): 149–75. Available at http://links.jstor.org.

Smith, George David. *From Monopoly to Competition: The Transformation of Alcoa, 1888–1986.* Cambridge: Cambridge University Press, 1988.

Smith, George Otis, ed. *The Strategy of Minerals: A Study of the Mineral Factor in the World Position of American in War and in Peace.* New York: Appleton, 1919. Available at http://books.google.com/books.

Smoot, Reed. "Why a Protective Tariff?" *Saturday Evening Post,* September 10, 1932. Available at http://web.ebscohost.com/.

Snowiss, Sylvia. "Presidential Leadership of Congress: An Analysis of Roosevelt's First Hundred Days." *Publius* 1, no. 1 (1971): 59–87.

Snyder, George H. "Domestic Tungsten Ore Supply Becomes Increasingly Important." *Steel* 100 (May 3, 1937): 79–80.

Sobel, Robert. *The Age of Giant Corporations: A Microeconomic History of American Business, 1914–1992.* 3rd ed. Westport Conn.: Praeger, 1993.

———. *The Big Board: A History of the New York Stock Market.* New York: Free Press, 1965.

Somers, Herman Miles. *Presidential Agency: OWMR, the Office of War Mobilization, and Reconversion.* Washington, D.C.: National Defense University Press, 1996.

Stabile, Donald R. "Herbert Hoover, the FAES, and the AF of L." *Technology and Culture* 27 (October 1986): 819–27. Available at http://links.jstor.org.

Stevens, Donald G. "World War II Economic Warfare: The United States, Britain, and Portuguese Wolfram." *Historian* 61 (Spring 1999): 539–55. Available at http://web.ebscohost.com/.

Stevenson, David. *Cataclysm: The First World War as Political Tragedy.* New York: Basic Books, 2004. Available at http://books.google.com/books.

Stevenson, Elizabeth. *Babbitts and Bohemians: The American 1920s.* New York: Macmillan, 1967.

Stoll, Richard J. "Steaming in the Dark? Rules, Rivals, and the British Navy, 1860–1913." *Journal of Conflict Resolution* 36 (June 1992): 263–83. Available at http://links.jstor.org.

Strachan, Hew. *The First World War*. Vol. 1, *To Arms*. Oxford: Oxford University Press, 2001. Available at http://books.google.com/books.

Strikwerda, Carl. "The Troubled Origins of European Economic Integration: International Iron and Steel and Labor Migration in the Era of World War I." *American Historical Review* 98 (October 1993): 1106–29. Available at http://web.ebscohost.com/.

Stromberg, Roland. "American Business and the Approach of War, 1935–1941." *Journal of Economic History* 13 (Winter 1953). Available at http://links.jstor.org.

———. "On Cherchez Le Financier: Comments on the Economic Interpretation of World War I." *History Teacher* 10 (May 1977): 435–43. Available at http://links.jstor.org.

Sturm, Hobert P. "Webb-Pomerene Associations." *Western Political Quarterly* 8 (March 1955): 82–89. Available at http://links.jstor.org.

Taussig, F. W. "Necessary Changes in Our Commercial Policy." *Foreign Affairs* 11 (April 1933). Available at http://links.jstor.org.

———. *Tariff History of the United States*. Pt. 1. 5th ed. New York: G. P. Putnam's Sons, 1910. Online edition prepared by Wm. Harshbarger, 2003, available at http://www.mises.org.

Terkel, Studs. *"The Good War": An Oral History of World War Two*. New York: Pantheon Books, 1984.

Thompson, William R., and Gary Zuk. "World Power and the Strategic Trap of Territorial Commitments." *International Studies Quarterly* 30 (September 1986): 249–67. Available at http://links.jstor.org.

Thomson, David. *Europe Since Napoleon*. New York: Alfred A. Knopf, 1960.

Toland, John. *The Rising Sun: The Decline and Fall of the Japanese Empire, 1936–1945*. Vol. 1. New York: Random House, 1970.

Travers, T. H. E. "The Offensive and the Problem of Innovation in British Military Thought, 1870–1915." *Journal of Contemporary History* 13 (July 1978): 531–53. Available at http://links.jstor.org.

Trebilcock, Clive. "British Armaments and European Industrialization, 1890–1914." *Economic History Review* 26, no. 2 (1973): 254–72. Available at http://links.jstor.org.

———. "'Spin-off' in British Economic History: Armaments and Industry, 1760–1914." *Economic History Review* 22 (December 1969): 474–90. Available at http://links.jstor.org.

Trubowitz, Peter. *Defining the National Interest: Conflict and Change in American Foreign Policy*. Chicago: University of Chicago Press, 1998.

Tweedale, Geoffrey. "Metallurgy and Technological Change: A Case Study of Sheffield Steel and America, 1830–1930." *Technology and Culture* 27 (April 1986): 189–222. Available at http://links.jstor.org.

———. *Sheffield Steel and America: A Century of Commercial and Technological Interdependence, 1830–1930*. Cambridge: Cambridge University Press, 1987.

———. "Sir Robert Abbot Hadfield F.R.S. (1858–1940) and the Discovery of Manganese Steel." *Notes and Records of the Royal Society of London* 40 (November 1985): 63–74. Available at http://links.jstor.org.

———. *Steel City: Entrepreneurship, Strategy, and Technology in Sheffield, 1743–1993.* Oxford: Clarendon Press, 1995.

Twight, Charlotte. "The Political Economy of the National Defense Stockpile." *Policy Studies Review* 8 (Summer 1989). Available at http://web.ebscohost.com/.

Vaccari, John A., et al. *Materials Handbook: An Encyclopedia for Managers, Technical Professionals, Purchasing and Production Managers, Technicians, and Supervisors.* 15th ed. 1929. Reprint, New York: McGraw-Hill, 2002. Available at http://books.google.com/.

Vernon, Raymond. *Two Hungry Giants: The United States and Japan in the Quest for Oil and Ores.* Cambridge: Harvard University Press, 1983.

Wall, Joseph Frazier. *Andrew Carnegie.* New York: Oxford University Press, 1970.

Wallace, Benjamin B. "Postwar Tariff Changes and Tendencies." *Annals of the American Academy of Political and Social Science* 94 (March 1921): 175–85. Available at http://links.jstor.org.

Wallace, Henry A. *America Must Choose: The Advantages and Disadvantages of Nationalism, of World Trade, and of a Planned Middle Course.* World Affairs Pamphlet no. 3. New York: Foreign Policy Association, 1934.

Wang, Kung-Ping. "Mineral Resources of China: With Special Reference to the Nonferrous Metals." *Geographical Review* 34 (October 1944): 621–35. Available at http://links.jstor.org.

Weintraub, Ruth G., and Rosalind Tough. "Federal Housing and World War II." *Journal of Land and Public Utility Economics* 18 (May 1942): 155–62. Available at http://links.jstor.org.

Whealey, Robert H. *Hitler and Spain: The Nazi Role in the Spanish Civil War.* Lexington: University Press of Kentucky, 1989. Available at http://books.google.com/books.

Wheeler, Douglas L. "The Price of Neutrality: Portugal, the Wolfram Question, and World War II." *Luso-Brazilian Review* 23 (Summer 1986): 107–27. Available at http://links.jstor.org.

White, Langdon. "The Iron and Steel Industry of the Pittsburgh District." *Economic Geography* 4 (April 1928): 115–39. Available at http://links.jstor.org.

Wilson, Joan Hoff. *Herbert Hoover: Forgotten Progressive.* Edited by Oscar Handlin. 1975. Reprint, New York: Harper Collins, 1992.

Wormser, Felix Edgar. "The Importance of Foreign Trade in Copper and Other Metals." *Annals of the American Academy of Political and Social Science* 84 (March 1921): 65–76. Available at http://links.jstor.org.

Wright, Chester W. "American Economic Preparations for War, 1914–1917 and 1939–1941." *Canadian Journal of Economic and Political Science* 8 (May 1942): 157–75. Available at http://links.jstor.org.

Wright, Gavin. "The Origins of American Industrial Success, 1879–1940." *American Economic Review* 80 (September 1990): 651–68. Available at http://links.jstor.org.

Wriston, Walter B. "Technology and Sovereignty." *Foreign Affairs* (Winter 1988–1989). Reprinted in *The American Encounter: The United States and the Making of the Modern World; Essays From 75 Years of Foreign Affairs*, edited by James F. Hogue and Fareed Zakaria. New York: Basic Books, 1997.

Wyman, Mark. *Hard Rock Epic: Western Miners and the Industrial Revolution, 1860–1910.* Berkeley and Los Angeles: University of California Press, 1989.

Yeager, Mary A. "Trade Protection as an International Commodity: The Case of Steel." *Journal of Economic History* 40 (March 1980): 33–42. Available at http://links.jstor.org.

Yoshpe, Harry B. "Economic Mobilization Planning Between the Two World Wars." *Military Affairs* 15 (Winter 1951): 199–204. Available at http://links.jstor.org.

Young, Arthur N. *China's Nation-Building Effort, 1927–1937: The Financial and Economic Record.* Stanford: Hoover Institution Press, 1971.

Young, George J. "The Tungsten Mining Industry in 1919." *Engineering and Mining Journal* 109 (January 17, 1920): 211–12.

Zanjani, Sally. *Goldfield: The Last Gold Rush on the Western Frontier.* Athens, Ohio: Swallow Press, 1992.

INDEX